ENVIRONMENTAL ENGINEERING

"If seven maids with seven mops
  Swept it for half a year,
Do you suppose," the Walrus said,
  "That they could get it clear?"
"I doubt it," said the Carpenter,
  And shed a bitter tear.
                    —Lewis Carroll

... it will not be enough to fashion the world's
largest mop in an attempt to clean up the human
habitat. Humanity (also) needs a world order.

                    Norman Cousins

ENVIRONMENTAL ENGINEERING

p. aarne vesilind
j. jeffrey peirce

ANN ARBOR SCIENCE
THE BUTTERWORTH GROUP

Library of Congress Catalog Card Number 81-70872
ISBN 0-250-40422-2

Butterworths, Ltd., Borough Green, Sevenoaks
Kent TN15 8PH, England

Preface

If any single characteristic can be used to describe environmental engineering, it is that *it is a profession where everything matters*. The natural sciences, the social sciences and the humanities can be just as important in the performance of environmental engineering as the engineering aspects. This fact provides great appeal and challenge for the professional, but also creates an extremely difficult task for authors of a textbook on environmental engineering.

If everything matters, what is to be included in a textbook designed to cover the rudiments of the profession in one semester? We respond to this challenge by organizing the text in five major areas: water resources; air quality; solid and hazardous waste; noise; and environmental impact. We begin by introducing the profession of environmental engineering in its historical context, and finish with a discussion of environmental ethics. There is some justification for introducing the ethical considerations early in the course, and we feel this may be done without loss in continuity. Otherwise, we attempt to carefully introduce concepts and definitions in an orderly manner. If chapters are to stand alone or be taken out of sequence, the instructor should first determine if additional background is needed.

We are grateful to our students at Duke University who have read and tested the manuscript. Special thanks go to Dr. Yakir Hasit of Drexel University and his students who reviewed and tested the manuscript. Fred Avent drafted many of the figures, and Judy Edwards typed the numerous changes and revisions, with a minimal number of "Oh, Lord!"s.

Finally, we would like to acknowledge our wives, Gail and Jodie. Their constant encouragement and cheer and their contributions to our professional careers have been invaluable.

P. A. Vesilind
J. J. Peirce

v

to
Gail and Jodie

Contents

Chapter 1

Environmental Engineering

Environmental engineering is a new profession with a long and honorable history.

This is not a contradiction. The descriptive title of "environmental engineer" was not used until the 1960s, when academic programs in engineering and public health schools broadened their scope, and required a more accurate title to describe their curricula and their graduates. The roots of this profession, however, go back as far as recorded history.

These roots reach into four major disciplines: civil engineering, public health, ecology and ethics. From each, the environmental engineering profession draws knowledge, skill, professionalism and concern for the greater good.

CIVIL ENGINEERING

Skills in agriculture spawned the development of a cooperative social fabric and the growth of communities. As farming efficiency increased, a division of labor was possible, and communities began to build permanent public and private structures. Defense of these structures and the land became paramount, and other structures were built purely for defensive purposes. In other societies, the conquest of neighbors also required the construction of the machines of war. Builders of these facilities became known as military engineers, and the term engineer continued to imply military involvement well into the eighteenth century.

In 1782, John Smeaton, builder of roads, structures and canals in England, recognized that since his profession was in the construction of public facilities, he should correctly be designated a *civil* engineer. This title was widely adopted by engineers engaged in public works.[1]

The first engineering curriculum in the United States was established at The United States Military Academy at West Point in 1802. The first engineering course outside the Academy was offered in 1821 at the American Literary, Scientific, and Military Academy, which later became Norwich University. The Rensselaer Polytechnic Institute conferred the first truly *civil* engineering degree in 1835. In 1852, the American Society of Civil Engineers was founded.[2]

Among the public facilities for which the civil engineer became responsible were water supply and wastewater drainage. The availability of water has remained a critical component of civilizations.* Ancient Rome, for example, had water supplied by nine different aquaducts, with cross sections from 2 m to 15 m (7 to 50 ft), up to 80 km (50 mi) long.

As cities grew, the demand for water became increasingly acute. In the eighteenth and nineteenth centuries, common people in cities lived under abominable conditions, with water supplies grossly polluted, expensive or nonexistent.

In London, the water supply was controlled by nine different private companies, and water was sold to the public. Often the poorer people could not afford to pay for the water and begged or stole it. During epidemics, the privation was so great that people went into plowed fields and drank water out of depressions in the ground. Droughts caused water supplies to be curtailed, causing great crowds to form at the public pumps, wanting their "turn" at the pump.[3]

In the New World, the first public water supply system consisted of wooden pipes, bored and charred, with metal rings shrunk on the ends to prevent splitting. The first such pipes were installed in 1652, and the first city-wide system was constructed in Winston-Salem, North Carolina, in 1776. The first American water works was built in the Moravian settlement of Bethlehem, Pennsylvania. A wooden waterwheel, driven by the flow of Monococy Creek, powered wooden pumps that lifted spring water to a wooden reservoir on top of a hill where it was distributed by

[1]Kirby, R. S., S. Withington, A. B. Darling and F. G. Kilgrour. *Engineering in History* (New York: McGraw-Hill Book Company, 1956).

[2]Wisely, W. H. *The American Civil Engineer, 1852-1974,* (The American Society of Civil Engineers, New York, 1974).

* A fascinating account of the importance of water supply to a community through the ages can be found in James Michener's *The Source.*

[3]Ridgeway, J. *The Politics of Ecology,* (E. P. Dutton & Co., 1970).

gravity.[4] One of the first major water supply undertakings was the Croton Aqueduct, started in 1835 and completed six years later. This engineering marvel brought clear water to Manhattan Island, which had an inadequate supply of groundwater.[5]

Although the quantities of water provided by municipal systems might have been adequate, the quality of water was often suspect. As one observer described it:[6]

> The appearance and quality of the public water supply were such that the poor used it for soup, the middle class dyed their clothes in it, and the very rich used it for top-dressing their lawns. Those who drank it filtered it through a ladder, disinfected it with chloride of lime, then lifted out the dangerous germs which survived and killed them with a club in the back yard.

The earliest known acknowledgment of the effect of impure water is found in *Susruta Samhitta,* a collection of fables and observations on health, dating back to 2000 B.C., which recommended the water be boiled before drinking. The filtration of water became commonplace only toward the middle of the nineteenth century. The first successful filter for water supply was in Parsley, Scotland, in 1804, and many less successful attempts to clarify water by filtration followed.[7]. A notable failure was the system for New Orleans, which was to filter water from the Mississippi River. The water proved to be so muddy that the filters clogged up too fast for the system to be workable. This problem was not alleviated until aluminum sulfate (alum) began to be used as a pretreatment. Although the use of alum to clarify water as a pretreatment to filtration was first proposed in 1757, it wasn't until 1885 that this was convincingly demonstrated. Disinfection of water with chlorine was started in 1902 in Belgium and followed in 1908 in Jersey City, New Jersey. The years 1900 to 1920 saw dramatic drops in deaths from infectious diseases, due in part to the effect of cleaner water supplies.

Human waste disposal in early cities also presented both a nuisance

---

[4] American Public Works Association. *History of Public Works in the United States 1776-1976* (Chicago: APWA, 1976).

[5] Lankton, L. D. "1842: The Old Croton Aquaduct Brings Water, Rescues Manhattan from Fire, Disease," *Civil Eng.* 47(10) (1977).

[6] Smith, G., *Plague on Us* (Oxford: Oxford University Press, 1941).

[7] Baker, M. N. *The Quest for Pure Water* (New York: American Water Works Association, 1949).

and a serious health problem. Often the method of disposal consisted of nothing more than flinging the contents of chamberpots out the window (Figure 1-1).

Figure 1-1   Human excreta disposal, from an old woodcut. [Source: Reyburn, W. *Flushed With Pride* (London: McDonald, 1969).]

Stormwater was considered the main "drainage" problem, and it was in fact illegal in many cities to discharge wastes into the ditches and storm sewers. Eventually, as water supplies developed,* the storm sewers became used for both sanitary waste and stormwater. Such "combined sewers" exist in some of our major cities even today.

The first system for urban drainage in America was constructed in Boston around 1700. There was a surprising amount of resistance to the construction of sewers for waste disposal. Most American cities even at the end of the nineteenth century had cesspools or vaults. The most economical means of waste disposal was to pump these out at regular intervals and cart the waste to a disposal site outside the town. The engineers argued that although sanitary sewers were capital intensive, they provided the best means of wastewater disposal in the long run. Their argument won, and there was a remarkable period of sewer construction between 1890 and 1900.

Actually, the first separate sewerage systems in America were built in the 1880s in Memphis, Tennessee and Pullman, Illinois. The Memphis system was a complete failure. It used small pipes which were to be flushed periodically. No manholes were constructed and thus cleanout became a major problem. The system was later removed and larger pipes, with manholes, were installed.[4]

Initially, all sewers emptied into the nearest watercourse, without any treatment. As a result, many of these lakes and rivers became grossly polluted, and as an 1885 Boston Board of Health report put it, "large territories are at once, and frequently, enveloped in an atmosphere of stench so strong as to arouse the sleeping, terrify the weak, and nauseate and exasperate everybody."

Wastewater treatment first consisted only of screening for the removal of the large floatables to protect sewage pumps. Screens had to be cleaned manually, and the wastes were buried or incinerated. The first mechanical screens were installed in Sacramento, California in 1915, and the first mechanical comminutor for grinding up the screenings was installed in Durham, North Carolina. The first complete treatment systems were operational by the turn of the century with land spraying of the effluent being a popular method of wastewater disposal.

The engineering for these facilities was the responsibility of the civil engineer, since these were constructed public works. There was, however, little appreciation of the broader perspectives of environmental pol-

---

*To hold down the quantity of wastewater discharge, the city of Boston in 1844 passed an ordinance prohibiting the taking of baths without doctor's orders.

lution control and management. These considerations historically have come from the public health professions, from the science of ecology, and perhaps most subtly, from the philosophy of ethics.

PUBLIC HEALTH

Life in cities during the middle ages and through the industrial revolution was difficult, sad and short. In 1842, the *Report from the Poor Law Commissioners on an Inquiry into the Sanitary Conditions of the Labouring Population of Great Britain* described the sanitary conditions in this manner:

> Many dwellings of the poor are arranged round narrow courts having no other opening to the main street than a narrow covered passage. In these courts there are several occupants, each of whom accumulated a heap. In some cases, each of these heaps is piled up separately in the court, with a general receptacle in the middle for drainage. In others a pit is dug in the middle of the court for the general use of all the occupants. In some the whole courts up to the very doors of the houses were covered with filth.

The great rivers in urbanized areas were in effect open sewers. The River Cam, like the Thames, was for many years grossly polluted. There is a tale of Queen Victoria being shown over Trinity by the Master, Dr. Whewell, and saying, as she looked down over the bridge: "What are all those pieces of paper floating down over the river?" To which, with great presence of mind, he replied: "Those, ma'am, are notices that bathing is forbidden."[8]

During the middle of the nineteenth century, medical knowledge was still primitive, and public health measures were inadequate and often counterproductive. The germ theory was not as yet appreciated, and great epidemics swept over the great cities of the world. Some intuitive measures taken by the public health agencies did, however, have a positive effect. Removal of corpses during epidemics and appeals for cleanliness undoubtedly helped the public health.

The 1850s witnessed what is now called the "Great Sanitary Awakening." Led by tireless public health advocates like Sir Edwin Chadwick, proper and effective measures began to evolve. Possibly the single most

---

[8]Raverat, G. *Period Piece,* as quoted in W. Reyburn's *Flushed with Pride* (London: McDonald, 1969).

important investigation of a public health problem was John Snow's classic epidemiological study of the 1849 cholera epidemic in London. By using a map of the area and identifying the residences of the people who contracted the disease, he was able to pinpoint the cause of the epidemic as the water from a public pump on Broad Street. Removal of the handle from the Broad Street pump eliminated the source of the cholera organism and the epidemic subsided.* Ever since, waterborne diseases have become one of the major concerns of the public health. The reduction of such diseases by providing safe and pleasing water to the public has been one of the dramatic successes of the public health profession.

Today, the concerns of public health encompass not only water, but all aspects of civilized life, including food, air, toxic materials, noise and other environmental insults.

ECOLOGY

The recognition that all life forms are inextricably dependent on one another and on the physical environment is a fairly recent phenomenon. The word "ecology" was not even invented until the mid-1800s, and the study of ecology as a branch of natural sciences was not widespread until a few decades ago.

The science of ecology defines "ecosystems" as groups of organisms which interact with each other and the physical environment, and which affect the population of the various species in the environment. For example, the simplest type of ecosystem consists of two animals such as the hare and the lynx. If the hare population in a specific locality is high, the lynx have an abundant food supply, procreate, and increase in population until they outstrip the availability of hares. As the lynx population decreases due to the unavailability of food, the hares increase since the number of predators is fewer, and the cycle repeats (Figure 1-2). Such a system is dynamic in that the numbers of each population are continually changing, but over a long time span is at a steady state condition, known as "homeostasis."

A slightly more complex example includes three species: the sea otter, the sea urchin and kelp. The kelp forests along the Pacific coast consist of 60-m (200-ft) streamers fastened to the ocean floor. Kelp is an economically valuable plant since it is the source of algin (used in foods,

---

*Interestingly, it wasn't until 1884 that Robert Koch proved that *Vibrio comma* was the microorganism responsible for cholera.

Figure 1-2   The hare and lynx homeostasis. [Source: MacLurich, D. A. "Fluctuations in the Numbers of the Varying Hare," University of Toronto Studies, Biological Sciences No. 43, Reproduced in Odum, G. *Fundamentals of Ecology,* 3rd ed. (Philadelphia: W. B. Saunders Co., 1971).]

paints, cosmetics, etc.) and its harvesting is regulated to preserve the forests.

A few years ago, kelp started to mysteriously disappear, leaving barren ocean floors. The puzzle was solved when it was recognized that sea urchins feed on the kelp, weaken the stems, and cause them to detach and float away. The sea urchin population had increased because the population of their predators, the sea otters, had been reduced drastically. The solution to the problem was to protect the sea otter and allow its population to increase, thus reducing the number of sea urchins, and save the kelp forests.

Some ecosystems are fragile, while others are resilient and able to withstand even serious perturbations. One characteristic of a resilient ecosystem is that more than one species fills a "niche" within that system. That is, many species perform the same function. For example, in the hare/lynx system, if the hare were the only source of food for the lynx (a niche), the destruction of the hares would result in the eventual eradication of the lynxes as well. If, however, the lynx used other small animals for food, the system might survive the loss of the hares.

Engineers must appreciate the fundamental principles of ecology, and design with the environment, so as to not impact adversely on especially fragile ecosystems. For example, one of the most fragile of all ecosystems is the deep oceans, yet ocean disposal of hazardous waste is seriously advocated by some engineers. The inclusion of ecological principles in engineering decisions is a major component of the environmental engineering profession.

ETHICS

Historically, engineering was devoid of the need for ethical decisions. Ethics as a framework for deciding what one ought to do was irrelevant, since the engineer did precisely what employer or client required. Indeed, even today, some engineers revel in the supposed freedom from concern for ethical questions since engineering is applied science, and science is value free.[10]

Unfortunately (or perhaps fortunately), this is a weak argument, and engineers are daily confronted with questions which demand responses for which purely technical tools are inadequate. This is especially true for engineers engaged in environmental and pollution control activities, since their work interfaces a newly emerging field of "environmental ethics."[11]

Among the interesting questions asked in the search for environmental ethics is the origin of our attitude toward the environment. This problem, and much of the following discussion, is covered in greater detail in the last chapter. Suffice for now to recognize that Western traditions encourage private ownership of land and other resources and the exploitation of the environment. This tradition is contrary to the view of the American Indians and other cultures which consider land and natural resources as being a public trust. European settlers, arriving in the New World from countries where all land was owned by wealthy aristocracies, considered it their right to similarly own and exploit land.* And indeed, the natural resources were so plentiful that there grew a "myth of super-abundance," where the likelihood of running out of any resource (including oil!) was considered remote.[12]

Some dissenting voices were raised during the nineteenth century rape of the environment. Henry David Thoreau, Ralph Waldo Emerson, and later John Muir, Gifford Pinchot and President Theodore Roosevelt all

---

[10]Florman, S. *The Existential Pleasures of Engineering* (New York: St. Martins Press, 1976).

[11]*Environmental Ethics*, a professional journal published quarterly by the University of Georgia, Athens, GA.

*An exception to this general rule is found in states within the boundaries of the Northwest Purchase, notably Wisconsin. Included in the Purchase agreement between France and America is the condition that state constitutions must ensure that the water and air resources must be held in trust by the state for the people "as long as the wind blows and the water flows." Witness the incalculable number of public accesses to lakes and rivers in Wisconsin.

[12]Udall, S. *Fundamentals of Ecology*, 3rd ed. (Philadelphia: W. B. Saunders Company, 1971).

contributed to the growth of awareness in environmental concern. The first explicit statement of a need for an environmental ethic was penned by Aldo Leopold in the 1930s. Since then, many people have contributed thoughtful and well reasoned arguments toward the development of a comprehensive and useful ethic for judging questions of conscience and value. This component of the environmental engineer's work will continue to grow in importance as increasingly difficult questions will demand answers.

ENVIRONMENTAL ENGINEERING AS A PROFESSION

The purpose of colleges and universities is to allow students to mature intellectually and socially and to prepare themselves for careers which will earn them a livelihood. One of the main considerations used by students, especially in engineering, in selecting a profession is that the skills learned be marketable. There are, however two other factors which should go into the equation by which career decisions are calculated.

First, the vocation should ideally be an avocation as well. It should be a job which is "fun," a job that is approached with enthusiasm even after many years in the profession. The people who are doing exactly what they would like to do, people who made a fortunate career decision, are surprisingly rare. Most engineers in fact get bored with their work, search around for other jobs, often leaving the profession altogether. In the selection of a career, therefore, the enjoyment factor is immensely important.

Secondly, it is necessary to be proud of what one has accomplished. Building a bridge to allow access to a previously isolated community can be a positive and personally satisfying undertaking. On the other hand, building a dam on a river which wipes out a wild whitewater stream and which creates a useless, highly eutrophied body of water cannot be a source of pride, regardless of what the boss says, or what political favors and deals are involved. In other cases, the social benefits from one's job are even more clearly lacking. Consider, for example, the people involved in the production of cigarette commercials. After completing a professional career on Madison Avenue, is it possible to note with pride that one wrote the world's best cigarette advertisements?

One of the fortunate characteristics of environmental engineering is that in the course of performing a professional function, one can enjoy the work and be proud of the results. It is satisfying to be part of the solution and not part of the problem.

SUGGESTED READING

History of the Water Pollution Control Federation, 1928–1977 (Washington, DC: Water Pollution Control Federation, 1977).

Civil Engineering, Special Issue on Civil Engineering History, 47(10) (1977).

Vesilind, P. A. *History of Environmental Sciences and Engineering,* Department of Environmental Sciences and Engineering, University of North Carolina, Chapel Hill (1972).

Petulla, J. M. *American Environmental History* (San Francisco: Boyd & Fraser Publishing Company, 1977).

Turk, A., et al. *Energy Pollution Environment* (Philadelphia: W. B. Saunders Co., 1972).

Boughey, A. S. *Strategy for Survival* (Menlo Park, CA: W. A. Benjamin Inc., 1976).

Barbour, I. G., Ed. *Western Man and Environmental Ethics* (Reading, MA: Addison-Wesley Publishing Company, 1973).

Leopold, A. *A Sand Country Almanac* (New York: Ballantine Books, Inc., 1966).

Chapter 2

Water Pollution

Although people intuitively relate filth with disease, the fact that pathogenic organisms can be transmitted by polluted water was not recognized until the middle of the nineteenth century. Probably the most dramatic demonstration that water can indeed transmit disease was the Broad Street pump-handle incident.

A public health physician named John Snow, assigned to attempt to control a cholera epidemic, realized that there seemed to be an extremely curious concentration of cholera cases in one part of London. Almost all of the people affected drew their drinking water from a community pump in the middle of Broad Street. Even more curious was the fact that the people who worked and lived in an adjacent brewery were not afflicted. Although this seemed to demonstrate the health benefits of beer, welcome news to most students, Snow recognized that the absence of cholera in the brewery might be because the brewery obtained its water from a private well and not the Broad Street pump.

Snow's evidence convinced the city council to ban the obviously polluted water supply, which was done by simply removing the pump handle, thus effectively preventing the people from the using the water. The source of infection was stopped, the epidemic subsided, and a new era of public health awareness related to water supplies began.

The concern with water pollution was, until recently, a concern about health effects. In many countries it still is. In the United States and other developed countries, however, water treatment and distribution methods have for the most part eradicated the transmission of bacterial waterborne disease. We now think of water pollution not so much in terms of health, but rather of conservation, aesthetics and the preservation of natural beauty and resources. Man has an inexplicable affinity for water, and the fouling of lakes, rivers and oceans is intrinsically unacceptable to the concerned citizen.

SOURCES OF WATER POLLUTION

The United States has more than 40,000 factories that use water, and their industrial wastes are probably the greatest single water pollution problem.

Organic wastes from industrial plants, at present-day treatment levels, are equal in polluting potential to the *untreated* raw sewage of the entire population of the United States. In most cases the organic wastes, as potent as they might be, are at least treatable, in or out of the plant. Inorganic industrial wastes are much trickier to control, and potentially more hazardous. Chromium from metal-plating plants is an old source of trouble, but mercury discharges have only recently received their due attention.

As important as these and other well-known "heavy metals" might be, many scientists are much more concerned with the unknown chemicals. Industry is creating a fantastic array of new chemicals each year, all of which eventually find their way to the water. For most of these, not even the chemical formulas are known, much less their acute, chronic or genetic toxicity.

Another industrial waste is heat. Heated discharges can drastically alter the ecology of a stream or lake. This alteration is sometimes called beneficial, perhaps because of better fishing or an ice-free docking area. The deleterious effects of heat, in addition to promoting modifications of ecological systems, include a lessening of dissolved oxygen solubility and increases in metabolic activity. Dissolved oxygen is vital to healthy aquatic communities, and the warmer the water the more difficult it is to get oxygen into solution. Simultaneously, the metabolic activity of aerobic (oxygen-using) aquatic species increases, thus demanding more oxygen. It is a small wonder, therefore, that the vast majority of fish kills due to oxygen depletion occur in the summer.

Municipal waste is a source of water pollution second in importance only to industrial wastes. Around the turn of the century, most discharges from municipalities received no treatment whatsoever. In the United States, sewage from 24 million people was flowing directly into our watercourses. Since that time, the population has increased, and so has the contribution from municipal discharges. It is estimated that presently the population equivalent* of municipal discharges to water-

---

* Population equivalent is the number of people needed to contribute a certain amount of pollution. For example, if a town has 10,000 people, and the treatment plant is 50% effective, then their discharge has a population equivalent of 5000 people. Similarly, if an industry discharges 1000 lb of solids per day, and if each person contributes 0.2 lb/day into domestic wastewater, the industrial waste can be expressed as being the equivalent of $1000/0.2 = 5000$ people.

courses is about 100 million. Even with the billions of dollars spent on building wastewater treatment plants, the contribution from municipal pollution sources has not been significantly reduced. We seem to be holding our own, however, and at least are not falling further behind.

One problem, especially in the older cities on the east coast of the United States, is the sewerage systems. When the cities were first built, the engineers realized that sewers were necessary for both stormwater and sanitary wastes, and they saw no reason why both stormwater and sanitary wastes should not flow in the same pipelines. After all, they both ended up in the same river or lake. Such sewers are now known as *combined sewers*.

As years passed and populations increased, the need for the treatment of sanitary wastes became obvious, and two-sewer systems were built, one to carry stormwater and the other, sanitary waste. Such systems are known as *separate sewers*.

Almost all of the cities with combined sewers have built treatment plants which can treat the *dry weather flow,* or in other words, sanitary wastes. As long as it doesn't rain, they can provide sufficient treatment. When the rains come, however, the flows swell to many times the dry weather flow and most of it must be bypassed directly into a river or lake. This overflow contains sewage as well as stormwater, and has a high polluting capacity. All attempts to capture this excess flow for subsequent treatment, such as storage in underground caverns and rubber balloons, are expensive. However, the alternative solution, separating the sewers, is estimated at a staggering $60 billion for the major cities in the United States.

In addition to industrial and municipal wastes, water pollution emanates from many other sources.

Agricultural wastes, should they all flow into a stream, would have a population equivalent of about 2 billion. Fortunately, very little of it does reach streams. The problems are intensifying, however, with the increase in the size and number of feed lots. Feed lots are cattle pens constructed for the purpose of fattening up the cattle before slaughter. These are usually close to slaughterhouses (hence cities) and a large number of animals are packed into a small space. Drainage from these lots has an extremely high pollutional strength.

Sediment from land erosion can also be classified as a pollutant. Sediment consists of mostly inorganic material washed into a stream as a result of farming, construction or mining operations. The detrimental effects of sediment include interference with the spawning of fish by covering gravel beds; interference with light penetration, thus making food more difficult to find; and direct damage to gill structures. In the long run, sediment could well be one of our most harmful pollutants.

The concern with pollution from petroleum compounds is relatively new, starting to a large extent with the Torrey Canyon disaster in 1967. Ignoring maps showing submerged rocks, the huge tanker loaded with crude oil plowed into a reef in the English Channel. Almost immediately, oil began seeping out, and both the French and British became concerned. Rescue efforts failed and the Royal Air Force attempted to set it on fire, with little success. Almost all of the oil eventually leaked out and splashed on the beaches of France and England. The French started the back-breaking chore of spreading straw on the beaches, allowing the straw to adsorb the oil, and then collecting and burning the oil-soaked straw. The English, being more sophisticated, used detergents to disperse the oil and then flushed the emulsion off the beaches. Time has shown the French way to be best, since the English detergents have now been shown to be potentially more harmful to coastal ecology than the oil would have been.

Although the Torrey Canyon disaster was the first big spill, many have followed it. It is estimated that there are no fewer than 10,000 serious oil spills in the United States every year. In addition, the contribution from routine operations such as flushing oil tankers may well exceed all the oil spills.

The acute effect of oil on birds, fish and microorganisms is reasonably well cataloged. What is not so well understood, and potentially more harmful, is the subtle effect on aquatic life. Salmon, for example, have been known to find their home stream by the specific smell (or taste) of the water, caused in large part by the hydrocarbons present. If man continues to pour (albeit unintentionally) hydrocarbons into salmon rivers, it is possible that the salmon will become so confused that they will refuse to enter their spawning stream.

Another form of industrial pollution, much of it willed to us by our ancestors, is acid mine drainage. The problem is caused by the leaching of sulfur-laden water from old abandoned mines (as well as some active mines). On contact with air, these compounds are soon oxidized to sulfuric acid, a deadly poison to all living matter.

It should be amply clear, therefore, that water can be polluted by many types of waste products.

Water pollution problems (and hence their solutions) can be best understood by first describing them in the context of an ecosystem, and then studying one specific aspect of that ecosystem: the biodegradation of organics.

ELEMENTS OF ECOLOGY

Plants and animals in their physical environment make up an *ecosystem*. The study of such ecosystems is *ecology*. Although we often draw lines around a specific ecosystem in order to be able to study it more fully (e.g., a farm pond) and in so doing assume that the system is totally self-contained, this obviously is not true, and we must remember that one of the tenets of ecology is that "everything is connected with everything else."

Within an ecosystem there exist three broad categories of actors. The *producers* take energy from the sun, nutrients such as nitrogen and phosphorus from the soil, and through the process of photosynthesis produce high-energy chemicals. The energy from the sun is thus stored in the chemical structure of their organic molecules. These organisms are often referred to as being in the first *trophic* level and are called *autotrophs*.

A second group of organisms are the *consumers* who use some of this energy by ingesting the high-energy molecules. These organisms are in the second trophic level in that they directly use the energy of the producers. There can be several more trophic levels as the consumers use the level above as a source of energy. A simplified ecosystem showing various trophic levels is shown in Figure 2-1, which also illustrates the progressive use of energy through the trophic levels.

The third group of organisms, the *decomposers* or decay organisms, use the energy in animal wastes and dead plants and animals, and in so doing convert the organic molecules to stable inorganic compounds. The residual inorganics then become the building blocks for new life, using the sun as the source of energy.

Ecosystems exhibit a flow of both energy and nutrients. Energy flow is in only one direction: from the sun and through each trophic level. Nutrient flow, on the other hand, is cyclic. Nutrients are used by plants to make high-energy molecules, which are eventually decomposed to the original inorganic nutrients, ready to be used again.

The entire food web, or ecosystem, stays in dynamic balance, with adjustments being made as required. Such a balance is called *homeostasis*. For example, a drought one year may produce little grass, thus exposing field mice to predators such as owls. The mice, in turn, spend more time in burrows, thus not eating as much, and allowing the grass to reseed for the following year. External perturbations, however, can upset and even destroy an ecosystem. In the previous example, the use of an herbicide to kill the grass might also destroy the field mouse population, and in turn diminish the number of owls. It must be recognized that although most ecosystems can absorb a certain amount of insult, a sufficiently large perturbation can cause irreparable damage.

The amount of perturbation a system is able to absorb without being destroyed is tied to the concept of the *ecological niche*. The combination of function and habitat of an organism in an ecological system is its niche. A niche is not a property of a type or organism or species, but is its

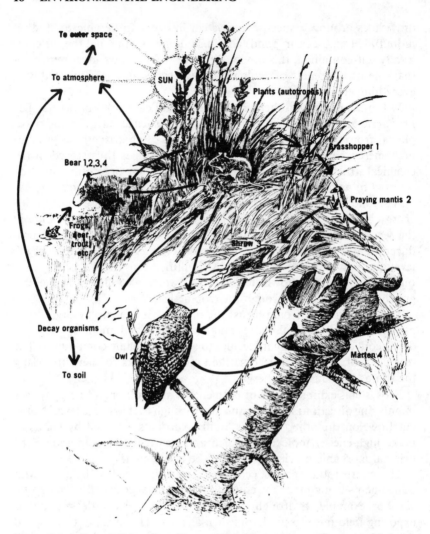

Figure 2-1 A typical land-based ecosystem. The numbers refer to trophic level, and the arrows show progressive loss of energy for life systems [From Turk, A. et al. *Environmental Science* (Philadelphia: W. B. Saunders Co., 1974).]

best accommodation with the environment. In the example above, the grass is a producer which acts as food for the field mouse, which in turn is food for the owl. If the grass were destroyed, both the mice and owls might eventually die out. But suppose there were *two* types of grass, each equally acceptable as mouse food. Now if one died out, the ecosystem would not be destroyed because the mice would still have food. This simple example demonstrates an important ecological principle: the stability of an ecosystem is proportional to the number of organisms capable of filling various niches. A jungle, for example, is a very stable ecosystem, whereas the tundra in Alaska is extremely fragile. Another fragile system is deep oceans—a fact which should be a consideration in the disposal of hazardous and toxic materials in deep ocean areas. Inland water courses tend to be fairly stable ecosystems, but certainly not totally resistant to destruction by outside forces. Other than the direct effect of toxic materials such as heavy metals and refractory organics,* the most serious effect of water pollution for inland waters is the depletion of dissolved (free) oxygen. All higher forms of aquatic life exist only in the presence of oxygen, and most desirable microbiologic life also requires oxygen. Generally, all natural streams and lakes are aerobic (containing dissolved oxygen). If a watercourse becomes anaerobic (absence of oxygen), the entire ecology changes to make the water unpleasant or unsafe.

Problems associated with pollutants which affect the dissolved oxygen levels cannot be appreciated without a fuller understanding of the concept of decomposition or biodegradation, part of the total energy transfer system of life.

BIODEGRADATION

Plant growth, or photosynthesis, can be represented by the equation

$$CO_2 + H_2O \xrightarrow[\text{\& nutrients}]{\text{sunlight}} HCOH + O_2$$

In this representation formaldehyde (HCOH) and oxygen are produced from carbon dioxide and water, with sunlight the source of energy.** If the formaldehyde and oxygen are combined and ignited, an explosion results. The energy which is released during such an explosion is stored in the carbon-hydrogen-oxygen bonds of formaldehyde.

As discussed above, plants (producers) use inorganic chemicals as nutrients and, with sunlight as a source of energy, build high-energy mol-

---

*Refractory organics are man-made organic materials, such as the pesticide DDT, which decompose very slowly in the environment.

**Of course, formaldehyde is not the end product of photosynthesis, but it is an organic molecule and happens to provide a simple equation.

ecules. The animals (consumers) eat these high-energy molecules and during their digestion process some of the energy is released and used by the animals. The release of this energy is quite rapid and the end products of digestion (excrement) consist of partially stable compounds. These compounds become food for other organisms and are thus degraded further but at a slower rate. After several such steps, very low-energy compounds are formed which can no longer be used by microorganisms for food. Plants then use these compounds to build more high-energy molecules and the process starts all over. The process is symbolically shown in Figure 2-2.

Figure 2-2   Energy loss in biodegradation.

It is important to realize that many of the organic materials responsible for water pollution enter watercourses at a high energy level. It is the biodegradation, or the gradual use of this energy, by a chain of organisms which causes many of the water pollution problems.

AEROBIC AND ANAEROBIC DECOMPOSITION

Decomposition, or biodegradation, can take place in one of two distinctly different ways: aerobic (using free oxygen) or anaerobic (in the absence of free oxygen).

The basic equation of aerobic decomposition is

$$\text{Complex Organics} + O_2 \rightarrow CO_2 + H_2O + \text{Stable Products}$$

Carbon dioxide and water are always two of the end products of aerobic decomposition. Both are stable, low in energy, and are used by plants in the process of photosynthesis. If sulfur compounds are involved in the reaction, the most stable end product is $SO_4^=$, the sulfate ion. Similarly, phosphorus ends up as $PO_4^=$, orthophosphate. Nitrogen goes through a series of increasingly stable compounds, finally ending up as nitrate. The progression is

Organic Nitrogen $\rightarrow$ NH_3 (ammonia) $\rightarrow$ NO_2^- (nitrite) $\rightarrow$ NO_3^- (nitrate)

Because of this distinctive progression, nitrogen has been in the past and to some extent is still used as an indicator of pollution.

A schematic representation of the aerobic cycle for carbon, sulfur and nitrogen compounds is shown as Figure 2-3. This figure illustrates only the basic facts, and is a gross simplification of the actual steps and mechanisms involved.

Figure 2-3　Aerobic nitrogen, carbon and sulfur cycles.

A second type of biodegradation is anaerobic, performed by a completely different set of microorganisms, to which oxygen is in fact toxic. The basic equation of anaerobic decomposition is

Complex Organics $\rightarrow$ CO_2 + CH_4 + other partially stable compounds

Note that many of the end products shown are biologically unstable. CH_4, for example, is methane, a high-energy gas commonly called marsh gas, physically stable but still able to be decomposed biologically. Nitrogen compounds stabilize only to ammonia (NH_3), and sulfur ends up as evil-smelling hydrogen sulfide (H_2S) gas. Figure 2-4 is a schematic representation of anaerobic decomposition. Note that the left half of the

Figure 2-4   Anaerobic nitrogen, carbon and sulfur cycles.

cycle, the photosynthesis by plants, is identical to the aerobic cycle in Figure 2-3.

Biologists often speak about various compounds as "hydrogen acceptors." The hydrogen atoms, torn from high-energy organic molecules, must be attached to various compounds. In aerobic decomposition oxygen serves this purpose and is thus known as the hydrogen acceptor. It accepts the hydrogen atoms to form water.

In anaerobic decomposition free oxygen is not available, and the next preferred hydrogen acceptor is nitrogen, thus forming ammonia, NH_3. If free oxygen is not available, ammonia cannot be converted to nitrites or nitrates. If nitrogen is not available, the next preferred hydrogen acceptor is sulfur, thus forming hydrogen sulfide, H_2S, the chemical responsible for the notorious rotten egg smell.

EFFECT OF POLLUTION ON STREAMS

When a high-energy organic material such as raw sewage is discharged to a stream, a number of changes occur downstream from the point of discharge. As the organics are decomposed, oxygen is used at a greater rate than before the pollution occurred, and the dissolved oxygen (DO) level drops. The rate of reaeration, or solution of oxygen from the air, also increases, but this is often not great enough to prevent a total depletion of oxygen in the stream. When this happens, the stream is said to become anaerobic. Often, however, the DO does not drop to zero, and the stream recovers without experiencing a period of anaerobiosis. Both of these situations are depicted graphically in Figure 2-5. The dip in DO is referred to as a *dissolved oxygen sag curve.*

Figure 2-5   Dissolved oxygen downstream from a source of organic pollution. Curve A depicts an oxygen sag without anaerobic conditions; curve B shows DO levels when pollution is sufficiently strong to create anaerobic conditions. D_a is the oxygen deficit in the stream after the stream has mixed with the pollutant.

The effect of a certain waste on a stream's oxygen level can be estimated mathematically. The basic assumption is that there is an oxygen balance at any point in the stream and that this level is dictated by: (1) how much oxygen is being used by the microorganisms, and (2) how much oxygen is supplied to the water through reaeration (O_2 dissolved into the water from the atmosphere). The rate of oxygen use, or oxygen depletion, can be expressed as:

$$\text{Rate of deoxygenation} = k_1' z$$

where Z = the amount of oxygen still required at any time t, or the bio-chemical demand for oxygen remaining in the water, mg/l*

k_1' = deoxygenation constant, a function of the type of waste material decomposing, temperature, etc., days^{-1}.

The value of k_1' is measured in the laboratory, as discussed in the next chapter.

Integrating the above equation yields

$$z = L_o e^{-k't}$$

where L_o is the *ultimate oxygen demand.* Since the long-term need for oxygen is L_o and the amount of oxygen still needed at any time t is Z, the amount of oxygen *used* at any time t must be

$$y = L_o - z$$

This relationship is shown in Figure 2-6. The term Y is defined later in this text as the *biochemical oxygen demand* (BOD) and expressed as

$$y = L_o (1 - e^{-k't})$$

Although y is commonly termed oxygen *demand,* it is more correctly described as the dissolved oxygen *used.* In this text we use the terms *demanded* and *used* synonymously.

The reoxygenation of a water course can be expressed as:

$$\text{Rate of reoxygenation} = k_2' D$$

where D = deficit in dissolved oxygen, or the difference between satura-tion (maximum dissolved oxygen the water can hold) and the actual DO, mg/l

k_2' = reoxygenation constant, days^{-1}.

---

*mg/l (milligrams per liter) is a common way of expressing the concentration of chemi-cals in water. It is a weight/volume measurement, with the stated milligrams of a chemical per liter of water.

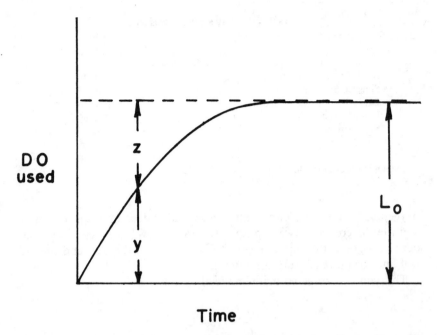

Figure 2-6   The dissolved oxygen used at any time t (y) plus what is still needed (z) is equal to the ultimate oxygen demand (L_o).

The value of k_2' is obtained by conducting a study on a stream using a tracer. If this cannot be done, a generalized formula can be used:

$$k_2' = \frac{3.9 \, v^{1/2}[(1.037)^{(T-20)}]^{1/2}}{H^{3/2}}$$

where   T = temperature of the water, °C
H = average depth of flow, m
v = mean stream velocity, m/s.

Alternatively, k_2' values may be estimated using generalized tables, such as Table 2-1.

If a stream is loaded with organic material, the simultaneous action of deoxygenation and reoxygenation forms the dissolved oxygen sag curve, first discussed by Streeter and Phelps in 1925.

The shape of the oxygen sag curve, as shown in Figure 2-5, is the result of adding the rate of oxygen use and rate of supply. If the rate of use is great, as in the stretch of stream immediately after the pollution, the DO level drops because the supply rate cannot keep up with it. As the rate of

Table 2-1.   Reaeration Constants

	k_2' at 20°Cª, (days^{-1})
Small Ponds or Backwaters	0.1–0.23
Sluggish Streams	0.23–0.35
Large Streams, Low Velocity	0.35–0.46
Large Streams, Normal Velocity	0.46–0.69
Swift Streams	0.69–1.15
Rapids	>1.15

ªFor temperatures other than 20°C, $k_{2_T}' = k_{2_{20}}' = 1.024^{T-20}$.

oxygen use decreases (fewer high-energy readily decomposable organics) and the difference between oxygen saturation level and actual DO (the deficit) is great, the supply begins to keep up with the use, and the DO will once again reach saturation levels.

This can be expressed mathematically as:

$$\frac{dD}{dt} = k_1'z - k_2'D$$

where all the terms are as defined above. The rate of change in the deficit (D) depends on the concentration of decomposable organic matter, or the need by the microorganisms for oxygen (z) and the deficit at any time t. The need for oxygen at any time t was previously expressed as:

$$z = L_o e^{-k_1't}$$

where L_o = ultimate oxygen demand, or the maximum oxygen required in mg/l. Substituting this into the above expression and integrating,

$$D = \frac{k_1'L_o}{k_2' - k_1'}(e^{-k_1't} - e^{-k_2't}) + D_a e^{-k_2't}$$

where D_a = the initial oxygen deficit, at the point of pollution, after the stream flow has mixed with the source of pollution, mg/l
         D = deficit at any time t, mg/l

Note that

$$D_a = \frac{D_s Q_s + D_p Q_p}{Q_s + Q_p}$$

where D_s = oxygen deficit in the stream directly upstream from the point of pollution, mg/l
Q_s = stream flow, m^3/s
D_p = oxygen deficit in the pollutant stream, mg/l
Q_p = flow rate of pollutant, m^3/s.

The deficit equation is often expressed in common logarithms, and since

$$e^{-k't} = 10^{-kt} \quad \text{when} \quad k = 0.434\,k'$$

then

$$D = \frac{k_1 L_o}{k_2 - k_1}(10^{-k_1 t} - 10^{-k_2 t}) + D_a(10^{-k_2 t})$$

The most serious concern of water quality is of course the point in the curve where the deficit is the greatest, or the dissolved oxygen concentration the least. By setting the $dD/dt = 0$, we can solve for the critical time as:

$$t_c = \frac{1}{k_2' - k_1'} \ln\left[\frac{k_2'}{k_1'}\left(1 - \frac{D_a(k_2' - k_1')}{k_1' L_o}\right)\right]$$

where t_c = time downstream when the dissolved oxygen is the lowest.

Example 2.1
Assume that a large stream has a reoxygenation constant k_2' of 0.4/day, a flow velocity of 5 miles per hour, and at the point of the pollution discharge the stream is saturated with oxygen at 10 mg/l. The wastewater flow rate is small compared to the stream flow, so the mixture is assumed to have a DO at saturation, and an oxygen demand of 20 mg/l. The deoxygenation constant is 0.2/day. What is the DO level 30 miles downstream?

Velocity = 5 mph, hence it takes 30/5 = 6 hr to travel 30 mi
v = 6/24 = 1/4 days
D_a = 0 since the stream is saturated

$$D = \frac{(0.2)(20)}{0.4 - 0.2}[e^{-0.2(0.25)} - e^{-0.4(0.25)}] = 20[e^{-0.05} - e^{-0.1}] = 1.0 \text{ mg/l}.$$

The DO is thus the saturation level minus the deficit, or 10 − 1.0 = 9.0 mg/l.

An actual dissolved oxygen sag curve is shown in Figure 2-7. Note that the stream goes anaerobic at about mile 3.5, recovers, then drops back to zero after receiving effluents from a city and a paper mill. Also shown on the figure is the expected DO sag if 95% of the demand for oxygen is removed from all discharges.

Stream flow is of course variable, and the critical dissolved oxygen levels can be expected to occur when the flow is the lowest. Accordingly, most state regulatory agencies base their calculations on a statistical low flow, such as a 7-day, 10-year low flow, or the seven consecutive days of lowest flow that can be expected to occur once during a 10-year interval. This is calculated by first estimating the lowest 7-day discharge for each year, assigning these rankings, m as m = 1 for the most severe (least flow) and m = n (where n is the number of years) for the least severe case. The probability of a flow equal to or more than a low flow occurring is calculated as $m/(n+1)$, and plotted versus the flow. Usually log-probability paper is used since this gives the best straight line fit. Then $m/(n+1) = 0.1$ is read as the 10-year low flow.

Example 2.2

Calculate the 10-year, 7-day low flow given the data below.

Year	Lowest flow 7 consecutive days (m^3/sec)	Ranking (m)	Lowest flow in order of severity (m^3/sec)	$\dfrac{m}{n+1}$
1965	1.2	1	0.4	0.071
1966	1.3	2	0.6	0.143
1967	0.8	3	0.6	0.214
1968	1.4	4	0.8	0.285
1969	0.6	5	0.8	0.357
1970	0.4	6	0.8	0.428
1971	0.8	7	0.9	0.500
1972	1.4	8	1.0	0.571
1973	1.2	9	1.2	0.642
1974	1.0	10	1.2	0.714
1975	0.6	11	1.3	0.785
1976	0.8	12	1.4	0.857
1977	0.9	13	1.4	0.928

These data are plotted on Figure 2-8 and the minimum 7-day, 10-year flow is read off as 0.5 m^3/sec.

Figure 2-7 Actual DO sag (going to zero; anaerobic conditions) and the projected DO sags if various types of wastewater treatment are provided (courtesy Camp Dresser & McKee, Boston, MA).

Figure 2-8   Plot of 10-year, 7-day low flows, for Example 2.2.

When the rate of oxygen usage overwhelms the rate at which the oxygen can be supplied, the stream may go anaerobic. An anaerobic stream is easily identifiable. Since oxygen is no longer around to act as the hydrogen acceptor, ammonia and hydrogen sulfide are formed, among other gases. Some of these gases will dissolve readily, but others will attach themselves as bubbles to hunks of black solid material, known as benthic deposits (or simply sludge), and buoy this material to the surface.

Anaerobic streams are thus recognized by floating sludge solids and the formation of gas which bubbles to the surface. In addition, the H_2S emitted will advertise the anaerobic condition for a considerable distance. Two other telltale signs are the color of the water, generally black, and the presence of long filamentous fungus growths which cling to rocks and gracefully wave slimy streamers downstream.

It is logical to assume that such outward changes can also be described by the effect on aquatic life. Indeed, the types and numbers of species change drastically downstream from the point of gross pollution. The increased turbidity, settled solid matter and low DO all contribute to a decrease in fish life. Fewer and fewer species of fish are able to survive, but those types of fish that do survive find that food is plentiful and they often multiply in numbers. Carp and catfish can survive in quite foul water, and can even gulp air from the surface if necessary. Trout, on the other hand, need very pure, cold water to survive and are notoriously intolerant of pollution.

The number of other types of aquatic life are also reduced. Such characters as sludge worms, bloodworms or rat-tailed maggots abound, and their numbers can be staggering—as many as 50,000 sludge worms per square foot. The variation of both the number of species and the total number of organisms downstream from a source of pollution is illustrated in Figure 2-9.

The diversity of species can be quantified by using an index, such as

Figure 2-9   The number of species and the total number of organisms downstream from a point of organic pollution.

$$\bar{d} = \sum_{i=1}^{s} \left(\frac{n_i}{n} \right) \log_2 \left(\frac{n_i}{n} \right)$$

where $\bar{d}$ = diversity index
         n_i = number of individuals in the ith species
         n = total number of individuals in all S species.

In one study, the diversity index was calculated above and below a sewage outfall, and the results were as shown in Table 2-2.

Table 2-2.   Diversity of Aquatic Organisms

Location	Diversity Index ($\bar{d}$)
Above the Outfall	2.75
Immediately Below the Outfall	0.94
Downstream	2.43
Further Downstream	3.80

It was mentioned earlier that nitrogen compounds can be used as indicators of pollution. The changes in the various forms of nitrogen are shown in Figure 2-10. The first transformation, both in aerobic and anaerobic decomposition, is the formation of ammonia, and thus the

Figure 2-10   Typical variations in nitrogen compounds downstream from a point of organic pollution.

concentration of ammonia increases as organic nitrogen decreases. Similarly, the concentration of nitrate nitrogen will finally increase to become by far the dominant form of nitrogen.

It is important to remember that the reactions of a stream to pollution outlined above occur when a rapidly decomposable organic material is wasted. The stream will react much differently to an inorganic waste, say from a metal-plating plant. If the waste is toxic to aquatic life, both the kind and total number of organisms will decrease below the outfall. The DO will not fall, and might even rise. There are many types of pollution, and a stream will react differently to each. Even more complicated is a situation where two or more wastes are involved.

EFFECT OF POLLUTION ON LAKES

The effect of pollution on lakes differs in several respects from the effect on streams. For one thing, light and temperature have significant influences on a lake and must be included in any limnological* analysis. Light is the source of energy in the photosynthetic reaction and the penetration of light into the lake water is important. This penetration is logarithmic. If, for example, at a 1-foot depth the light is 10,000 footcandles,** at 2 feet it might be 1000, at 3 feet 100, and at 4 feet only 10 footcandles. Only the top few feet of a lake generally experience light penetration, and hence all photosynthetic reactions occur in that zone.

Temperature often has a profound effect on a lake. Water is at a maximum density at 4°C (water both colder and warmer is lighter; ice, therefore, floats). Water is also a poor conductor of heat and retains heat quite well.

Lakes usually undergo a seasonal variation in water temperature. These temperature-depth relationships are illustrated in Figure 2-11. During the winter, assuming the lake does not freeze, the temperature is often constant with depth. As warmer weather approaches the top layers begin to warm up. Since water is a poor conductor of heat and warmer water is lighter, a distinct temperature gradient is formed known as *thermal stratification*. These strata are often very stable and last through the summer months. The top layer is called the *epilimnion,* the middle the *metalimnion,* and the bottom the *hypolimnion.* The inflection point in the curve is called the *thermocline.* Circulation of water occurs only

---

*Limnology is the study of lakes.
** The intensity of light is measured in footcandles.

Temperature (°C)

Temperature (°C)

Figure 2-11   Typical temperature-depth relationships in a lake.

within a zone, and thus there is only limited transfer of biological or chemical material (including dissolved oxygen) across the boundaries.

As the colder weather approaches, the top layers begin to cool, become more dense, and sink. This creates circulation throughout the lake, a condition known as *fall turnover*. Often a spring turnover also occurs.

The biochemical reactions in a natural lake can be represented schematically as in Figure 2-12. A river feeding the lake would contribute carbon, phosphorus and nitrogen, either as high-energy organics or as low-energy compounds. The *phytoplankton* or *algae* (microbial free-floating plants) take C, P and N and using sunlight as a source of energy make high-energy compounds. Algae are eaten by *zooplankton* (tiny aquatic animals), which are in turn eaten by larger aquatic life such as fish. All of these forms of life defecate, thus contributing to a pool of *dissolved organic carbon*. This pool is further fed by the death of aquatic life. Bacteria utilize dissolved organic carbon and produce CO_2, in turn used

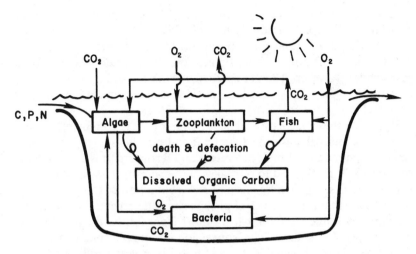

Figure 2-12   Schematic representation of lake ecology.

by the algae. Additional CO_2 is provided from the respiration of the fish and zooplankton as well as the CO_2 dissolved directly from the air.

In an unpolluted lake the supply of incoming C, P and N is sufficiently small to limit the production of algae, and the productivity of the entire ecological system is limited. But what happens when an excessive amount of C, P and N are introduced to the lake?

The plentiful supply of nutrients promotes the uncontrolled growth of algae. When the algae, along with the zooplankton and fish, die, they drop to the bottom and become another source of carbon for the bacteria. Aerobic bacteria will use all available dissolved oxygen in decomposing this material and may use enough oxygen to deplete all available DO, thus creating anaerobic conditions.

As more and more algae are produced in the epilimnion, the bacteria in the lower portions of the lake will utilize more and more oxygen, and the metalimnion might also become anaerobic. All of the aerobic biological activity would thus concentrate in the upper few feet of the lake, the epilimnion. All this activity causes turbidity, thus decreasing light penetration, and in turn limiting algal activity to the surface layers. The amount of DO contributed by the algae is therefore decreased. Eventually, when the epilimnion is also anaerobic, all aerobic aquatic life disappears and the algae, because of limited light penetration, concentrate on the surface of the lake, forming large green mats or *algal blooms*. These algae will also die, and eventually fill up the lake, producing what we now know as a peat bog.

This entire process is called *eutrophication*. It is a continually occurring natural process. But what may have taken thousands of years to occur naturally can be accomplished in only a decade if enough nutrients are introduced into a lake as a result of human activities.

It is worth repeating that eutrophication is a natural process. It is not man-made. However, it can be speeded up with the addition of nutrients, specifically phosphorus, which is considered the chemical limiting algal growth.*

Where do these nutrients originate? One source is excrement, since all human and animal wastes include C, P and N. This source, however, is small compared to synthetic detergents and fertilizers. It is estimated that of the total P discharged to our lakes, 1/2 comes from agricultural run-off, 1/4 comes from detergents and 1/4 from all other sources. It seems unfortunate that the presence of phosphates in detergents has received so much unfavorable attention when runoff from fertilized land is a much more important source. The conversion to nonphosphate detergents would be of limited value if other sources are not controlled.

It is generally believed that a P concentration between 0.01 and 0.1 mg/l is sufficient to promote accelerated eutrophication. Effluents from sewage treatment plants often contain from 5 to 10 mg/l of P. A river flowing through farm country might carry from 1 to 4 mg/l of P. In moving streams this high P concentration is not a problem since streams are continually flushed out and the algae do not have time to accumulate. Eutrophication can thus occur only in lakes, ponds, estuaries, and sometimes in very slowly moving rivers.

Incidentally, it has been estimated that only about 15% of the people in the United States should be concerned about phosphate discharges from their municipalities. About 85% of the wastewater flows into moving streams and rivers where eutrophication is not a problem.

There is also some question as to the validity of blaming phosphorus for accelerated eutrophication. Generally a P:N:C ratio of 1:16:100 is required for growth. It takes 16 parts N and 100 parts C for every part P for algae to grow. If there are more than 16 parts N and 100 parts C for every P, then N and C are said to be in excess, and thus P must be limiting growth. Although this is the case with most lakes, there are those which have very high P levels and have few problems with algal blooms. Conversely, some lakes have experienced serious algal problems while carrying extremely low P levels. In addition, recent evidence suggests that

---

*Phosphorus "limits" algal growth in that it is a constraint against unlimited growth, much like the gas pedal on your car "limits" the speed of your car. The car is able to go faster—all the components are there and could function at a higher rate—but the gas pedal is a constraint against higher speed. Dumping excess phosphorus into a lake is like tromping down on the gas pedal.

nitrogen is limiting algal growth in brackish waters such as bays and estuaries. Suffice it to say, the blame for the accelerated eutrophication of many of our lakes cannot at this time be placed on any one chemical or product. An interaction among the many pollutants involved is the probable answer.

Actual profiles in a lake for a number of parameters are shown in Figure 2-13. Based on the above discussion, it should be clear why the temperature is warmer on the top, why the DO drops to zero, and why there is high phosphorus and nitrogen concentration at the deeper areas, while the phytoplankton (algae) are all on the surface.

EFFECT OF POLLUTION ON OCEANS

Not many years ago, the oceans were considered infinite sinks; the immensity of the seas and oceans seemed impervious to assault. This is no longer the case. We now recognize seas and oceans as fragile environments and are able to measure the detrimental effect of our actions.

The water in the oceans is the most complicated chemical solution imaginable, and there is evidence that it has changed very little over millions of years. Because of this constancy, however, marine organisms have become highly specialized and intolerant to environmental change. Oceans are thus fragile ecosystems, quite susceptible to pollution.

A relief map of the ocean bottom reveals that there are two major areas of what we think of ocean; the continental shelf and the deep oceans. The continental shelf, and especially the areas near major estuaries, are the most productive in terms of food supply. These also receive the greatest pollutional load. Many major estuaries have become so badly polluted they have been closed to commercial fishing. Some large areas, such as the Baltic and Mediterranean, are also in danger of becoming permanently damaged.

Although ocean disposal of wastewater is severely restricted in the United States, many major cities all over the world still discharge all of their untreated sewage into the oceans. This is usually done by pipelines running considerable distances from the shore and discharging through diffusers to achieve maximum dilution. The controversy continues as to the wisdom of using the oceans for wastewater disposal, and what the long-term consequences might be.

CONCLUSION

The sources, causes and effects of water pollution are legion. We have discussed only a limited number. The effects of pollution on oceans or

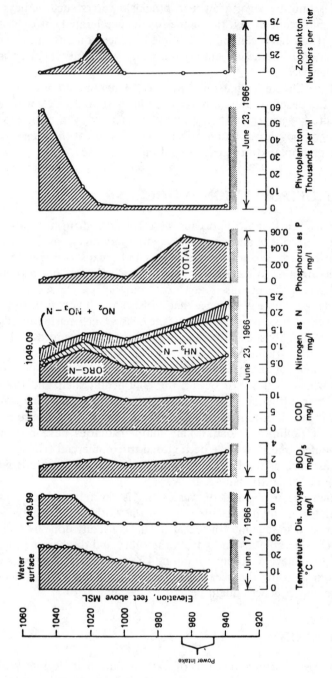

Figure 2-13   Actual water quality profiles for a water supply reservoir. (From Berthouex, P. and D. Rudd, *Strategy of Pollution Control* (New York: John Wiley & Sons, 1977).]

groundwater, or the effects of inorganic poisons such as mercury and cadmium on aquatic life also deserve attention but are ignored here because of the lack of space, and not because they are not important. The list of suggested reading contains further information on the water pollution problem.

PROBLEMS

2.1 The aerobic cycle, shown as Figure 2-2, is only for nitrogen, carbon and sulfur. Phosphorus should also have been included in the cycle since it exists as organic phosphorus in living and dead tissue, decomposes to polyphosphates (such as $(P_2O_7)^{-4}$, $P_3O_{10})^{-5}$, etc.) and finally to the inorganic orthophosphate, PO_4^{-3}. Draw a phosphorus cycle similar to Figure 2-2.

2.2 Draw typical oxygen sag curves for the following wastes:
1. potato waste (high BOD)
2. electroplating waste (Cr, Co, etc.)
3. brick waste (clay).

Assume that the stream has a DO of 5 mg/l at the point of the pollution, and a temperature of 20°C.

2.3 Some researchers have suggested that the empirical analysis of some algae has the following chemical composition (by weight):

$$C_{106}H_{181}O_{45}N_{16}P$$

Suppose an analysis of a lake water yields the following:

$$C = 62 \text{ mg/l}$$
$$N = 1.0 \text{ mg/l}$$
$$P = 0.01 \text{ mg/l}$$

Which element would be limiting the growth of algae in this lake? (Show your calculations.)

2.4 A stream feeding a lake has an average flow of 1 cubic foot per second and a phosphate concentration of 10 mg/l. The water leaving the lake has a phosphate concentration of 5 mg/l.
a. How much phosphate is deposited in the lake every year?
b. Where does this phosphorus go? (The outflow is less than the inflow, so it has to go somewhere.)

c. Would you expect the average phosphate concentration to be higher near the surface or the bottom?

d. Would you expect this lake to have accelerated eutrophication problems? Why?

2.5 Suppose you have a 55-gallon drum full of distilled water, and you add one frog weighing 0.2 pound. What is the concentration of frogs in terms of mg/l? Show all your calculations.

2.6 What could be done to maximize the silting in a reservoir?

2.7 Given a make-believe high-energy molecule H_2CSNO_3, show how it decomposes anaerobically and how these end products in turn decompose aerobically to stabilized sulfur and nitrogen compounds.

2.8 The temperature soundings for a lake are as follows:

Depth (ft)	Water Temperature (°F)
surface	80
4	80
8	60
16	40
24	40
30	40

Draw a graph of depth vs temperature and label the hypolimnion, epilimnion and thermocline.

2.9 If an industrial plant discharges solids at a rate of 5000 lb/day, and if each person contributes 0.2 lb/day, what is the population equivalent of the waste?

2.10 Draw the DO sag curves you would expect in a stream from the following wastes. Assume the stream flow equals the flow of wastewater. (Do not calculate.)

Waste	Source	Demand for oxygen, BOD (mg/l)	Suspended Solids, SS (mg/l)	Phosphorus (mg/l)
A	Dairy	2000	100	40
B	Brick mfg.	5	100	10
C	Fertilizer mfg.	25	5	200
D	Plating plant	0	100	10

2.11 Starting with nitrogenous dead organic matter, follow N around the aerobic and anaerobic cycles by writing down all the various forms of nitrogen.

2.12 Why does eutrophication rarely occur in a stream?

2.13 Name six ways you can recognize an anaerobic stream (without instrumentation).

2.14 Given a make-believe molecule $(C_2H_8)_3S$, show how it would degrade under anaerobic conditions (use as many oxygens as needed).

2.15 Suppose a stream with a velocity of 1 ft/sec, flow of 10 mgd and an ultimate BOD of 5 mg/l was hit with treated sewage at 5 mgd with an ultimate BOD (L_o) of 60 mg/l. The temperature of the stream water is 20°C, at the point of the sewage discharge the stream is 90% saturated with oxygen and the wastewater is at 30°C and has no oxygen. Measurements show the deoxygenation constant $k_1' = 0.5$ and the reoxygenation $k_2' = 0.6$ both as days^{-1}. Calculate: (a) the oxygen deficit 1 mile downstream, (b) the minimum DO (the lowest part of the sag curve), and (c) the minimum DO (or maximum deficit) if the ultimate BOD of the treatment plant's effluent was 10 mg/l. Use a computer program if possible.

2.16 An industry discharges sufficient quantities of organic wastes to depress the oxygen sag curve to 2 mg/l 5 miles downstream, at which point the DO begins to increase. This DO is too low, and the state regulatory agency wants to fine them for noncompliance with standards. One solution is to build a wastewater treatment plant. This is expensive, and so the industry looks around for other means of resolving their problem.

A salesman tells the plant engineer about freeze-dried bacteria, which come to life in water and are especially adapted to the kind of organic waste the plant is discharging. The salesman suggests that the plant engineer dump the bacteria into the river at the point of the industrial waste discharge, and that this will solve the problem.

Will it? Why or why not? Draw the DO sag curve before and after he adds the bacteria.

2.17 An industry discharges into a river with a velocity of 0.1 m/sec and a temperature of 26°C and average depth of 4 m. The ultimate oxygen demand (L_o) immediately below the effluent discharge point where the waste has mixed with the river water is 14.5 mg/l and the DO is 5.4 mg/l. Laboratory studies show that $k_1 = 0.11$ day^{-1} at 25°C. Find the point of critical DO and the minimum DO.

LIST OF SYMBOLS

D = deficit in dissolved oxygen, mg/l
D_a = initial dissolved oxygen deficit, mg/l
$\bar{d}$ = diversity index
H = depth of stream flow, m
k_1 = deoxygenation constant, $(\log_{10})$ sec^{-1}
k_1' = deoxygenation constant, $(\log_e)$ sec^{-1}
k_2 = reoxygenation constant, $(\log_{10})$ sec^{-1}
k_2' = reoxygenation constant, $(\log_e)$ sec^{-1}
L_o = ultimate biochemical oxygen demand, mg/l
m = rank assigned to low flows
n = number of individuals in all S species
n = number of years in low flow records
n_i = number of individuals in species i
T = temperature, °C
t = time, sec
t_c = critical time, time when the minimum DO occurs, sec
v = velocity, m/sec
y = oxygen used, mg/l
z = oxygen required for decomposition, mg/l

SUGGESTED READING

Mitchell, R. *Water Pollution Microbiology* (New York: John Wiley & Sons, 1972).

Warren, C. E. *Biology and Water Pollution Control* (Philadelphia, PA: W. B. Saunders Co., 1971).

U.S. Department of the Interior. *The Practice of Water Pollution Biology* (1969).

McGaughey, P. H. *Engineering Management of Water Quality* (New York: McGraw-Hill, 1968).

Kittrell, F. W. U.S. Department of the Interior. *A Practical Guide to Water Quality Studies of Streams* (1969).

Masters, G. M. *Introduction to Environmental Science and Technology* (New York: John Wiley & Sons, 1974).

Turk, A. et al. *Environmental Science* (Philadelphia, PA: W. B. Saunders Co., 1974).

Pettyjohn, W. A., Ed. *Water Quality in a Stressed Environment* (Minneapolis, MN: Burgess, 1972).

Nemerow, N. L. *Scientific Stream Pollution Analysis* (New York: McGraw-Hill, 1974).

McCaull, J., and J. Crossland. *Water Pollution* (New York: Harcourt Brace Jovanovich, 1974).

Chanlett, E. *Environmental Protection* (New York: McGraw-Hill, 1973).

Velz, C. J. *Applied Stream Sanitation* (New York: John Wiley & Sons, 1970).

Chapter 3

Measurement of Water Quality

Quantitative measurements of pollutants are obviously necessary before water pollution can be controlled. Measurement of these pollutants is, however, fraught with difficulties.

The first problem is that the specific materials responsible for the pollution are sometimes not known. The second difficulty is that these pollutants are generally at low concentrations, and very accurate methods of detection are therefore required.

Only a few of the many analytical tests available to measure water pollution are discussed in this chapter. A complete volume of analytical techniques used in water and wastewater engineering is compiled as *Standard Methods*.[1] This volume, now in its 15th edition, is the result of a need for standardizing test techniques. It is considered definitive in its field and has the weight of legal authority.

Many of the pollutants are measured in terms of milligrams of the substance per liter of water (mg/l). This is a weight/volume measurement. In many older publications pollutants are measured as parts per million (ppm), a weight/weight parameter.* If the liquid involved is water these two units are identical, since 1 milliliter of water weighs 1 gram. Because of the possibility of some wastes not having the specific gravity of water, the ppm measure has been scrapped in favor of mg/l.

A third commonly used parameter is percent, a weight/weight relationship. Obviously 10,000 ppm = 1 percent and this is equal to 10,000 mg/l only if 1 ml = 1 g.

---

[1] American Public Health Association. *Standard Methods for the Examination of Water and Wastewater,* 15th ed., WPCF, AWWA, (1980).

* In air pollution, however, ppm is in volume/volume.

SAMPLING

Some tests require the measurement to be conducted in the stream since the process of obtaining a sample may change the measurement. For example, if it is necessary to measure the dissolved oxygen in a stream, the measurement should be conducted right in the stream, or the sample must be extracted with great care to assure that no transfer of oxygen between the air and water (in or out) has occurred.

Most tests can be performed on a water sample taken from the stream. The process by which that sample is obtained, however, can greatly influence the result.

There are basically three types of samples:

1. grab
2. composite
3. flow weighed composite

The grab, as the name implies, simply measures a point. Its value is that it represents accurately the water quality at the moment of sampling, but obviously says nothing about the quality before or after the sampling.

The composite sample is obtained by taking a series of grab samples and mixing them together. The flow weighed composite is obtained by taking each sample so that the volume of the sample is proportional to the flow at that time. The last method is especially useful when daily loadings to wastewater treatment plants are calculated.

Whatever the technique or method, however, it is necessary to recognize that the analysis can only be as accurate as the sample, and often the sampling methodology is far more sloppy than the analytical determination.

DISSOLVED OXYGEN

Probably the most important measure of water quality is the dissolved oxygen. Oxygen, although poorly soluble in water, is fundamental to aquatic life. Without free dissolved oxygen, streams and lakes become uninhabitable to most desirable aquatic life. Yet, the maximum oxygen that can possibly be dissolved in water at normal temperatures is about 9 mg/l, and this saturation value decreases rapidly with increasing water temperature, as shown in Table 3-1. The balance between saturation and depletion is therefore tenuous.

The amount of oxygen dissolved in water is usually measured either by

Table 3-1.   Solubility of Oxygen

Temperature of Water (°C)	Saturation Concentration of Oxygen in Water (mg/l)
0	14.6
2	13.8
4	13.1
6	12.5
8	11.9
10	11.3
12	10.8
14	10.4
16	10.0
18	9.5
20	9.2
22	8.8
24	8.5
26	8.2
28	8.0
30	7.6

an oxygen probe or the old standard wet technique, the Winkler Dissolved Oxygen Test. The Winkler test for dissolved oxygen, developed more than 80 years ago, is the standard to which all other methods are compared.

Chemically simplified reactions in the Winkler test are as follows:

1. Manganese ions added to the samples combine with the available oxygen

$$Mn^{++} + O_2 \rightarrow MnO_2 \downarrow$$

forming a precipitate.

2. Iodide ions are added, and the manganous oxide reacts with the iodide ions to form iodine

$$MnO_2 + 2I^- + 4H^+ \rightarrow Mn^{++} + I_2 + 2H_2O$$

3. The quantity of iodine is measured by titrating with sodium thiosulfate, the reaction being

$$I_2 + 2S_2O_3^= \rightarrow S_4O_6^= + 2I^-$$

Note that all of the dissolved oxygen combines with Mn^{++}, so that the quantity of MnO_2 is directly proportional to the oxygen in solution. Similarly, the amount of iodine is directly proportional to the manganous oxide available to oxidize the iodide. Although the titration measures iodine, the quantity of iodine is thereby directly related to the original concentration of oxygen.

The Winkler test has obvious disadvantages, such as chemical interferences and the necessity to either carry a wet laboratory to the field or bring the samples to the laboratory and risk the loss (or gain) of oxygen during transport. All of these disadvantages are overcome by using a dissolved oxygen electrode, often called a probe.

The simplest (and historically the first) probe is shown in Figure 3-1. The principle of operation is that of a galvanic cell. If lead and silver electrodes are put in an electrolyte solution with a microammeter between, the reaction at the lead electrode would be

$$Pb + 2OH^- \rightarrow PbO + H_2O + 2e^-$$

Figure 3-1   Schematic of a galvanic cell, oxygen probe.

At the lead electrode, electrons are liberated which travel through the microammeter to the silver electrode where the following reaction takes place:

$$2e^- + \tfrac{1}{2}O_2 + H_2O \rightarrow 2OH^-$$

The reaction would not go unless free dissolved oxygen is available, and the microammeter would not register any current. The trick is to construct and calibrate a meter in such a manner that the electricity recorded is proportional to the concentration of oxygen in the electrolyte solution.

In the commercial models the electrodes are insulated from each other with nonconducting plastic and are covered with a permeable membrane with a few drops of an electrolyte between the membrane and electrodes. The amount of oxygen that travels through the membrane is proportional to the DO concentration. A high DO in the water creates a strong push to get through the membrane, while a low DO would force only limited O_2 through to participate in the reaction and thereby create electrical current. Thus the current registered is proportional to the oxygen level in solution.

BIOCHEMICAL OXYGEN DEMAND

Perhaps even more important than the determination of dissolved oxygen is the measurement of the rate at which this oxygen is used. A very low rate of use would indicate either clean water or that the available microorganisms are uninterested in consuming the available organics. A third possibility is that the microorganisms are dead or dying. (Nothing decreases oxygen consumption by aquatic microorganisms quite so well as a healthy slug of arsenic.)

The rate of oxygen use is commonly referred to as *biochemical oxygen demand* (BOD). It is important to understand that BOD is not a measure of some specific pollutant. Rather, it is a measure of the amount of oxygen required by bacteria and other microorganisms while stabilizing decomposable organic matter.

The BOD test was first used for measuring the oxygen consumption in a stream by filling two bottles with stream water, measuring the DO in one and placing the other in the stream. In a few days the second bottle was retrieved and the DO measured. The difference in the oxygen levels was the BOD, or oxygen demand, in milligrams of oxygen used per liter of sample.

This test had the advantage of being very specific for the stream in question since the water in the bottle is subjected to the same environmental factors as the water in the stream, and thus the result was an accurate measure of DO usage in that stream. It was impossible, however, to compare the results in different streams, since three very important variables were not constant: temperature, time and light.

Temperature has a pronounced effect on oxygen uptake (usage), with metabolic activity increasing significantly at higher temperatures. The time alloted for the test is also important, since the amount of oxygen used increases with time. Light is also an important variable since most natural waters contain algae and oxygen can be replenished in the bottle if light is available. Different amounts of light would thus affect the final oxygen concentration.

The BOD test was finally standardized by requiring the test to be run in the dark at 20°C for five days. This is defined as five-day BOD, or BOD_5, or the oxygen used in the first five days. Although there appear to be some substantial scientific reasons why five days was chosen, it has been suggested that the possibility of preparing the samples on a Monday and taking them out on Friday, thus leaving the weekend free, was not the least important of these reasons.

The BOD test is almost universally run using a standard BOD bottle (about 300 ml volume) as shown in Figure 3-2. It is of course also possible to have a 2-day, 10-day, or any other day BOD. One measure used in some cases is *ultimate BOD* or the O_2 demand after a very long time.

Figure 3-2   A BOD bottle.

If we measure the oxygen in several samples every day for five days, we may obtain curves such as Figure 3-3. Referring to this figure, sample A had an initial DO of 8 mg/l, and in five days this has dropped to 2 mg/l. The BOD therefore is $8 - 2 = 6$ mg/l.

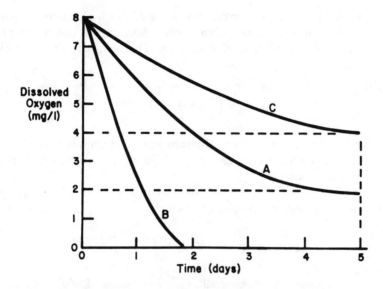

Figure 3-3   Typical oxygen uptake (use) curves in a BOD test.

Sample B also had an initial DO of 8 mg/l, but the oxygen was used up so fast that it dropped to zero. If after five days we measure zero DO, we know that the BOD of sample B was more than $8 - 0 = 8$ mg/l, but we don't know how much more since the organisms might have used more DO if it were available. For samples containing more than about 8 mg/l, dilution of the sample is therefore necessary.

Suppose sample C shown on the graph is really sample B diluted by $1:10$. The BOD of sample B is therefore

$$\frac{8 - 4}{0.1} = 40 \text{ mg/l}$$

It is possible to measure the BOD of any organic material (e.g., sugar) and thus estimate its influence on a stream, even though the material in its original state might not contain the necessary organisms. *Seeding* is a process in which the microorganisms which create the oxygen intake are added to the BOD bottle.

Suppose we used the water previously described by the A curve as seed water since it obviously contains microorganisms (it has a 5-day BOD of 6 mg/l). We now put 100 ml of an unknown solution into a bottle and add 200 ml of seed water, thus filling the 300-ml bottle. Assuming that the initial DO of this mixture was 8 mg/l and the final DO was 1 mg/l, the total oxygen used was 7 mg/l. But some of this was due to the seed water, since it also has a BOD, and only a portion was due to the decomposition of the unknown material. The DO uptake due to the seed water was

$$6 \times \left(\frac{2}{3}\right) = 4 \text{ mg/l}$$

since only 2/3 of the bottle was the seed water, which has a BOD of 6 mg/l. The remaining oxygen uptake ($7-4=3$ mg/l) must have been due to the unknown material.

If the seeding and dilution methods are combined, the following general formula is used to calculate the BOD:

$$\text{BOD (mg/l)} = \frac{(I-F) - (I'-F')(X/Y)}{D}$$

where
    I = initial DO of bottle with sample and seeded dilution water
    F = final DO of bottle with sample and seeded dilution water
    I' = initial DO of bottle with seeded dilution water
    F' = final DO of bottle with seeded dilution water
    X = ml seeded dilution water in sample bottle
    Y = ml in bottle with only seeded dilution water
    D = dilution of sample.

Example 3.1

Calculate the BOD_5, given the following data:
    Temperature of sample, 20°C
    Initial DO is saturation
    Dilution is 1:30, with seeded dilution water
    Final DO of seeded dilution water is 8 mg/l
    Final DO bottle with sample and seeded dilution water is 2 mg/l
    Volume of BOD bottle is 300 ml.

From Table 3-1, at 20°C saturation is 9.2 mg/l; hence this is the initial DO. Since the BOD bottle contains 300 ml, a 1:30 solution would have 10 ml of sample and 290 ml of seeded dilution water,

$$BOD_5 = \frac{(9.2-2) - (9.2-8)(290/300)}{0.033} = 181 \text{ mg/l}$$

If, instead of stopping the test after five days, we allowed the reactions to proceed and measured the DO each day, we might get a curve like Figure 3-4. Note that some time after five days the curve takes a sudden jump. This is due to the exertion of oxygen demand by the micro-

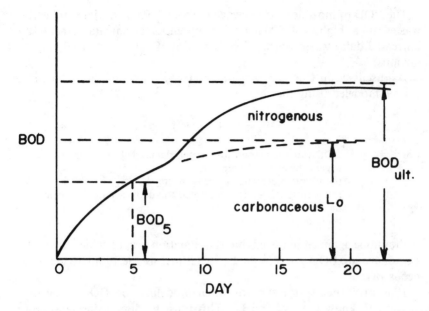

Figure 3-4   Long-term BOD. Note that BOD_{ult} includes nitrogenous as well as the ultimate carbonaceous BOD (L_o).

organisms which decompose nitrogenous organics and which are con-verting these to the stable nitrate, NO_3^-. The curve is thus divided into nitrogenous and carbonaceous BOD areas. Note also the definition of the ultimate BOD.*

It is important to remember that BOD is a measure of oxygen use, or potential use. An effluent with a high BOD can be harmful to a stream if the oxygen consumption is great enough to cause anaerobic conditions. Obviously, a small trickle going into a great river will have negligible effect, regardless of the mg/l of BOD involved. Similarly, a large flow into a small stream can seriously affect the stream even though the BOD might be low. Accordingly, engineers often talk of "pounds of BOD," a value calculated by multiplying the concentration by the flow rate, with a conversion factor, so that

$$\text{lb BOD/day} = [\text{mg/l BOD}] \times \left[\begin{array}{c} \text{flow in million} \\ \text{gallons per day} \end{array} \right] \times 8.34$$

---

*This, however, is different from the definition of ultimate BOD on p. 26. The latter would more accurately be termed *ultimate carbonaceous BOD*.

The BOD of most domestic sewage is about 250 mg/l. Many industrial wastes run as high as 30,000 mg/l. The potential detrimental effect of an untreated dairy waste which might have a BOD of 20,000 mg/l is quite obvious.

As discussed in Chapter 2, the carbonaceous part of the BOD curve can be modeled using the equation

$$y = L_0 (1 - e^{-k_1't})$$

where y = BOD_t, or the amount of oxygen demanded by the micro-organisms at some time t, mg/l
L_0 = the ultimate demand for oxygen, mg/l
k_1' = rate constant, previously termed the deoxygenation constant, days^{-1}
t = time, days.

It should be again emphasized that this equation is applicable only to the carbonaceous BOD curve, and L_0 is defined as the ultimate carbonaceous BOD.

It is often necessary, such as when modeling the DO profile in a stream, to know both k_1' and L_0. This must be done using laboratory BOD tests.

There are a number of techniques for calculating k_1' and L_0. One of the simplest is a method devised by Thomas.[2] Starting with the equation

$$y = L_0 (1 - 10^{-k_1't})$$

we can rearrange it to read

$$\left(\frac{t}{y}\right)^{1/3} = (2.3 k_1 L_0)^{-1/3} + \left(\frac{k_1^{2/3}}{3.43 L_0^{1/3}}\right) t$$

This equation is in the form of a straight line

$$x = a + bt$$

where x = $(t/y)^{1/3}$
a = $(2.3 k L_0)^{-1/3}$
b = $k_1^{2/3}/(3.43 L_0^{1/3})$

Thus plotting x vs t, the slope (b) and intercept (a) can be obtained, and

---

[2]Thomas, H. A. Jr. "Graphical Determination of BOD Curve Constants," *Water and Sewer Works,* 97:123 (1950).

$$k_1 = 2.61 \ (b/a)$$

$$L_o = 1/(2.3 \, k_1 a^3)$$

Example 3.2

The BOD vs time data for the first five days of a BOD test were obtained as follows:

time (*days*)	BOD, *y* (*mg/l*)
2	10
4	16
6	20

Calculate the k_1 and L_o.

The $(t/y)^{1/3}$ values 0.585, 0.630 and 0.669 are plotted as shown in Figure 3-5. The intercept $a = 0.545$ and the slope $b = 0.021$. Thus

$$k_1 = 2.61 \left(\frac{0.021}{0.545} \right) = 0.10 \ \text{day}^{-1}$$

$$(\text{note: } k_1' = 0.23 \ \text{day}^{-1})$$

$$L_o = \frac{1}{2.3(0.10)(0.545)^3} = 26.8 \ \text{mg/l}$$

For streams and rivers with travel times greater than about five days, the ultimate demand for oxygen must include the nitrogenous demand. Although the use of BOD_{ult} (carbonaceous plus nitrogenous) in dissolved oxygen sag calculations is not strictly accurate, it is often assumed that the ultimate BOD can be calculated as

$$BOD_{ult} = a(BOD_5) + b(KN)$$

where   KN = Kjeldahl nitrogen (organic plus ammonia mg/l)
a and b = constants.

The state of North Carolina, for example, uses $a = 1.2$ and $b = 4.0$ for calculating the ultimate BOD, which is then substituted for L_o in the dissolved oxygen sag equation.

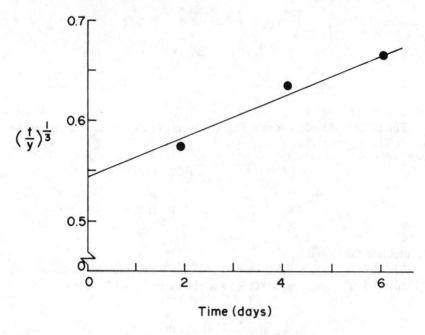

Figure 3-5   Calculation of k_1 and L_o for Example 3.2

CHEMICAL OXYGEN DEMAND

Among the many drawbacks of the BOD test the most important is that it takes five days to run. If the organics were oxidized chemically instead of biologically, the test could be shortened considerably. Such oxidation is accomplished with the chemical oxygen demand (COD) test. Because nearly all organics are oxidized in the COD test and only some are decomposed during the BOD test, COD values are always higher than BOD values. One example of this is wood pulping wastes where compounds such as cellulose are easily oxidized chemically (high COD) but are very slow to decompose biologically (low BOD).

Potassium dichromate is generally used as an oxidizing agent. It is an inexpensive compound which is available in very pure form. A known amount of this compound is added to a measured amount of sample and the mixture boiled. The reaction, in unbalanced form is

$$C_X H_Y O_Z + Cr_2 O_7^= + H^+ \xrightarrow{\Delta} CO_2 + H_2O + Cr^{+++}$$
(organic)   (dichromate)

After boiling with an acid, the excess dichromate (not used for oxidizing) is measured by adding a reducing agent, usually ferrous ammonium sulfate. The difference between the chromate originally added and the chromate remaining is the chromate used for oxidizing the organics. The more chromate used, the more organics were in the sample, and hence the higher the COD.

TOTAL ORGANIC CARBON

Since the ultimate oxidation of organic carbon is to CO_2, the total combustion of a sample will yield some significant information on the amount of organic carbon present in a wastewater sample. Without elaboration, this is done by allowing a little bit of the sample to be burned in a combustion tube and measuring the amount of CO_2 emitted. This test is not widely used at present, mainly because of the expensive instrumentation required. It has significant advantages over the BOD and COD tests, however, and will undoubtedly be more widely used in the future.

TURBIDITY

Water which is not clear but is "dirty," in the sense that light transmission is inhibited, is known as turbid water. Turbidity can be caused by many materials, some of which are discussed in Chapter 2. In the treatment of water for drinking purposes, turbidity is of great importance first because of the aesthetic considerations, and second because pathogenic organisms can hide on (or in) the tiny colloidal particles.

The standard method of measuring turbidity is with the Jackson Candle Turbidimeter first developed in 1900. It consists of a long flat-bottomed glass tube under which a candle is placed. The turbid water is poured into the glass tube until the outline of the flame is no longer visible. Obviously, with very turbid water, only a little is required. The centimeters of water in the tube are then measured and compared to the standard turbidity unit, which is

$$1 \text{ mg/l } SiO_2 = 1 \text{ unit of turbidity}$$

Recent years have seen the development of electronic devices which measure light scatter or transmittance. These have completely replaced the antiquated Jackson Candle in the laboratory.

COLOR AND ODOR

Color and odor are both important measurements in water treatment. Along with turbidity they are called the *physical* parameters of drinking water quality. Color and odor are important from the standpoint of aesthetics. If a water looks colored or smells bad, people will instinctively avoid using it, even though it might be perfectly safe from the public health aspect. Both color and odor can be and often are caused by organic substances such as algae or humic compounds.

Color is measured by comparison with standards. Colored water made with potassium chloroplatinate when tinted with cobalt chloride closely resembles the color of many natural waters. Where multicolored industrial wastes are involved, such color measurement is meaningless.

Odor is measured by successive dilutions of the sample with odor-free water until the odor is no longer detectable. This test is obviously subjective and depends entirely on the olfactory senses of the tester.

pH

The pH of a solution is a measure of hydrogen ion concentration. An abundance of hydrogen ions makes a solution acid, while a dearth of H^+ ions makes it basic. A basic solution has, instead, an abundance of hydroxide ions OH^-.

Water molecules HOH (commonly written H_2O) have the ability to dissociate, or ionize, very slightly. In a perfectly neutral water (neither acidic nor basic) equal concentrations of H^+ and OH^- are present.

If the concentration of both H^+ and OH^- ions is measured in moles (molecular weights in grams) per liter of water, the product of H^+ and OH^- concentrations is always constant.

$$[H^+] \times [OH^-] = K_w = 1 \times 10^{-14}$$

where K_w is the *ion-product constant of water.** If the hydrogen ion concentration is 10^{-4}, the OH^- concentration must be $10^{-14}/10^{-4} = 10^{-10}$. Since 10^{-10} is smaller than 10^{-4}, the solution is acidic.

The hydrogen ion concentration is so important in aqueous solutions that an easier method of expressing it has been devised. Instead of speak-

---

*The use of square brackets around a chemical symbol is shorthand for denoting the concentration of the chemical.

ing in terms of moles per liter, we can take the negative logarithm of $[H^+]$ so that

$$pH = -\log [H^+] = \log \frac{1}{[H^+]}$$

The pH value is the negative power to which 10 must be raised to equal the hydrogen ion concentration, or

$$[H^+] = 10^{-pH}$$

For a neutral solution $[H^+]$ is 10^{-7}, or pH = 7. Obviously for greater hydrogen ion concentrations the pH is lower, and for greater hydroxide ion concentrations (or lower hydrogen ion concentrations since the products, as you recall, are constant) the pH drops. The pH range is from 0 (very acidic) to 14 (very alkaline).

The measurement of pH is now almost universally by electronic means. Electrodes which are sensitive to hydrogen ion concentration (strictly speaking, the hydrogen ion activity) convert the signal to electrical current.

pH is important in almost all phases of water and wastewater treatment. Aquatic organisms are sensitive to pH changes, and biological treatment requires either pH control or monitoring. In water treatment pH is important in ensuring proper chemical treatment as well as in disinfection and corrosion control. Mine drainage often involves the formation of sulfuric acid (high H^+ concentration) which is extremely detrimental to aquatic life.

SOLIDS

One of the main problems with wastewater treatment is that so much of the wastewater is actually solid. The separation of these solids from the water is in fact one of the primary objectives of treatment.

Strictly speaking, in wastewater anything other than water would be classified as a solid. The usual definition of solids, however, is the residue on evaporation at 103°C (slightly higher than the boiling point of water). The solids thus measured are known as *total solids*. Total solids can be divided into two fractions: the *dissolved solids* and the *suspended solids*. If we put a teaspoonful of common table salt in a glass of water, the salt will dissolve. The water will not look any different, but the salt will remain behind if we evaporate the water. A spoonful of sand, however, will not dissolve and will remain as sand drains in the water. The salt is an example of dissolved solids while the sand would be measured as a suspended solid.

The separation of suspended solids from dissolved solids is by means of a special crucible, called a *Gooch crucible*. As shown in Figure 3-6 the Gooch crucible has holes on the bottom on which a glass fiber filter is placed. The sample is then drawn through the crucible with the aid of a

Figure 3-6   The Gooch crucible for suspended solids (with filter) and the evaporating dish for total solids.

vacuum. The suspended material is retained on the filter while the dissolved fraction passes through. If the initial dry weight of the crucible and filter are known, the subtraction of this from the total weight of crucible, filter and the dried solids caught on the filter yields the weight of suspended solids, expressed as mg/l.

Solids can be classified in another way: those which are volatilized at a high temperature and those which are not. The former are known as *volatile solids,* the latter as *fixed solids.* Usually volatile solids are organic. Obviously, at 600°C, the temperature at which the combustion takes place, some of the inorganics are decomposed and volatilized, but this is not considered a serious drawback.

The relationship between the total solids and the total volatile solids can best be illustrated by an example.

Example 3.3

Given the following data:

- weight of dish (such as shown in Figure 3-6) = 48.6212 g
- 100 ml of sample is placed in dish and evaporated. Weight of dish and dry solids = 48.6432 g
- Dish is placed in 600°C furnace, then cooled. Weight = 48.6300 g.

Find the total, the volatile and fixed solids.

$$\text{Total Solids} = \frac{(\text{dish} + \text{dry solids}) - (\text{dish})}{\text{volume of sample}}$$

$$= \frac{(48.6432) - (48.6212)}{100}$$

$$= 220 \times 10^{-6} \, \text{g/ml}$$

$$= 220 \times 10^{-3} \, \text{mg/ml}$$

$$= 220 \, \text{mg/l}$$

$$\text{Fixed Solids} = \frac{(\text{dish} + \text{unburned solids}) - (\text{dish})}{\text{volume of sample}}$$

$$= \frac{(48.6300) - (48.6212)}{100}$$

$$= 88 \, \text{mg/l}$$

$$\text{Volatile Solids} = \text{Total Solids} - \text{Total Fixed Solids}$$

$$= 220 - 88 = 132 \, \text{mg/l}$$

It is often necessary to measure the volatile fraction of suspended material, since this is a quick (if gross) measure of the amount of micro-organisms present. The *volatile suspended solids* are determined by simply burning the Gooch crucible and weighing again. The loss in weight is interpreted as volatile suspended solids.

NITROGEN

Recall from Chapter 2 that nitrogen is an important element in biological reactions. Nitrogen can be tied up in high-energy compounds such as amino acids and amines. In this form the nitrogen is known as organic nitrogen. One of the intermediate compounds formed during biological metabolism is ammonia. Together with organic nitrogen, ammonia is considered an indicator of recent pollution. These two forms of nitrogen are often combined in one measure, known as *Kjeldahl nitrogen,* after the scientist who first suggested the analytical procedure.

Aerobic decomposition eventually leads to the nitrite (NO_2^-) and finally nitrate (NO_3^-) nitrogen forms. A high-nitrate and low-ammonia nitrogen therefore suggests that pollution has occurred, but quite some time ago.

All of the above forms of nitrogen can be measured analytically by *colorimetric techniques.* The basic idea of colorimetry is that the ion in question will combine with some compound and form a color, and that the intensity of this color is proportional to the original concentration of the ion. For example, ammonia can be measured by adding a compound called Nessler reagent to the unknown sample. This reagent is a solution of potassium mercuric iodide, K_2HgI_4, and combines with ammonium ions to form a yellow-brown colloid. Since Nessler reagent is in excess, the amount of colloid formed is proportional to the concentration of ammonium ions in the sample.

The color is measured either by visual comparison (in long tubes called Nessler tubes) or photometrically. The basic workings of a photometer, illustrated in Figure 3-7, consist of a light source, a filter, the sample and a photo cell. The filter allows only certain wavelengths of light to pass

| Light Source | Filter | Sample | Photocell | Ammeter |

Figure 3-7   Elements of a filter photometer.

through, thus lessening interferences and increasing the sensitivity of the photocell, which converts light energy to electrical current. An intense color will allow only a limited amount of light to pass through, and thus create little current. On the other hand, a sample containing very little of the chemical in question will be clear, allow almost all of the light to pass through, and set up a substantial current. If the color intensity (and hence light absorbance) is directly proportional to the concentration of the unknown ion, the color formed is said to obey Beer's law.

A photometer can be used to measure ammonia concentration by measuring the absorbance of light by samples containing known ammonia concentration and comparing the absorbance of the unknown sample to these standards.

Example 3.4

Several known samples and an unknown sample were treated with Nessler reagent and the color measured with a photometer. Find the ammonia concentration of the unknown sample.

Sample	% Absorbance
Standards: 0 mg/l ammonia (Distilled water)	0
1 mg/l ammonia	6
2 mg/l ammonia	12
3 mg/l ammonia	18
4 mg/l ammonia	24
Unknown sample	15

A plot (Figure 3-8) of ammonia concentration of the standards vs percent absorbance results in a straight line (Beer's Law is adhered to). We can then enter this chart at 15% absorbance (the unknown) and read the concentration of ammonia in our unknown as 2.5 mg/l.

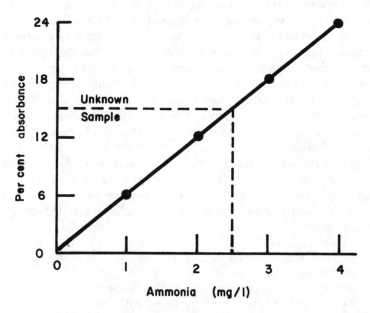

Figure 3-8   Calculation using colorimetric standards.

PHOSPHATES

The importance of phosphorus compounds in the aquatic environment is discussed in Chapter 2. Phosphorus in wastewater can be either inorganic or organic. Although the greatest single source of inorganic phosphorus is synthetic detergents, organic phosphorus is found in food and human waste as well. All phosphates in nature will, by biological action, eventually revert to inorganic forms to be again used by the plants in making high-energy material.

Ever since phosphates were indicted as one culprit in lake eutrophication, the measurement of total phosphate has assumed considerable importance. Total phosphates can be measured by first boiling the sample in acid solution, which converts all the phosphates to the inorganic forms. From that point the test is colorimetric, using a chemical which when combined with phosphates produces a color directly proportional to the phosphate concentration.

BACTERIOLOGICAL MEASUREMENTS

From the public health standpoint the bacteriological quality of water is as important as the chemical quality. A number of diseases can be transmitted by water, among them typhoid and cholera. However, it is one thing to declare that water must not be contaminated by pathogens (disease-causing organisms) and another to determine the existence of these organisms. First, there are many pathogens. Each has a specific detection procedure and must be screened individually. Second, the concentration of these organisms can be so small as to make their detection impossible. It is a perfect example of the proverbial needle in a haystack. And yet only one or two organisms in the water might be sufficient to cause an infection.

How then can we measure for bacteriological quality? The answer lies in the concept of indicator organisms. The indicator most often used is a group of microbes called *coliforms* which are organisms normal to the digestive tracts of warm-blooded animals. In addition to that attribute, coliforms are:

- plentiful, hence not difficult to find
- easily detected with a simple test
- generally harmless except in unusual circumstances
- hardy, surviving longer than most known pathogens.

Coliforms have thus become universal indicator organisms. But the presence of coliforms does not prove the presence of pathogens. If a large number of coliforms are present, there is a good chance of recent pollution by wastes from warm-blooded animals, and therefore the water *may* contain pathogenic organisms.

This last point should be emphasized. The presence of coliforms does not mean that there are pathogens in the water. It simply means that there *might* be. A high coliform count is thus suspicious and the water should not be consumed (although it may be perfectly safe).

There are three ways of measuring for coliforms. The simplest is to filter a sample through a sterile filter, thus capturing any coliforms. The filter is then placed in a petri dish containing a sterile agar which soaks into the filter and promotes the growth of coliforms while inhibiting other organisms. After 24 or 48 hr of incubation the number of shiny black dots, indicating coliform colonies, is counted. If we know how many milliliters were poured through the filter, the concentration of coliforms can be expressed as coliforms/ml. The equipment used for such filter tests is shown in Figure 3-9.

The second method of measuring for coliforms is called the most probable number (MPN), a test based on the fact that in a lactose broth coliforms will produce gas and make the broth cloudy. The production of gas is detected by placing a small tube upside down inside a larger tube (Figure 3-10) so as not to have air bubbles in the smaller tube. After incubation, if gas is produced, some of it will become trapped in the smaller tube and this, along with a cloudy broth, will indicate that the tube had been inoculated with at least one coliform.

And here is the trouble. One coliform can cause a positive tube just as easily as a million coliforms can. Hence it is not possible to ascertain the concentration from just one tube. We get around this problem by inoculating a series of tubes with various quantities of sample, the reasoning being that a 10-ml sample would be more likely to contain a coliform than a 1-ml sample.

For example, if we take three different inoculation amounts, 10, 1 and 0.1 ml of sample, and inoculate three tubes with each amount, after incubation we might have an array such as follows. The plus signs indicate a positive test (cloudy broth with gas formation) and the minus sign represent tubes where no coliforms were found.

Amount of sample, ml put in test tube	Tube Number		
	1	2	3
10	+	+	+
1	−	+	−
0.1	−	−	+

Figure 3-9   Millipore Filter apparatus for the measurement of coliforms (courtesy Millipore Inc.).

1.   Put membrane filter on filter stand.
2.   Replace funnel.
3.   Pour in measured amount of sample.
4.   Turn on vacuum.
5.   Remove membrane filter.
6.   Center membrane filter on pad in petri dish containing agar.
7.   Place petri dish in oven at 35°C.
8.   After 8 hr, count the coliform colonies.

Figure 3-10   Test tubes used for MPN coliform test.

Based on these data we would suspect that there is at least 1 coliform per 10 ml, but we still have no firm number.

The solution to this dilemma lies in statistics. It can be proven statistically that such an array of positive and negative results will occur most probably if the coliform concentration was 75 coli/100 ml. A higher concentration would most probably have resulted in more positive tubes while a lower concentration would most probably have resulted in more negative tubes. This is how MPN is established.

A third way of measuring coliforms is by a proprietary device called a "Coli-Count." A sterile pad with all the necessary nutrients is dipped into the water sample, incubated, and the colonies counted. The pad is designed to adsorb exactly 1 ml of sample water so that the colonies counted give a coliform concentration per ml.

VIRUSES

Because of their minute size and extremely low concentrations, and the need to culture them on living tissues, pathogenic (or animal) viruses are fiendishly difficult to measure. Because of this problem, there are as yet no standards for viral quality of water supplies as there are for pathogenic bacteria.

One possible method of overcoming this difficulty is to use an indicator organism, much like the coliform group is used as an indicator for bacterial contamination. This can be done by using a bacteriophage, or a virus that attacks only a certain type of bacterium. For example, coliphages attack coliform organisms, and because of their association with wastes from warm-blooded animals, seem to be an ideal indicator. The test for coliphage is performed by inoculating a petri dish containing an ample supply of a specific type of coliform with the wastewater sample. Coliphages will attack the coliforms, leaving visible spots or plaques which can be counted and an estimate made of the number of coliphages per known volume.

CONCLUSION

Discussed above are only some of the most important tests used in water pollution control. *Standard Methods,* for example, contains no fewer than 239 analytical procedures, many of which can be performed only with special equipment and skilled technicians.

Understanding this, and realizing the complexity, variation and objectives of some of the measurements of water pollution, how would you answer someone who brings a jug of water to your office, sets it on your desk and asks, "Can you tell me how much pollution is in this water?"

PROBLEMS

3.1 Given the following BOD_5 test results:
Initial DO   8 mg/l
Final DO   0 mg/l
Dilution   1/10
what can you say about
a. BOD_5?
b. BOD ultimate?
c. COD?

3.2 If you had two bottles full of lake water and kept one dark and the other in daylight, which one would have a higher DO after a few days? Why?

3.3 Name three substances you would need to seed if you wanted to measure their BOD.

3.4 The following data were obtained for a sample:
total solids $= 4000$ mg/l

suspended solids = 5000 mg/l
volatile suspended solids = 2000 mg/l
fixed suspended solids = 1000 mg/l.
Which of these numbers is questionable and why?

3.5 A water has a BOD_5 of 10 mg/l. The initial DO in the BOD bottle was 8 mg/l and the dilution was 1 to 10. What was the final DO in the BOD bottle?

3.6 If the BOD_5 of a waste is 100 mg/l, draw a curve showing the effect of adding progressively higher doses of chromium (a toxic chemical) on the BOD_5.

3.7 Consider the following data from a BOD test:

Day	DO (mg/l)	Day	DO (mg/l)
0	9	5	6
1	9	6	6
2	9	7	4
3	8	8	3
4	7	9	3

What is the: (a) BOD_5, (b) carbonaceous BOD, and (c) ultimate BOD? Why was there no oxygen used until the third day? If the sample were "seeded," would the final DO have been higher or lower and why?

3.8 An industry discharges 10 million gallons a day of a waste which has a BOD_5 of 2000 mg/l. How many pounds of BOD_5 are discharged?

3.9 If you dumped a half gallon of milk every day into a stream, what would be your discharge in lb BOD_5/day?

3.10 Given the same standard ammonia samples as in Example 3.3, if your unknown measured 20% absorbance, what is the ammonia concentration in the unknown?

3.11 Suppose you ran a multiple tube coliform test and got the following results: 10 ml samples, all 3 positive; 1 ml samples, all 3 positive; 0.1 samples, all 3 negative. Use the table in *Standard Methods* to estimate the concentration of coliforms.

3.12 If coliform bacteria are to be used as an indicator of viral pollution as well as an indicator of bacterial pollution, what attributes must the coliform organisms have (relative to viruses)?

3.13 Draw a typical BOD curve. Label the: (a) carbonaceous BOD, (b) nitrogenous BOD, (c) ultimate BOD, and (d) 5-day BOD. On the same graph, plot the BOD curve if the test had been run at 30°C instead of at the usual temperature. Also plot the BOD curve if a substantial mount of toxic materials were added to the sample.

3.14 Consider the BOD data below (no dilution, no seed).
   a. What is the ultimate carbonaceous BOD?
   b. What might have caused the lag in the beginning of the test?
   c. Calculate k_1, the reaction constant.

Day	DO	Initial DO
0	–	8
1	8	
3	7	
5	6	
7	6	
15	4	
20	4	

3.15 Given that the BOD test for a waste diluted 1:10 produced the following curve:

   a. What is the BOD_5 of the waste?
   b. What is the ultimate carbonaceous BOD?
   c. What is the ultimate BOD due to nitrification alone?
   d. The curve doesn't follow the "typical" first-order reaction curve. Comment on what happened during the first few days.

3.16 A wastewater is estimated to have a BOD of 200 mg/l.
   a. What dilution is necessary for this BOD to be measured by the usual technique?
   b. If the initial and final DO of the test thus conducted was 9.0 and 4.0, and the dilution water has a BOD of 1 mg/l, what is the actual BOD of the sample?

LIST OF SYMBOLS

BOD = biochemical oxygen demand, mg/l
BOD_5 = five-day BOD
   D = dilution (volume of sample/total volume)
   F = final BOD of sample, mg/l
   F′ = final BOD of seeded dilution water, mg/l
   I = initial BOD of sample, mg/l
   I′ = initial BOD of seeded dilution water, mg/l
   X = seeded dilution water in sample bottle, ml
   Y = volume of BOD bottle, ml

SUGGESTED READING

Sawyer, C. N., and P. L. McCarty. *Chemistry for Sanitary Engineers,* (New York: McGraw-Hill, 1978).

Environmental Protection Agency. *Methods for Chemical Analysis of Water and Wastes* (1971).

Chapter 4

Water Supply

A supply of water is critical to the survival of life as we know it. Man needs water to drink, animals need water to drink, and plants need water to drink. Additionally, basic functions of a society require water: cleaning for public health, consumption for industrial processes, and cooling for electricity generation. In this chapter, we discuss the supply of water in terms of:

- the hydrologic cycle and water availability
- groundwater supplies
- surface water supplies
- water transmission.

The direction of our discussion is that sufficient water supplies exist for the world, and the nation as a whole, but many areas are water poor while others are water rich. The trick in water supply is to engineer the supply and transmission from one area to another. In the following chapter, we address treatment methods available to clean up the water once it reaches the areas of demand.

THE HYDROLOGIC CYCLE AND WATER AVAILABILITY

The concept of the hydrologic cycle is a useful starting point to begin our study of water supply. Illustrated in Figure 4-1, this cycle is the precipitation of water from clouds, infiltration into the ground or runoff into surface watercourses, followed by evaporation and transpiration of the water back into the atmosphere.

The rates of precipitation and evaporation/transpiration help define the baseline quantity of water that is available for human consump-

Figure 4-1   The hydrologic cycle.

tion. *Precipitation* is the term applied to all forms of moisture origi-
nating in the clouds and falling to the ground, and a range of instruments
and technologies has been developed for measuring the amount and
intensity of rain, snow, sleet and hail. The average depth of precipitation
over a given region, either on a storm, seasonal or annual basis, is
required in many water availability studies. Any open receptacle with
vertical sides has application as a rain gauge, but varying wind and splash
effects must be considered if amounts collected by different gauges are to
be compared.

Evaporation and *transpiration* help define the baseline quantity of
water that is available. The same meteorological factors which influence
evaporation are at work in the transpiration process: solar radiation,
ambient air temperature, humidity and wind speed, as well as the amount
of soil moisture available to the plants impact the rate of transpiration.
Many measurements for transpiration are made with a phytometer, a
large vessel filled with soil and potted with selected plants. The surface of
the soil is hermetically sealed to prevent evaporation; thus the only
escape of moisture is through transpiration, which can be determined by
weighing the plant-vessel system at time intervals up to the life of the
plant. Because it is impossible to simulate natural conditions, the results
of phytometer tests are of limited value. However, they can be used as an
index of water demand by a crop under field conditions, and thus relate
to the calculations which help an engineer determine the water supply
requirements for that crop. Generally speaking, the terms evaporation
and transpiration are linked by engineers into evapotranspiration, or the
total water loss to the atmosphere by the two processes.

GROUNDWATER SUPPLIES

Groundwater is both an important direct source of supply which is
tapped by wells and a significant indirect source of supply since a large
portion of the flow to streams is derived from subsurface water.

Near the surface of the earth in the *zone of aeration* soil pore spaces
contain both air and water. This zone, which may have a zero thickness
in swamplands and be several hundred feet thick in mountainous regions,
contains three types of moisture. *Gravity water* is in transit after a storm
through the larger soil pore spaces. *Capillary water* is drawn through
small pore spaces by capillary action, and is available for plant up-take.
Hygroscopic moisture is water held in place by molecular forces during
all except the driest climatic conditions. Moisture from the zone of
aeration cannot be tapped as a water supply source.

On the other hand, the *zone of saturation* offers water in a quantity that is directly available. In this zone, located below the zone of aeration, the pores are filled with water, and this is what we consider *groundwater*. The stratum which contains a substantial amount of groundwater is called an *aquifer*. At the surface between the two zones, labeled the *water table* or *phreatic surface,* the hydrostatic pressure in the groundwater is equal to atmospheric pressure. An aquifer may extend to great depths, but because the weight of overburden material generally closes pore spaces, little water is found at depths greater than 600 m (2000 ft). The water readily available from an aquifer is that which will drain by gravity. Each soil type thus has a *specific yield,* defined as the volume of water, expressed as a percent of the total volume of water in the aquifer, which will drain freely from the aquifer.

Although Hagen and Poiseville were the first to propose that the velocity of flow of water through capillary tubes is proportional to the slope of the hydraulic gradient,[1, 2] it was Henri Darcy who in 1856 applied their theory to groundwater to obtain:[3]

$$v = ks$$

where v = the velocity of flow
s = the scope of the hydraulic gradient
k = a coefficient having the units of v.

The water discharged from an aquifer, q (in m^3/sec or gallons per minute, gpm), is the product of the cross-sectional area A and velocity. Only the "effective area" is of concern, i.e., the total area of the pores in the ground, so we must correct the total cross-sectional area by the porosity p of the soil. Porosity K_p is defined as the ratio of pore volume to the total volume of a soil formation. After combining terms, we get:

$$q = K_p As$$

Here, q is in gallons per day while all other terms are in feet. K_p is defined as the coefficient of permeability and in common units is expressed as the

[1]Hagen, G. "Uber die Bewegung des Wassers in engen cylindrischen Rohren," *Ann. Physik und Chemie,* 46:423 (1839).

[2]Poiseville, G. L. M. "Recherches experimentales sur le mouvement des liquides dans les tubes de tres petit diametre," *Roy. Acad. Sci. Insti. France, math. Phys. Sci. Man.* 9:433 (1846).

[3]Darcy, H. *Les fontanines publiques de la Ville de Dijon,* Paris, France (1856).

discharge in gallons of water per day through an area one foot square
under a gradient of one foot per foot at 60°F. In SI units, K_p is expressed
as m^3/day through an area $1\ m^2$ under a gradient of $1\ m/m$. Table 4-1
lists some average porosities, specific yields and permeabilities of
selected materials.

Table 4-1.   Estimate of Average Permeability and Porosity
Values for Selected Materials[a]

Materials	Porosity (%)	Specific Yield (%)	Permeability K_p	
			(gal/day · ft²)	(m³/day · m²)
Clay	45	3	1	0.04
Sand	35	25	800	32
Gravel	25	22	5000	200
Sandstone	15	8	700	28
Granite	1	0.5	0.1	0.004

[a] Adapted from *Elements of Hydrology,* by R. K. Linsley and J. B. Franzini. Copyright
1958 by McGraw-Hill Book Company. Used with the permission of the McGraw-Hill
Book Company.

Values for the coefficient of permeability can be determined for other
temperatures, K_{P_T}, by multiplying by the ratio of the kinematic
viscosities; ν_{60} and ν_T:

$$K_{P_T} = K_{P_{60}}\ \frac{\nu_{60}}{\nu_T}$$

Also, it is useful to define the transmissibility T to represent the rate of
flow through a section of aquifer of a unit thickness under a unit head
(slope = 1 foot per foot). In SI units, this would be $m^3/day \cdot m$ whereas in
common units T is defined in terms of gallons/day through a section
1-foot thick under a unit head (1 foot/1 foot slope). Thus,

$$q = TBS$$

where B is the thickness of the aquifer.

Permeameters are instruments used in the laboratory for measuring
the permeability of sample materials. The sample is placed under a water

column and the flows through the sample during given time periods are measured. The applicability of such tests is, however, limited because of the difficulty associated with placing the samples into the permeameter in their natural state; in addition, flows in large-solution cavities (such as would occur in limestone aquifers) and fractures cannot be simulated in this bench-scale test.

Groundwater supplies are often tapped by drilling wells into the ground and pumping the water up and out. If a flow of water is to occur to the well, a gradient must exist. Figure 4-2 illustrates this *cone of depression* in the water table resulting from pumping activities at a sample well site. In 1935, Theis developed a model to relate drawdown to T, the transmissibility of the aquifer:[4]

$$Z_r = \frac{q}{4\pi T} \int_u^\infty \frac{e^{-u}}{u} \, du$$

where   q = the discharge
          Z_r = the drawdown at an observation well a distance r from the water supply well

$$u = \frac{1.87\, r^2 S_c}{Tt}$$

where   t = the time from pumping start-up at the water supply well
          S_c = the storage constant for the given aquifer (the volume of water removed from a column of aquifer 1 unit dimension square when the water table is lowered 1 unit length, essentially the specific yield of the aquifer)
          r = the distance from the water supply well to the observation well
          u = a constant calculated for each well-aquifer system

When the pumping activity at the water supply well has proceeded for a relatively long time, i.e., as t is large u approaches zero, it is possible to modify the Theis model to:[5]

$$T = \frac{2.3\, q}{4\pi \Delta Z} \log_{10} \frac{t_2}{t_1}$$

---

[4]Theis, C. V. "The Relation between the Lowering of the Piezometric Surface and the Rate and Duration of Discharge of a Well Using Groundwater Storage," *Trans. Am. Geophys. Union* 16:519–524 (1935).

[5]Jacob, C. E. "Drawdown Test to Determine the Effective Radius of Artesian Well," *Trans. ASCE* 112:1047–1070 (1947).

Figure 4-2 Cone of depression.

where ΔZ is the drawdown at the observation well between times t_1 and t_2.

Example 4.3

When the drawdown values for a sample well system are plotted versus the time, on log-linear paper, we get results seen in Figure 4-3. We can simplify our calculations when we take ΔZ during a single log cycle. When between 4 hours and 40 hours, $\log_{10}(t_2/t_1) = 1$, and we can easily calculate T from the recorded value of discharged water q:[6]

$$T = \frac{2.3\,q}{4\pi\Delta Z}$$

If $q = 200$ gpm (288,000 gpd), then:

$$T = \frac{2.3\,(288,000)}{4(3.14)(5.2)}$$

$$T = 10,142 \text{ gpd per foot width}$$

Thus far, we have relied on the existence of perfectly symmetrical cones of depression. In the field, symmetry is not often the case. If several wells are close together, cones of depression may overlap. If a cone of depression intersects a surface stream or lake, the gradient from the stream to the well causes water to move out of the stream and the cone's surface will also be far from symmetrical. Variation in soil materials also impacts the shape of these drawdown cones.

Many factors help determine the "safe yield" of a given well or a selected aquifer:

- precipitation (P)
- net groundwater flow to the area (G_N)
- evapotranspiration (E_T)
- surface flow out of the area (S_o)
- change in groundwater storage (G_s)
- change in surface water storage (S_s)
- cost of pumping.

Any notion of "safe yield" must consider each factor, and must reference a given time period. If we focus on the quantity of water available for a year, for example, and neglect cost, we can define safe yield S_F as:

---

[6] *Ibid.*, Linsley, p. 138.

$$S_F = P + G_N - E_T - S_o - S_s - G_s$$

Other factors in the determination of safe yields include the transmissibility of the aquifer (can the aquifer transmit the water to the well(s) at a rate great enough to sustain the demand) and the location of contaminated bodies of water (including salt water intrusion possibilities from the oceans and chemical migrations from disposal or spill sites).

Besides offering a source of well water, aquifers also combine with precipitation to feed surface water courses. Streams, rivers and lakes also offer options for water supply, and they are addressed next.

SURFACE WATER SUPPLIES

Surface water supplies are not as reliable as groundwater sources since quantities often fluctuate widely during the course of a year or even a week, and qualities are restricted by various sources of pollution. If a river has an average flow of 10 cubic feet per second (cfs), it means often

Figure 4-3   Application of the modified Theis method, with sample data.

the flow is less than 10 cfs. If a community wishes to use this for a water supply, its demands for the water should be considerably less than 10 cfs in order to be assured of a reasonably dependable supply.

The variation in the river flow can be so great that even a small demand cannot be met during dry periods in many parts of the nation, and storage facilities must be constructed to hold the water during wet periods so it can be saved for the dry ones. The objective is to build these reservoirs sufficiently large to have dependable supplies.

One method of arriving at the proper reservoir size is by using the *mass curve*. Basically, the total flow in a stream at the point of a proposed reservoir is summed and plotted against the time. On the same curve the water demand is plotted and the difference between the total water flowing in and the water demanded is the quantity that the reservoir must hold if the demand is to be met. The method is illustrated by the following example problem:

Example 4.2

A reservoir is needed to provide a constant flow of 15 cfs. The monthly stream flow records, in total cubic feet, are:

Month	J	F	M	A	M	J	J	A	S	O	N	D
Cubic feet of water ($\times 10^6$)	50	60	70	40	32	20	50	80	10	50	60	80

We can calculate the storage requirement by plotting the cumulative water as in Figure 4-4. Note that for January, we plot 50 million cubic feet, for February we add 60 to that and plot 110 million cubic feet, and so on.

The demand for water is constant at 15 cfs, or

$$15 \; \frac{\text{cubic feet}}{\text{sec}} \times 60 \; \frac{\text{sec}}{\text{min}} \times 60 \; \frac{\text{min}}{\text{hr}} \times 24 \; \frac{\text{hr}}{\text{day}} \times 30 \; \frac{\text{days}}{\text{month}}$$

$$= 38.8 \times 10^6 \text{ cubic feet/month}$$

This can be represented as a sloped line in Figure 4-4, and plotted on the curved supply line. Note that the stream flow in May was lower than the demand, and this was the start of a drought lasting into June. The demand slope was greater than the supply, and thus the reservoir had to make up the deficit. In July the rains came and the supply increased until the reservoir could be filled up again, late in August. The reservoir capacity needed to get through that particular drought was 60×10^6 cubic feet. A second drought, starting in September, lasted into November and required 35×10^6 cubic feet of capacity. If the municipality therefore had a reservoir with at least 60×10^6 cubic feet capacity, they could have drawn water from it throughout the year.

A mass curve such as Figure 4-4 is actually of little use if only limited stream flow data are available. One year's data yield very little informa-

Figure 4-4   Mass curve for determining required reservoir capacity.

tion about long-term variations. For example, was the drought in the above example that required 60 million cubic feet of storage the worst drought in 20 years, or was the year shown actually a fairly wet year?

To get around this problem, it is necessary to predict statistically the recurrence of events such as droughts and then to design the structures according to a known risk. Water supplies are often designed to meet demands 19 out of 20 years. In other words, once in 20 years the drought will be so severe that the reservoir capacity will not be adequate. If running out of water once every 20 years is not acceptable, the people can choose to build a bigger reservoir and thus expect to be dry only once every 50 years. The question really is one of an increasing investment of capital for a steadily smaller added benefit.

Example 4.3

Suppose the need was to build a reservoir which could supply the water demand 9 out of 10 years. The reservoir capacities, which would be required in order to prevent running out of water, as calculated from mass curves, follow:

Year	Required Reservoir Capacity, mil cu ft	Year	Required Reservoir Capacity, mil cu ft
1951	60	1961	53
1952	40	1962	62
1953	85	1963	73
1954	30	1964	80
1955	67	1965	50
1956	46	1966	38
1957	60	1967	34
1958	42	1968	28
1959	90	1969	40
1960	51	1970	45

These data must now be ranked, with highest required capacity getting rank 1, the next highest 2, and so on.

Rank	Capacity		Rank	Capacity	
1	90	(1959)	11	51	
2	85		12	46	
3	80		13	45	
4	73		14	42	
5	67		15	40	
6	62		16	40	
7	60		17	38	
8	60		18	34	
9	55		19	30	
10	53		20	28	(1968)

These data are now plotted on a graph in Figure 4-5 showing the capacity versus the percent of time this capacity requirement was equaled or exceeded. For example, the highest capacity was equaled or exceeded once out of 20 years, or 5% of the time. The next highest capacity was exceeded or equaled twice out of 20 years, or 10% of the time, and so on. Using the curve in Figure 4-5 we can pick the 1 in 10 years drought (10%) as 85 million cubic feet.

Figure 4-5   Frequency analysis of reservoir capacity.

This procedure is known as a *frequency analysis* of recurring natural events such as droughts. In the above example we selected a "10-year drought," or the drought which on the average occurs once every 10 years. Recognize that there is no guarantee that it would indeed occur once every 10 years. In fact, it could happen 3 years in a row, and then not again for 50 years. On the average, however, a ten-year recurrence interval is a reliable estimate for this example.

WATER TRANSMISSION

Water can be transported from either a ground or surface supply source directly to a community or, if water quality considerations indicate, initially to a water treatment facility, by different types of conduits, including:

- Pressure conduits: tunnels, aqueducts and pipelines
- Gravity-flow conduits: grade tunnels, grade aqueducts and pipelines

The location of the well field or river reservoir defines the length of the conduits, while the topography indicates whether the conduits are designed to carry the water in open-channel flow or under pressure. The profile of a water supply conduit must generally follow the hydraulic grade line to take advantage of the forces of gravity and thus minimize pumping costs.

Service reservoirs are also necessary in the transmission system to help level out peak demands. In practice, intermediate reservoirs close to the city, or water towers, are sized to meet three design constraints:

- hourly fluctuations in water consumption within the service area
- short-term shutdown of the supply network for servicing
- back-up water requirements to control fires.

These distribution reservoirs are most often constructed as open or covered basins, elevated tanks, or, in the past, standpipes. If the service reservoirs are adequately designed to meet these capacity considerations, then the supply conduits leading to them generally must only be designed to carry approximately 50% in excess of the average daily demand of the system or subsystem.

Elements of Closed-Conduit Hydraulics

Before discussing flow, it might first be useful to review some important fluid properties.

The *density* of a fluid is its mass per unit volume. In common units, density is expressed as slugs per cubic foot, or as $lb \cdot sec^2/ft^4$. Density in the metric system is in terms of grams per cubic centimeter. The density of water under standard conditions is 1.94 in common units and unity in the metric system.

Specific weight represents the force exerted by gravity on a unit volume of fluid and therefore must be in terms of force per unit volume, such as pounds per cubic foot. The specific weight is related to density as

$$w = \rho g$$

where w = specific weight lb/ft^3
  ρ = density, lb·sec^2/ft^4
  g = gravitational constant ft/sec^2.

The specific weight of water is 62.4 pounds per cubic foot.

Specific gravity of a liquid is the ratio of its density to that of pure water at a standard temperature. In the metric system the density of water is one gram per cubic centimeter and hence the specific gravity has the same numerical value as the density.

The *viscosity* of a fluid is a measure of its resistance to shear or angular deformation and is defined as the proportionality constant relating the shear stress τ to the rate of deformation du/dy. This proportionality constant is usually written as

$$\tau = \mu \frac{du}{dy}$$

The assumption inherent in this definition is that the shear stress is directly proportional to the rate of deformation. Such a definition holds for many fluids, which are known as Newtonian fluids. Fluids for which the shear stress is not proportional to the shear rate are known as non-Newtonian fluids. An example of a non-Newtonian fluid is biological sludge. The term viscosity, when applied to biological sludge, is therefore significant only if either the rate of deformation or the shear stress is also specified.

In the metric system, a unit of viscosity is a poise, with units of grams per centimeter-second. Most fluids have low viscosity and a more convenient unit is the centipoise or 0.01 poise. The viscosity of water at 20°C (68.4°F) is 1 centipoise. In the English system the unit of viscosity is pound seconds per square foot. One lb·sec/ft^2 equals 479 poise.

Kinematic viscosity is defined as the absolute viscosity, μ, divided by the density of the fluid, or $\nu = \mu/\rho$. The dimensions of kinematic viscosity are square centimeters per second.

Fluid viscosity is a function of temperature. Some representative values of the viscosity of water are shown in Table 4-2.

Table 4-2. Viscosity of Water

Temperature		Viscosity	
°F	°C	lb · sec/ft²	Centipoise
40	4.4	3.1×10^{-5}	1.5
50	10.0	2.7×10^{-5}	1.3
60	15.5	2.3×10^{-5}	1.1
68.4	20.0	2.1×10^{-5}	1.0
70	21.0	2.0×10^{-5}	0.96
80	26.6	1.8×10^{-5}	0.86
90	32.2	1.6×10^{-5}	0.77

Closed-Conduit Flow

One of the fundamental principles of hydraulics states that in a system, the total energy of a perfect liquid under ideal conditions does not change as it flows from point to point. Total energy is the sum of the position energy, pressure energy and velocity energy. These energies are usually stated in terms of meters or feet of fluid, so that position energy is static head Z, pressure energy is fluid pressure P/w, called pressure head, and the velocity energy is velocity head $v^2/2g$. These terms are defined as:

Z = elevation, m (ft)
P = pressure, kg/m² (lb/ft²)
w = specific weight of the fluid, kg/m³ (lb/ft³)
v = velocity, m/s (ft/sec)
g = acceleration due to gravity, m/sec² (ft/sec²).

Consider a system such as is shown in Figure 4-6. Water is flowing through this pipe from a reservoir with constant surface elevation, and assuming there are no losses in this system, it is obvious that the total energy or the total head of water remains constant, although the form of the energy or head of water is converted from one type of energy to another. For example, at point 1, at the surface of the reservoir, all the energy is static head, while at points 3, 4, and 5, the energy is distributed among static, pressure and velocity head. At point 6, the energy of the jet, as it enters the atmosphere, is made up of only the velocity head and

Figure 4-6   Hydraulics of ideal closed-conduit flow.

the static head. The total energy, however, is constant at all points in the system.

This is the constant energy principle, commonly known as Bernoulli's Theorem, which if written between any two points in a system is:

$$Z_1 + \frac{P_1}{w} + \frac{v_1^2}{2g} = Z_2 + \frac{P_2}{w} + \frac{v_2^2}{2g}$$

This equation holds for an ideal fluid, in which no losses occur. In real systems, however, the flow of a fluid is always accompanied by a loss of energy. Taking into account losses occurring in any system between two points, the equation is:

$$Z_1 + \frac{P_1}{w} + \frac{v_1^2}{2g} = Z_2 + \frac{P_2}{w} + \frac{v_2^2}{2g} + h_L$$

in which h_L represents the energy losses within the system.

The energy loss may occur in many places in a system, such as valves, bends and sudden changes in pipe diameter. One of the major losses of energy is in the friction between the moving fluid and the pipe wall.

Because of the practical problems involved in the application of more elegant and theoretically sound friction loss equations, engineers have resorted to fitted exponential equations for flow calculations. Among these, the Hazen-Williams formula is most widely used in the United States for flow in pressure pipes and the Manning formula for flow in open channels or pipes not flowing full. These formulas are limited to turbulent flow of water and to common ambient temperatures.

The Hazen-Williams formula is

$$v = 1.318\,C\,r^{0.63}\,s^{0.54}$$

where   v = the mean velocity of flow in feet per second
        r = the hydraulic radius (area divided by wetted perimeter) in feet
        s = the slope of the hydraulic grade line
        C = the Hazen-Williams friction coefficient.

Combined with the continuity equation, $Q = Av$, the formula for discharge, Q, in gallons per minute for circular pipe of diameter D, in inches, is

$$Q = 0.285\,C\,D^{2.63}\,s^{0.54}$$

In metric units, these two equations are

$$v = 0.849\,C\,r^{0.63}\,s^{0.54}$$

$$Q = 0.278\,C\,D^{2.63}\,s^{0.54}$$

where D and r are in meters and v and Q are m/s and m³/s respectively. The nomographs shown as Figures 4-7 and 4-8 are solutions of the above equation in both common and metric units. Table 4-3 summarizes Hazen-Williams coefficients for various pipe materials.

Example 4.4

The pressure drop through a 6-in. asbestos-cement pipe, 3000 ft in length is 20 psi. What is the flow rate? 20 psi × 2.31 ft of water/psi = 46.2 ft of head loss (h_L). 46.2 ft ÷ 3000 ft = 15.4 ft/1000 ft(s). From Table 4-3 the Hazen-Williams coefficient is 140 and from the nomograph, Figure 4-7, the discharge is 460 gpm. If the flow had been given, the head loss could be found in reverse manner.

Energy or head is lost at the entrance to a pipe or conduit, at valves, meters, fittings, and other irregular features, at enlargements and at flow contractions. These losses, called *minor losses,* are in excess of friction losses over the same length of straight pipe or conduit, and can be expressed as:

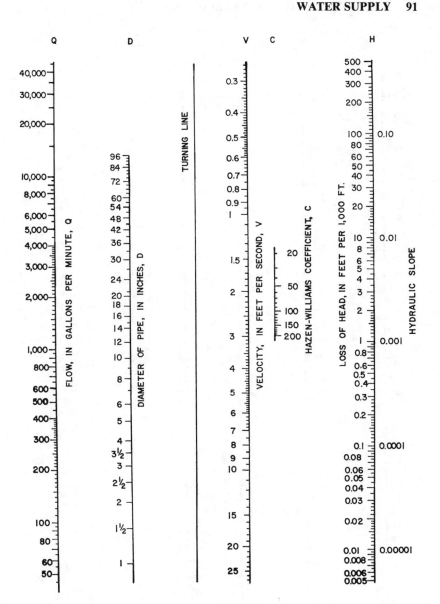

Figure 4-7   Hazen-Williams equation nomograph, common units.

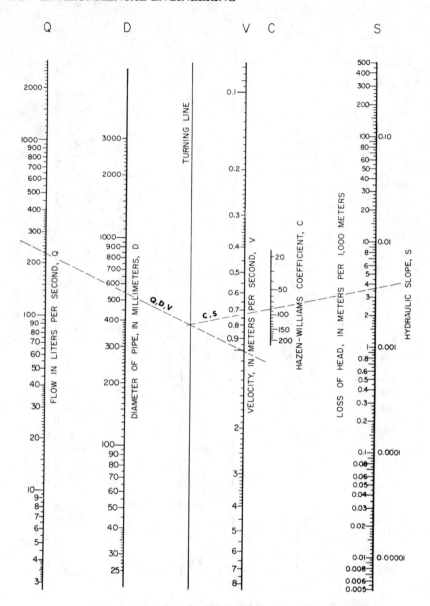

Figure 4-8   Hazen-Williams equation nomograph, metric units.

Table 4-3.   Values of "C" for Hazen-Williams Formula

Type of Pipe	C
Asbestos Cement	140
Brass	130–140
Brick Sewer	100
Cast Iron	
New, unlined	130
Old, unlined	40–120
Cement-lined	130–150
Bitumastic enamel-lined	140–150
Tar-coated	115–135
Concrete or Concrete-Lined	
Steel forms	140
Wooden forms	120
Centrifugally spun	135
Copper	130–140
Fire Hose (Rubber-Lined)	135
Galvanized Iron	120
Glass	140
Lead	130–140
Masonry Conduit	120–140
Plastic	140–150
Steel	
Coal-tar enamel-lined	145–150
New unlined	140–150
Riveted	110
Tin	130
Vitrified	100–140
Wood Stave	120

$$h_L = K\,\frac{v^2}{2g}$$

for which values of K may be estimated using Table 4-4.

Example 4.5

The loss for a flow of 1.0 cfs through a given 6-in. main with a gate valve wide open is 20 ft. Find the head loss with the gate valve 3/4 closed.

From Table 4-4, the increase in K value when the valve is partially closed is $24.0 - 0.2 = 23.8$ (difference in K for 3/4 closed and fully open).

Table 4-4.   Minor Losses of Head

Nature of Special Resistance	Loss (K)	Nature of Special Resistance	Loss (K)
Angle Valve		45° Elbow	
Wide open	2–5	Use ¾ of loss for 90°	
Butterfly Valve		Bend of same radius	
$\theta = 10°$	1	22 ½° Elbow	
$\theta = 40°$	10	Use ½ of loss for 90°	
$\theta = 70°$	320	bend of same radius	
Check Valves		Entrance Losses	
Horizontal lifts	8–12	Pipe projecting into tank	0.8–1.0
Ball	65–70	End of pipe flush with tank	0.5
Swing	0.6–2.5	Slightly rounded	0.23
Gate Valves		Bell-mouthed	0.04
Wide open	0.2	Outlet Losses	
¼ closed	1.2	From pipe into still water	
½ closed	5.6	or atmosphere	1.0
¾ closed	24.0	Sudden Contraction	
Globe Valves		$d/D = ¼$	0.42
Wide open	10	$d/D = ½$	0.33
90° Elbow		$d/D = ¾$	0.19
Regular flanged	0.21–0.30	Sudden Enlargement	
Long radius flanged	0.14–0.23	$d/D = ¼$	0.92
Short radius screwed	0.9	$d/D = ½$	0.56
Medium radius screwed	0.75	$d/D = ¾$	0.91
Long radius screwed	0.60		

$$v = \frac{Q}{A} = \frac{1.0}{0.2} = 5 \text{ ft/s}$$

$$h_L = h_o + K(v^2/2g)$$

$$= 20 + 23.8(5^2/64.4)$$

$$= 29.2 \text{ ft}$$

Two pipes, two systems of pipes, or a single pipe and a system of pipes are said to be equivalent when their losses of head for equal rates of flow are equal (or flow is equal for equal loss of head). Compound pipes, whether in parallel or in series (Figures 4-9 and 4-10) can be reduced to single, equivalent pipes. The following examples are illustrative:

PIPE NO. 1
D = 8"
L = 1300'

A   ACTUAL
         PIPES   B

PIPE NO. 2
D = 6"
L = 1400'

A   EQUIVALENT PIPE   B
D = 8.8"
L = 1000'

NOTE: C = 100 FOR ALL PIPES

Figure 4-9   Equivalent pipes, parallel.

Example 4.6

For the parallel pipes as shown in Figure 4-9, find the diameter of equivalent pipe (length = 1000 ft assumed).

1. Loss of head through pipe 1 must always equal loss of head through pipe 2 between points A and B.
2. Assume any arbitrary head loss, say 10 ft.
3. Calculate head loss in feet per 1000 feet for pipes 1 and 2.
   Pipe 1: $(10/1300) \times 1000 = 7.7$ ft/1000 ft
   Pipe 2: $(10/1400) \times 1000 = 7.1$ ft/1000 ft
4. Use Figure 4-7 to find flow in gallons per minute.
   Pipe 1: D = 8 in., s = 0.0077,
      Q = 495 gpm
   Pipe 2: D = 6 in., s = 0.0071,
      Q = 220 gpm
   Total Q through both pipes = 715 gpm.
5. Using Figure 4-7 with s = 0.010 and Q = 715 gpm, equivalent pipe size is found to be 8.8 in. in diameter.

Example 4.7

For the pipes in series as shown in Figure 4-10, find the diameter of equivalent pipe (length = 1000 ft assumed).

1. Quantity of water flowing through pipes 3 and 4 is the same.
2. Assume any arbitrary flow through pipes 3 and 4, say 500 gpm.
3. Using Figure 4-7, find head loss for pipes 3 and 4.
   Pipe 3: D = 8 in., L = 400 ft,
   $Q = 500$ gpm
   $h_L = 0.4 \times 8.0 = 3.2$ ft
   Pipe 4: D = 6 in., L = 600 ft,
   $Q = 500$ gpm
   $h_L = 0.6 \times 33.0 = 19.8$ ft
   Total head loss in both pipes = 23.0 ft.
4. Using Figure 4-7, with head loss = 23.0 ft/1000 and Q = 500 gpm, the equivalent pipe size is found to be 6.5 in. in diameter.

More complex systems can be reduced to a single equivalent pipe by piecework conversion of real and equivalent pipes.

Although the calculations become tedious, solution of network flow problems is dependent on the same basic physical principles as for single pipes, that is, the principles of energy conservation and continuity must be satisfied throughout the network. Examples of complex pipe networks are beyond the scope of this text.

Pumps and Pumping*

Pumps are mechanical devices for converting other forms of energy to hydraulic energy. When interposed in a pipe, they add energy to the liquid passing through the pipes. Nearly always, it is pressure energy. Pumps, like motor vehicles, are not individually designed for public works projects, except for very large or unusual installations. Rather, they are selected from predesigned and manufactured units readily available for a wide range of applications.

Economical selection requires that attention be given to: (1) The normal pumping rate, and also the minimum and maximum rate that the pump will ever be called on to deliver; (2) the total head capacity to meet flow requirements; (3) suction head (or lift); (4) pump characteristics including speed, number, power source, and spatial, environmental and other special requirements; and (5) the nature of the liquid to be pumped.

---

*This discussion on pumps and pumping is adapted from *Practical Hydraulics for Public Works Engineers,* F. E. McJunkin and P. A. Vesilind, published as a separate issue by *Public Works Magazine* (1968).

PIPE NO. 3
D = 8"
L = 400'

PIPE NO. 4
D = 6"
L = 600'

C   D   E
ACTUAL PIPES

C   EQUIVALENT PIPE   E
D = 6.5"
L = 1000'

NOTE: C = 100 FOR ALL PIPES

Figure 4-10   Equivalent pipes, series.

There are many types of pumps, but the two most often encountered in environmental engineering are *roto-dynamic* and *displacement* pumps. Roto-dynamic pumps impart energy to liquids with a rotating element, or *impeller*, shaped to force water outward at right angles to its axis (*radial flow*), to give the liquid an axial as well as a radial velocity (*mixed flow*), or to force the liquid in the axial direction alone (*axial flow*). Radial flow and mixed flow machines are commonly referred to as *centrifugal pumps* and axial flow machines are called *propeller pumps*. Displacement pumps include the *reciprocating* type, in which a piston draws water into a cylinder on one stroke and forces it out on the next, and the *rotary* type, in which two cams or gears mesh together and rotate in opposite directions to force water continuously past them. In addition there are *jet pumps* (*ejectors*), *air-lift pumps, hydraulic rams, diaphragm pumps,* and others which may be useful under special circumstances.

Pump hydraulics are depicted in Figure 4-11. The *static suction head* on a pump is the vertical distance, i.e., elevation difference, from a free liquid surface* on the intake to the pump centerline. This static head may be positive or negative (the latter sometimes called *suction lift*). The *net suction head* h_s is the static suction head minus friction head losses, including entrance loss, for the capacity under consideration. The *static discharge head* is the vertical distance from the pump centerline to the free surface (or discharge) on the discharge line. Similarly, the *net discharge head* h_d is the static discharge head plus the friction head losses.

___

*If no free surface exists, the gauge pressure at the pump flange (suction or discharge) for zero discharge, corrected to centerline pump elevation, should be used in lieu of the static head (suction and/or discharge as appropriate).

Figure 4-11   Pump hydraulics defined.

The total head H is the difference between the net discharge and suction heads, or,

$$H = h_d - h_s$$

In pumping liquids, the pressure at any point in the suction line should not be reduced to the vapor pressure of the liquid, if air binding, loss of prime and cavitation are to be avoided. The available energy that can be utilized to move liquid through the suction line to the impeller is thus the net suction head h_s, plus any pressure (a vacuum being a negative pressure) existing in the suction supply line, and less the vapor pressure of the liquid at the pumping temperature. This energy or head is known as the *net positive suction head* (NPSH).

The useful work done by a pump is the product of the weight of liquid pumped multiplied by the head developed by the pump. The power or work per unit time required is the *water horsepower* (*WHP*). By definition:

$$WHP \propto QHw$$

where   Q = pump discharge
         H = total head
         w = specific weight.

For water at 68°F, Q in gallons per minute (gpm) and H in feet:

$$WHP = \frac{QH}{3960}$$

The total power required to drive a pump is the *brake horsepower* (*BHP*). The ratio of the water horsepower to the brake horsepower is the pump efficiency, η:

$$\eta = \frac{WHP}{BHP} \times 100$$

Therefore:

$$BHP = \frac{100\,QH}{3960\,\eta}$$

for water at 68°F. BHP for other water temperature and other liquids can be determined by correcting for the change in specific weight. For natural waters, correction of specific weight for temperature is usually negligible.

Pump Characteristic Curves

For a given pump, the total head developed, the power required to drive it, and the resulting efficiency vary with discharge. These interrelations of head, power, efficiency and capacity are commonly known as the *characteristics of the pump*. They are best shown graphically, as *pump characteristic curves*. The usual practice is to plot head, power, and efficiency against capacity at constant impeller speed as shown in Figure 4-12.

The curve H–Q, showing the relation between capacity and total head is called the *head-capacity curve*; and pumps are often classified on the basis of the shape of their head-capacity curves.

The curve η–Q shows the relation between efficiency and capacity and is called simply the *efficiency curve*. The efficiency curve usually exhibits a maximum or peak efficiency.

Figure 4-12   Pump characteristic curves.

The curve P–Q showing the relation between power input and pump capacity is generally termed the *power curve.*

System Head Curves

The head loss in a pumping system increases of course with increasing flow through the system. This can be shown graphically as a *system head curve.* Figure 4-13 is an example. The system head loss for any flow rate is the sum of friction head loss at the rate plus the total static head in the system.

The total static head is the difference in elevation between the discharge liquid level and the suction liquid level (see Figure 4-11). Inasmuch as the static head is present, whether the pump is operating or not, it is plotted as the lower portion of the system head curve.

Friction losses, including minor losses, are determined as outlined in the previous section of this chapter. The system piping and fittings may be converted to one equivalent pipe and head losses for several flow rates determined readily from the Hazen-Williams nomograph. Alternatively the head loss through all pipe and fittings may be computed for a single

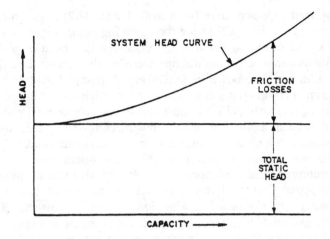

Figure 4-13 System head curve.

flow rate and losses for other flow rates determined from the relationship:

$$\frac{Q_1}{Q_2} = \left(\frac{H_1}{H_2}\right)^{0.54}$$

Total head for zero discharge will be equal to the total static head. This point plus several computed points will suffice to plot the curve. Quite commonly, the static head in a system will vary as tanks and reservoirs are filled and lowered. In such cases, system curves may readily be constructed for minimum and maximum heads, thereby enabling prediction of system pumping capacity for the entire range of possible static head conditions.

By separation of static and friction head in preparing the system head curve, the reasonableness of friction losses can be checked. If piping to be used is too small, then the pump and motor capital and operating costs due to the higher head capacity required will be excessive. As a rule of thumb, the friction losses should not be more than 10–20% of the static head for comparatively short pipelines. For elaborate and costly installations, an economic analysis of the tradeoff between pumping and piping costs may be justified.

Operating Head and Discharge

The usual design condition is that a system will be given and the proper pump must be selected. If on the system head curve (such as Figure 4-13), a pump head capacity curve (such as in Figure 4-12) is superimposed, then their intersection will locate the *operating point*. Figure 4-14 is an example. The operating point is the discharge and head at which the given system and given pump will operate. This also locates the efficiency of operation and the power requirements. A pump should be selected that has an operating point at or near its peak efficiency. Many other factors, some previously outlined, must also be considered in selecting from the multitude of options available, but meeting the system discharge requirement (for a given system, satisfying discharge automatically meets the head requirement) at reasonable efficiency is paramount.

The pumping head developed is doubled if two identical pumps are placed in series. Conversely, two pumps in parallel will double the pumping capacity (see Figure 4-15). This does not mean, however, that the head or capacity of a *system* is doubled. Superimposing a system curve on the head-capacity curves for two pumps in parallel, it is clear (from Figure 4-16) that the added capacity results in a greater friction head loss, the relationship being:

$$h_L = Q^2$$

and that the system capacity with two pumps in parallel is not double its capacity with one pump. Similarly, pumping in series will double neither head nor discharge.

Figure 4-14   Sample determination of pump operating point.

SINGLE PUMP   TWO PUMPS   TWO PUMPS
                     IN PARALLEL   IN SERIES

Figure 4-15   Pump characteristic curves with two pumps in parallel and in series.

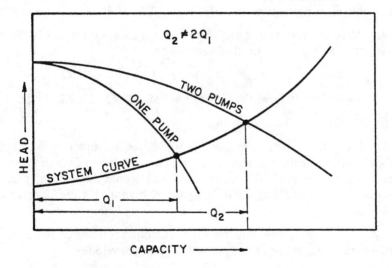

Figure 4-16   System head curve with two pumps in parallel.

CONCLUSION

As the hydrologic cycle indicates, water is a renewable resource because of the driving force of the sun. Thus, the earth is not running out of water, it may just be running out of clean water in general and sufficient water in isolated areas as climatic changes take place.

Both groundwater and surface water supplies are available, though to varying degrees, across the face of the earth. The trick is to develop these supplies using sound engineering design procedures for wells and dams, and to protect these supplies using sound engineering and ethical judgments. In the next chapter, we address methods of preparing and treating water for distribution and consumption once the supply has been provided.

PROBLEMS

4.1 Given the following data for monitoring well drawdown, calculate the transmissibility of the sample aquifer using the modified Theis method. Assume the observation well is 300 ft from a well pumped at 350 gpm.

Time of pumping (hr)	1.9	2.8	5.5	9.1	19.0	57.0
Drawdown at observation well (ft)	1.9	2.3	3.4	4.2	5.6	7.5

4.2 A storage reservoir is needed to ensure a constant flow of 20 cfs to a city. The monthly stream flow records are:

Month	J	F	M	A	M	J	J	A	S	O	N	D
ft³ of water × 10⁶	60	70	85	50	40	25	55	85	20	55	70	90

Calculate the storage requirement.

4.3 If a faucet drips at a rate of 2 drops per second and it takes 25,000 drops to equal 1 gallon of water, how much water is lost each day? each year? If water costs $1.20/1000 gal, how long would the lost water take to equal the cost of fixing the leak if you did it yourself (free labor!) for 10¢ in parts.

4.4 A 24-in. pipe with centrifugally spun concrete lining carries a flow of 10 mgd. Find: (a) head loss per unit length, (b) velocity.

4.5 A 6-in. main is carrying a flow of 1.0 cfs. What head loss per unit length is produced by two regular flanged 45° elbows and a gate valve one-half closed?

4.6 Three pipes, each 1500 ft long, are arranged in parallel. Two are 6-in. diameter; the other is 10-in. All three have a Hazen-Williams "C" value of 100. Find diameter of the equivalent pipe which is also 1500 ft long.

4.7 A pump delivers 900 gpm against a total head of 145 ft. Assuming a pump efficiency of 90%, what is the brake horsepower?

4.8 Given two identical pumps with characteristics shown in Figure 4-12, find discharge for a head of 100 feet if: (a) both pumps are operated in series and (b) both pumps are operated in parallel.

LIST OF SYMBOLS

γ = kinematic viscosity, cm^2/sec

ρ = density, g/cm^3 ($lb \cdot sec^2/ft^4$)

μ = viscosity, poise ($lb \cdot sec/ft^2$)

η = pump efficiency

τ = shear stress

A = cross-sectional area, m^2 (ft^2)

B = thickness of aquifer, m (ft)

BHP = brake horsepower

C = Hazen-Williams friction coefficient

cfs = cubic feet per second

E_T = evapotranspiration, m^3 (gal)

G_N = net groundwater flow, m^3 (gal)

G_s = change in groundwater storage, m^3 (gal)

g = acceleration due to gravity, m/sec^2 (ft/sec^2)

H = total head

h_d = net discharge head

h_L = head loss

h_s = net suction head, m (ft)

k = proportionality constant

K_p = coefficient of permeability, m^3/d (gal/day)

$NPSH$ = net positive suction head

P = pressure, kg/m^2 (lb/ft^2)

P' = precipitation, m^3 (gal)

Q = discharge, m^3/sec (gal/min)

q = discharge through a section of aquifer of unit thickness under a unit head, m^3/sec (gal/min)

r = radius, m (ft)

s = hydraulic gradient

S_c = storage constant, m^3 (gal)

S_F = safe yield, m^3 (gal)

S_o = surface flow, m^3 (gal)

S_s = change in surface water storage, m^3 (gal)

T = transmissibility, m^3/day

t = time

V_d = vapor pressure at dew point, inches of Hg
V_p = vapor pressure, inches of Hg
v = velocity of flow, m/sec (ft/sec)
w = specific weight, kg/m³ (lb/ft³)
WHP = water horsepower
Z = elevation, m (ft)
Z_r = drawdown, m (ft)

SUGGESTED READING

Jeppson, R. W. *Analysis of Flow in Pipe Networks* (Ann Arbor, MI: Ann Arbor Press, 1977).

Linsley, R. K., M. A. Kohler and J. L. H. Paulhus. *Hydrology for Engineers* (New York: McGraw-Hill, 1958).

McGauhey, P. H. *Engineering Management of Water Quality* (New York: McGraw-Hill, 1968).

McJunkin, F. E., and P. A. Vesilind. *Practical Hydraulics for Public Works Engineers,* Public Works Magazine (1968).

National Lime Association, *Water Supply and Treatment,* Washington, DC (1962).

Simon, A. *Practical Hydraulics* (New York: John Wiley & Sons, 1976).

Valentine, H. R. *Water in the Service of Man,* (Baltimore, MD: Penguine Books, 1967).

Chapter 5

Water Treatment

Many aquifers and isolated surface waters are high in water quality and may be pumped from the supply and transmission network directly to any number of end uses, including human consumption, irrigation, industrial processes or fire control. However, such clean water sources are the exception to the rule in many regions of the nation, particularly regions with dense populations or regions that are heavily agricultural. Here, the water supply must receive varying degrees of treatment prior to distribution.

Impurities enter the water as it moves through the atmosphere, across the earth's surface, and between soil particles in the ground. These background levels of impurities are often supplemented by man's activities. Chemicals from industrial discharges and pathogenic organisms of human origin, if allowed to enter the water distribution system, can cause health problems. Excessive silt and other solids can make the water both unsightly and aesthetically unpleasing. Water can be contaminated by many routes. For example, heavy metal pollution, including lead, zinc and copper, can be caused by corrosion of the very pipes which carry the water from its source to the consumer.

The method and degree of water treatment are important considerations for environmental engineers. Generally speaking, the characteristics of raw water determine the method of treatment. Because most public supply systems are relied on for drinking water, as well as industrial and fire consumption, the highest level of use, human consumption, defines the degree of treatment. Thus, we focus only on treatment technologies which produce potable water.

A typical water treatment plant is diagrammed in Figure 5-1. These plants are designed to remove odors, color and turbidity as well as bacteria and other contaminants from surface water. Raw surface water

KEY:

1: Chemical mixing basin ⎫
2: Flocculation basin   ⎬ Coagulation and Flocculation
3: Settling tank   ⎭
4: Rapid sand filter
5: Disinfection with chlorine
6: Clean water storage basin (Clear well)
7: Pump

Figure 5-1   Movement of water through a typical water treatment plant.

entering a water treatment plant usually has significant turbidity caused by tiny colloidal clay and silt particles. These particles have a natural electrostatic charge which keeps them continually in motion and prevents them from colliding and sticking together. Chemicals such as alum (aluminum sulfate) are added to the water, first to neutralize the charge on the particles and then to aid in making the tiny particles "sticky" so they can coalesce and form large particles called *flocs*. This process is called *coagulation* and *flocculation* and is represented by Stages 1 and 2 in Figure 5-1.

COAGULATION AND FLOCCULATION

Naturally occurring silt particles suspended in water are difficult to remove because they are very small, often colloidal in size, and they possess negative charges and thus are prevented from coming together to form larger particles which could more readily be settled out. The removal of these particles by settling requires first that their charges be neutralized, and second that the particles be encouraged to collide with each other. The charge neutralization is commonly termed coagulation, and the building of larger flocs from the small particles is called flocculation.

A fairly simple but not altogether satisfactory explanation of coagulation is available in the double-layer model. The charge on a particle is represented in Figure 5-2. The solid particle is negatively charged and it attracts to it counterions, positively charged. Some of these ions are so strongly attached to the particle that they travel with the particle, thereby forming a slippage plane. Around this inner layer is an outer layer of ions consisting mostly of counterions, but these are loosely attached and can easily slip off. The charge on the particle as it moves through a fluid is the negative charge, diminished in part by the positive ions in the inner layer. This charge is called the zeta potential.

If the net negative charge is considered a repulsive charge (since the neighboring particles are also so charged) the charge can be pictured as in Figure 5-3a. In addition to this repulsive charge, however, all particles have an attractive electrostatic charge, known as *Van der Waals force*, which is due to the molecular structure of the particle. This attractive charge is also shown in Figure 5-3a. Combining these two charges, we find a net repulsive charge, or an energy hill, which prevents the particles from coming together. The objective of coagulation is to reduce this energy hill to zero so that particles would not be repulsed by each other.

One means of accomplishing this end is to add trivalent cations to the

Figure 5-2   Charges on a suspended particle, as explained by the double-layer theory.

water. These ions would snuggle up to the negatively charged particle, and because they possess a stronger charge displace the monovalent cations. The effect of this would be to reduce the net negative charge and thus lower the replusive force. This is shown schematically in Figure 5-3b. In this condition, the particles will not repulse each other, and upon colliding, will stick together. A stable colloidal suspension has thus been made into an unstable colloidal suspension.

Figure 5-3   The reduction of the net charge on a particle as the result of the addition of trivalent counterions.

The usual source of trivalent cations in water treatment is alum (aluminum sulfate). Alum has an additional advantage in that some fraction of the aluminum may form aluminum oxides/hydroxides, represented simply as:

$$Al^{+++} + 3OH^- \rightarrow AlOH_3 \downarrow$$

These complexes are sticky and heavy, and will greatly assist in the clarification of the water in the settling tank if the unstable colloidal particles can be made to come into contact with the floc. This process is enhanced through an operation known as flocculation.

A flocculator introduces velocity gradients into the water so that particles in a fast-moving stream can catch up with and collide with slow-moving particles. Such velocity gradients are commonly introduced by rotating paddles. The power required for moving a paddle through the water is stated as

$$P = \frac{C_D \, A \, \rho \, v^3}{2}$$

where P = power, N/sec (ft·lb/sec)
         A = paddle area, m² (ft²)
         ρ = fluid density, kg/m³ (lb·sec²/ft⁴)
         v = velocity of the paddle relative to the fluid, m/sec (ft/sec)
         C_D = drag coefficient.

The velocity gradient produced as a result of a power input in a given volume of water is

$$G = \left(\frac{P}{\mu \forall} \right)^{1/2}$$

where G = velocity gradient, sec⁻¹
         μ = viscosity, Ns/m² (lb·sec/ft²)
         $\forall$ = tank volume, m³ (ft³)

Generally accepted design standards require G to be between 30 and 60 sec⁻¹.

Time is also an important variable in flocculation, and the term $G\bar{t}$ is often used in design, where $\bar{t}$ = hydraulic retention time in the flocculation basin. Typically, $G\bar{t}$ values range from 1×10^4 to 1×10^5.

Example 5.1

Assume the Durham, North Carolina, water plant is designed for 30 mgd. The flocculator dimensions are 100 ft long, 50 ft wide, 16 ft deep.

Revolving paddles attached to four horizontal shafts rotate at 1.7 rpm. Each shaft supports four paddles which are 6 in. wide and 48 ft long. Paddles are centered 6 ft from shaft. Assume $C_D = 1.9$, and mean velocity of water is 35% of paddle velocity. Find the velocity differential between the paddles and the water. Also calculate the value of G and the time of flocculation.

The rotational velocity is

$$v_r = \frac{2\pi rn}{60}$$

where r = radius, ft
n = rpm

$$v_r = \frac{2\pi \times 6 \times 1.7}{60} = 1.07 \text{ ft/sec}$$

The velocity differential between paddles and fluid is therefore

$$v = 0.65 \times v_r = 0.65 \times 1.07 = 0.70 \text{ ft/sec}$$

Total power input is

$$P = \frac{-C_D A \rho v^3}{2} = \frac{1.9 \times 384 \times 1.94 \times (0.70)^3}{2} = 243 \text{ ft·lb/sec}$$

where C_D = drag coefficient = 1.9, obtained from any fluid mechanics handbook
A = paddle area = $4 \times 4 \times 48 \times (6/12) = 384 \text{ ft}^2$
v = velocity differential = 0.70 ft/sec
ρ = density of water = 1.94 at 50°F

$$G = \left(\frac{P}{\Psi \mu}\right)^{1/2} = \sqrt{\frac{243}{(8 \times 10^4) \times (2.73 \times 10^{-5})}} = 10.5 \frac{\text{ft/sec}}{\text{ft}}$$

where Ψ = $100 \times 50 \times 16 = 80,000 \text{ ft}^3$
μ = viscosity = 2.73×10^{-5} at 50°F.

The value of $G = 10.5 \text{ sec}^{-1}$ is a little low.
Time of flocculation

$$\bar{t} = \frac{\Psi}{Q} = \frac{80,000 \times 7.48 \times 24 \times 60}{30 \times 10^6} = 28.7 \text{ min}$$

This is within the accepted range.

SETTLING

When the flocs have been formed they must be separated from the water. This is invariably done in gravity settling tanks which simply allow the heavier-than-water particles to settle to the bottom. Settling tanks are designed so as to approximate uniform flow and to minimize all turbulence. Hence the two critical elements of a settling tank are the entrance and exit configurations. Figure 5-4 shows one type of entrance and exit configuration used for distributing the flow entering and leaving the water treatment settling tank.

Alum sludge is not highly biodegradable and thus will not decompose at the bottom of the tank. After some time, usually several weeks, the accumulation of alum sludge at the bottom of the tank is such that it has to be removed. Typically, the sludge exits through a *mud valve* at the bottom and is either wasted into a sewer or to a sludge holding/drying pond. We will see in Chapter 7 that sludges collected in wastewater treatment plants can remain in the bottom of the settling tanks only a matter of hours before starting to produce odoriferous gases, thus floating some of the solids.

The water leaving a settling tank is essentially clear. The polishing is performed with a rapid sand filter.

FILTRATION

In Chapter 4 we discussed movement of water into the ground and through soil particles, and alluded to the cleansing action the particles have on contaminants in the water. Picture the extremely clear water that bubbles up from "underground streams" in hills and valleys across the nation. The soil particles definitively help filter the groundwater, and through the years environmental engineers have learned to apply this natural process in water treatment and supply systems and developed what we now know as the rapid sand filter. The actual process of separating impurities from a carrying liquid by rapid sand filtration involves two phases: filtration and backwashing.

A slightly simplified version of the rapid sand filter is illustrated in a cut-away in Figure 5-5. Water from the settling basins enters the filter and seeps through the sand and gravel bed, through a false floor and out into a clear well which stores the finished water. During filtration valves A and C are open.

Eventually the rapid sand filter becomes clogged and must be cleaned. This cleaning is performed hydraulically. The operator first shuts off the

Figure 5-4   Schematic of a common settling tank.

Figure 5-5   A schematic drawing of a rapid sand filter.

flow of water to the filter (closing valves A and C), then opens valves D and B which allow wash water (clean water stored in an elevated tank or pumped from the clear well) to enter below the filter bed. This rush of water forces the sand and gravel bed to expand and jolts individual sand particles into motion, rubbing against their neighbors. The light colloidal material trapped within the filter is released and escapes with the wash water. After a few minutes, the wash water is shut off and filtration is resumed.

The solid impurities in the water are removed by numerous processes, the most important of which are straining, sedimentation, interception and diffusion (see Figure 5-6). Straining, possibly the most important mechanism in general, takes place almost exclusively in the first few centimeters of the filter medium. As the filtering process begins, straining removes only particles in the water which are large enough to get caught in the pores (A in Figure 5-6). After a time, these trapped particles themselves begin to form a screen which has smaller openings than the original filter medium. Smaller particles suspended in the water are trapped by this mat, and immediately begin acting as part of the screen. Thus, removal efficiency due to screening tends to increase in some proportion to the time of the filtration phase.

A second removal mechanism is sedimentation in which larger and heavier particles do not follow the fluid streamline around the sand grain and settle on the grain (B in Figure 5-6).

Interception occurs with particles that do follow the streamline, but are too large and are caught because they brush up against the sand grain (C in Figure 5-6).

Finally, very small particles are experiencing Brownian motion and may by chance hit the sand grain. This last process is called diffusion (D in Figure 5-6.

The first three mechanisms are most effective for larger particles, whereas diffusion can occur only for subcolloidal-sized particles. This is represented by a typical removal efficiency curve for different-sized particles (Figure 5-7) where the efficiency of removal is high for both large and small particles, and substantially reduced for midsized (about 1-micron) particles. Unfortunately, many viruses, bacteria and fine clay particles are about 1 micron in size, and thus the filter may not be an effective tool for the removal of these particles.

Filter beds are often classified as single medium, dual media or trimedia. The latter two are often utilized in the treatment of wastewater because they permit solids to penetrate into the bed, have more storage capacity, and thus increase the required time between backwashings. Also, multimedia filters tend to spread head loss buildup over time and further permit longer filter runs.

One of the primary considerations in filter design is the head loss through the sand. Obviously, as the sand gets progressively dirtier the head loss increases. Figure 5-8 shows a simplified representation of head

Figure 5-6   Mechanisms of solids removal in a filter: (A) straining, (B) sedimentation, (C) interception, (D) diffusion.

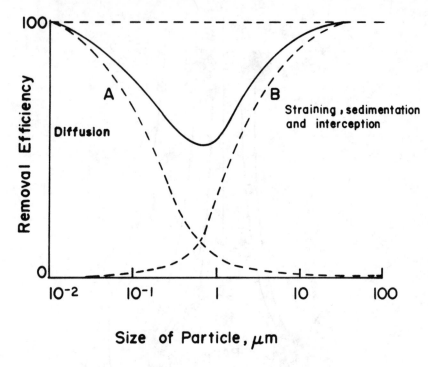

Figure 5-7   Removal of various-sized particles in a filter.

loss in a filter. Although it is not possible to predict without experimentation what head losses will be experienced in a given application, the head loss in clear sand can be estimated using several different equations. One of the earliest and still most widely used method is called the Carman-Kozeny equation.

A head loss in clean sand while filtering can be estimated by first considering the filter to be a mass of pipes, in which case the Darcy head loss equation holds;

$$h_L = f \frac{L}{D} \frac{v^2}{2g} \qquad (5\text{-}1a)$$

where   L = depth of filter
D = "pipe" diameter
v = velocity in "pipes"
f = friction factor.

But of course the "pipes", or channels through the sand, are not straight, and D varies, so we can substitute D = 4R where

Figure 5-8 Head loss in a filter as measured by a manometer. Note: h_L is head loss in a clean filter and H_L is head loss in a dirty filter.

$$R = \frac{area}{wetted\ perimeter} = \frac{\pi D^2/4}{\pi D} = \frac{D}{4} = \text{the hydraulic radius}$$

We thus have

$$h_L = f\, \frac{L}{4R}\, \frac{v^2}{2g} \qquad\qquad (5\text{-}1b)$$

The velocity of the water approaching the sand is

$$v_a = Q/A' \qquad\qquad (5\text{-}2a)$$

and the velocity through the bed is

$$v = \frac{v_a}{e} \qquad\qquad (5\text{-}2b)$$

where e = porosity, fraction of open spaces in the sand
          Q = flow rate
          A' = surface area of the sand bed.

The total channel volume is the porosity of bed times the total volume, or $e\Psi$. The total solids volume is $(1 - e)$ times the total volume, which is also equal to the number of particles times the volume of the particles. Thus total volume =

$$\frac{N\mathrm{V}_p}{1-e}$$

where N = number of particles
V_p = volume occupied by each particle.

The total channel volume is $e([N\mathrm{V}_p]/[1-e])$, and since the total wetted surface is NA_p (where A_p = surface area of each particle), the hydraulic radius

$$R = \frac{\text{Area}}{\text{wetted perimeter}} \propto \frac{\text{Volume}}{\text{Area}} = \frac{e([N\mathrm{V}_p]/[1-e])}{NA_p} = \left(\frac{e}{1-e}\right)\left(\frac{\mathrm{V}_p}{A_p}\right)$$

For spherical particles, $\mathrm{V}_p/A_p = d/6$, but for particles which are not true sphere $\mathrm{V}_p/A_p = \phi(d/6)$ where ϕ = shape factor. For example, $\phi = 0.95$ for Ottawa sand, a common filter sand. We thus have

$$R = \left(\frac{e}{1-e}\right)\left(\phi\,\frac{d}{6}\right) \qquad (5\text{-}3)$$

Substituting Equations 5-2 and 5-3 into 5-1b,

$$h_L = f_1 \left(\frac{L}{\phi d}\right)\left(\frac{1-e}{e^3}\right)\left(\frac{v_a^2}{g}\right) \qquad (5\text{-}4)$$

We can approximate the fraction factor by the equation

$$f = 150\left(\frac{1-e}{R}\right) + 1.75$$

where R = Reynolds number = $\phi\left(\dfrac{\rho v_a d}{\mu}\right)$.

The above discussion is applicable to a filter bed made of only one size particle. For a bed made of nonuniform sand grains,

$$d = \frac{6}{\phi}\left(\frac{\mathrm{V}}{A}\right)_{avg} \qquad d = \frac{6}{\phi}\left(\frac{\mathrm{V}_{avg}}{A_{avg}}\right) \qquad (5\text{-}5)$$

where V_{avg} = average volume of all particles
A_{avg} = average area of all particles.

Substituting Equation 5-5 into 5-4

$$h_L = f\left(\frac{L}{6}\right)\left(\frac{1-e}{e^3}\right)\left(\frac{v_a^2}{g}\right)\left(\frac{A}{V}\right)_{avg}$$

and $(A/V)_{avg}$ can be approximated as

$$\frac{6}{\phi}\sum\frac{x}{d'}$$

where   x = weight fraction of particles retained between any two sieves
         d' = geometric mean diameter between these sieves.

Thus

$$h_L = f_1\left(\frac{L}{\phi}\right)\left(\frac{1-e}{e^3}\right)\left(\frac{v_a^2}{g}\right)\sum\frac{x}{d'} \qquad (5\text{-}6)$$

This equation holds for nonstratified beds, such as are found in a slow sand filter, since the friction factor does not vary with depth. Where the particles are stratified, as in a rapid sand filter, we can write Equation 5-6 as

$$\frac{dh_L}{dL} = \left[\frac{1}{\phi}\left(\frac{1-e}{e^3}\right)\left(\frac{v_a^2}{g}\right)\right]f\frac{1}{d'}$$

The total head loss is thus

$$h_L = \int_0^{h_L}dh_L = \left[\frac{1}{\phi}\left(\frac{1-e}{e^3}\right)\left(\frac{v_a^2}{g}\right)\right]\int_0^L\frac{f}{d'}\,dL$$

Since dL = Ldx, where dx is the proportion of particles of size d

$$h_L = \left[\frac{1}{\phi}\left(\frac{1-e}{e^3}\right)\left(\frac{v_a^2}{g}\right)\right]L\int_{x=0}^{x=1}f\frac{dx}{d}$$

and if the particles between adjacent sieve sizes are assumed uniform,

$$h_L = \frac{L}{\phi}\left(\frac{1-e}{e^3}\right)\left(\frac{v_a^2}{g}\right)\sum\frac{fx}{d'}$$

Example 5.2
    A sand consisting of the following sizes is used[1]

[1] Adapted from *Water Supply and Pollution Control,* 3rd ed. by J. W. Clark, W. Viessman and M. J. Hammer. (New York: Thomas Y. Crowell Company, Inc., 1977).

Sieve Number	Percentage of sand retained on sieve, $\times 10^2$	Geometric mean size of sand, $ft \times 10^{-3}$
14–20	1.10	3.28
20–28	6.60	2.29
28–32	15.94	1.77
32–35	18.60	1.51
35–42	19.10	1.25
42–48	17.60	1.05
48–60	14.30	0.88
60–65	5.10	0.75
65–100	1.66	0.59

The filter bed measures 20×20 ft and is 2 ft deep. The sand has a porosity of 0.40 and a shape factor of 0.95. The filtration rate is 4 gal/min·ft². Assume the viscosity is 3×10^{-5} lb·sec/ft². Find the head loss through the clean sand.

The solution is shown in tabular form:

Reynolds Number	Friction factor, f	$\dfrac{x}{d}$	$f\dfrac{x}{d}$
1.80	51.7	3.4	1.74
1.37	67.4	28.8	1,941
1.06	86.6	90.1	7,802
0.91	100.6	123.2	12,394
0.75	121.7	152.8	18,595
0.63	144.6	167.6	24,235
0.53	171.5	162.5	27,868
0.45	201.7	68.0	13,715
0.35	258.8	28.1	7,272

Column 1: The approach velocity is

$$v_a = 4 \ \frac{gal}{min \cdot ft^2} \times \frac{1}{7.481} \ \frac{ft^3}{gal} \times \frac{1}{60} \ \frac{min}{sec} = 8.9 \times 10^{-3} \ ft/sec$$

For the first sized particle, $d = 3.28 \times 10^{-3}$ ft

$$\mathbf{R} = \phi \ \frac{\rho v_a d}{\mu} = \frac{(0.95)(1.94)(89 \times 10^{-3})(3.28 \times 10^{-3})}{3 \times 10^{-5}} = 1.8$$

Column 2:

$$f = 150 \left(\frac{1-e}{R} \right) + 1.75$$

$$= 150 \left(\frac{1-0.4}{1.8} \right) + 1.75 = 51.75$$

Columns 3 and 4:
for the first size, $x = 1.10\%$ and $d = 3.28 \times 10^{-3}$

$$f \frac{x}{d} = \frac{51.75(0.011)}{3.28 \times 10^{-3}} = 174$$

The last column is summed: $\Sigma f(x/d) = 113{,}977$

$$h_L = \frac{L}{\phi} \left(\frac{1-e}{e^3} \right) \left(\frac{v_a^2}{g} \right) \Sigma f \frac{x}{d}$$

$$= \frac{2}{0.95} \left(\frac{1-0.4}{0.4^3} \right) \left(\frac{(8.9 \times 10^{-3})^2}{32.2} \right) 113{,}977$$

$$= 5.78 \text{ ft.}$$

The deposition of material during the filtering process increases the head loss through the filter. A method of predicting head loss takes advantage of this limitation, predicting head loss

$$H_L = h_L + \sum_{i=1}^{n} (h_{L_i})_t$$

Here H_L is the total head loss through the filter, h_L is the clear water head loss at time zero, and $(h_{L_i})_t$ is the head loss in the ith layer of media in the filter at time t. (All values are in meters.) Head loss within an individual layer, $(h_{L_i})_t$, is related to the amount of material caught by that layer:

$$(h_{L_i})_t = x(q_i)_t^y$$

Here, $(q_i)_t$ is the amount of material collected in the ith layer at time t, in mg/cm^3, and x and y are experimental constants. Head loss data for sand and anthracite are summarized in Figure 5-9.

A filter run is the time a filter operates before it has to be cleaned. The end of a filter run is indicated in one of two ways:

Figure 5-9   Increased head loss in a filter due to materials deposits. Adapted from *Wastewater Engineering* by Metcalf and Eddy, Inc. (New York:McGraw-Hill Book Company, 1979).

- excessive head loss
- excessive turbidity of the filtered water.

If either of these occurs the filter must be washed.

Example 5.3

The filter described in Example 5.2 is run so that it reduces the suspended solids from 55 mg/l to 3 mg/l during a filter run of 4 hours. Assume this material is captured in the first 6 in. of sand, where the sand train size is adequately described as 0.8 mm. What is the head loss of the dirty filter?

The total water through the filter is

$$4 \ \frac{gal}{min \cdot ft^2} \times (20 \ ft \times 20 \ ft) \times 4 \ hr \times 60 \ \frac{min}{hr} = 384{,}000 \ gal$$

$$384{,}000 \ gal \times 3.785 \ \frac{liter}{gal} = 1.45 \times 10^6 \ liter$$

Solids removal is $55 - 3 = 52$ mg/l

$$52 \ mg/l \times 1.45 \times 10^6 \ l = 75.4 \times 10^6 \ mg$$

Effective filter volume $= 20 \ ft \times 20 \ ft \times \frac{1}{2} \ ft = 200 \ ft^3$

$$200 \ ft^3 \times 0.02832 \ \frac{m^3}{ft^3} \times 10^6 \ \frac{cm^3}{m^3} = 5.65 \times 10^6 \ cm^3$$

Suspended solids removed per unit filter volume is

$$\frac{75.4 \times 10^6 \ mg}{5.65 \times 10^6 \ cm^3} = 13.3 \ mg/cm^3$$

From Figure 5-8, the $h_L \cong 5m = 16.45$ ft, so the total head loss of the dirty filter is

$$5.78 \ ft + 16.45 \ ft = 22.23 \ ft.$$

Note that in the preceding example, the head loss of the dirty filter exceeds the total head available (if we assume that the filter is perhaps 10 ft deep). Obviously, a very dirty filter, as in the above example, has a region of negative pressure in the filter bed. This is not recommended since the negative pressure can cause gases to come out of solution and thus further impede the effectiveness of filtration.

The flow rate in most filters is controlled by a valve which allows only a given volume of water to pass through, regardless of the pressure. Such *rate controllers* allow the filter to operate at a constant rate, as shown by curve A in Figure 5-10. An alternative method of filter operation is to allow the water to flow through at a rate governed by the head loss, as shown by curve B in Figure 5-10. The relative advantages of these two methods of filter operation are still being debated in the profession.

DISINFECTION

After filtration, the finished water is often disinfected with chlorine (Step 5 in Figure 5-1). Disinfection is the process of killing the remaining

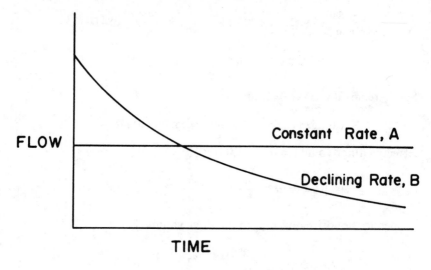

FLOW

Constant Rate, A

Declining Rate, B

TIME

Figure 5-10   Flow through a filter if operated at constant rate or declining rate.

microorganisms in the water, some of which may be pathogenic. Chlorine from bottles or drums is fed in correct proportions to the water in order to obtain a desired level of chlorine in the finished water. When chlorine comes in contact with any organic matter, including microorganisms, it oxidizes this material and is in turn reduced to inactive chlorides.

Chlorine gas is very soluble in water and rapidly forms hypochlorous acid through hydrolysis:

$$Cl_2 + H_2O \rightleftharpoons HOCl + H^+ + Cl^-$$

The hypochlorous acid itself ionizes:

$$HOCl \rightleftharpoons OCl^- + H^+$$

At temperatures found in water supply systems, the hydrolysis of chlorine is generally complete in a matter of seconds, while the ionization of hypochlorous acid is instantaneous. Both HOCl and OCl$^-$ are effective disinfectants and are called free available chlorine in water. This free available chlorine kills pathogenic bacteria and thus disinfects the water.

Many water plant operators prefer to maintain a residual of chlorine in the water, thus assuring that if within the distribution system organic matter, such as bacteria, enters the water there is sufficient chlorine present to eliminate this potential health hazard.

It must be noted that the secondary effects of chlorine addition are poorly understood. Chlorine may combine with synthetic organics in the water to possibly produce dangerous halogenated compounds, and the long-term health effects of such compounds as trihalomethane and chloroform are unknown.

In addition to chlorine, a growing number of municipalities now have fluoride added to the water. At proper concentration fluoride will do much to prevent dental decay in children and young adults.

From the clear well (Step 6 in Figure 5-1) the water is pumped to the distribution system. This is a closed network of pipes, all under pressure. In most cases water is pumped to an elevated storage tank which not only serves to equalize pressures but provides storage for fires and other emergencies as well.

CONCLUSION

Water treatment is often necessary if surface water supplies, and sometimes groundwater supplies, are to be available for human consumption. Because the vast majority of cities across the nation use one water distribution system for households, industries and fire control, large quantities of water often must be available to satisfy the "highest use," which is usually drinking water. Does it make sense to go to the trouble and expense of producing high-quality water, only to use it for lawn sprinkling? This is not an easy question, and the problems of future water supply have prompted the serious consideration of dual water supplies: one of high quality and one of a lower quality, perhaps reclaimed from wastewater. Many engineers are convinced that the next major environmental engineering concern will be the availability and production of potable water. The job, therefore, is far from done.

PROBLEMS

5.1 A water treatment system must be able to deliver 15 mgd of water to a city of 150,000 people. Estimate: (1) the diamteter of three equally sized mixing basins 10 ft deep, with detention times of 2 min, (2) length, width and corresponding surface area of three flocculator basins that are 10 ft deep, (3) surface area of three settling basins that are 10 ft deep with 2-hr detention periods, and (4) the required area of each of 15 rapid sand filters rated at 2 gpm/ft^2.

5.2 An engineer suggests the following design parameters for a city's proposed rapid sand filter: flow rate $= 0.6$ m³/sec, loading rate to filter $= 125.0$ m³/day·m². How much surface area is required for the filter? Select the number of equally sized filters, and size these filters assuming a width to length ratio of 1.0 to 2.5 with a maximum surface area of each filter tank of 75 m².

5.3 A flocculator is designed to treat 25 mgd. The flocculator basin is 100 ft long, 50 ft wide and 20 ft deep, and is equipped with 12-in. paddles (48 ft long). The paddles are attached to four horizontal shafts, two paddles per shaft, their center line is 8.0 ft from the shaft, and they rotate at 2.5 rpm. Assume the velocity of the water is 25% of the velocity of the paddles and water temperature is 50°F.

Calculate: (1) the time of flocculation, (2) the velocity differential between the water and the paddle, (3) the hydraulic power (and subsequent energy consumption) if the drag coefficient is 1.7 and the velocity differential between water and paddles is 70% of the linear velocity of the blades, and (4) the value of G and Gt̄

5.4 Discuss utilization and disposal options for water treatment sludges collected in the settling tanks following flocculation basins.

LIST OF SYMBOLS

ϕ = shape factor

ρ = fluid density kg/m³ (lb· sec²/ft⁴)

μ = fluid viscosity (lb· sec/ft²)

V = tank volume, m³ (ft³)

V_p = volume occupied by each particle, m³ (ft³)

A = paddle area, m² (ft²)

A' = surface area of the sand bed, m² (ft²)

A_p = surface area of the sand particles, m² (ft²)

C_D = drag coefficient

D = diameter of "pipe," m (ft)

d = diameter of sand particle, m (ft)

d' = geometric mean diameter between sieve sizes, m (ft)

e = porosity fraction of open spaces in the sand

f = friction factor

G = velocity gradient, sec⁻¹

g = acceleration due to gravity, m/sec² (ft/sec²)

H_L = total head loss through a filter, m (ft)

h_L' = clear water head loss through a filter, m (ft)

L = depth of filter, m (ft)

n = revolutions per minute

$(h_{L_i})_t$ = head loss in ith layer of media at time t

P = power N/sec (ft· lb/sec)

Q = flow rate, m^3/sec (gal/min)

R = hydraulic radius, m (ft)

R = Reynolds number

r = radius, m (ft)

$\bar{t}$ = time of flocculation, min

V = velocity of the paddle relative to the fluid, m/sec (ft/sec)

v = velocity of water through sand bed, m/sec (ft/sec)

v_a = velocity of water approaching sand, m/sec (ft/sec)

x = weight fraction of particles retained between two sieves

SUGGESTED READING

American Water Works Association, *Water Quality and Treatment* (New York: McGraw-Hill Book Company, 1972).

Fair, G. M., J. C. Geyer and D. A. Okun. *Water and Wastewater Engineering, Vol. 2,* (New York: John Wiley & Sons, Inc., 1968).

Office of Technology Transfer, *Process Design Manual for Carbon Adsorption,* U. S. Environmental Protection Agency, Washington, DC (1973).

Chapter 6

Collection of Wastewater

"The Shambles" is both a street and an area in London, and during the eighteenth and nineteenth centuries was a highly commercialized area, with meat packing as a major industry. The butchers in those days would throw all of their wastes into the street where it was washed away by rainwater into drainage ditches. The condition of the area was so bad that it contributed its name to the English language.

In old cities, drainage ditches like the ones at the Shambles were constructed for the sole purpose of moving stormwater out of the cities. In fact, it was illegal in London to discard human excrement into these ditches. Eventually, these ditches were covered over and became what we now know as *storm sewers.*

As water supplies developed and the use of the indoor water closet increased, the need for transporting domestic wastewaters, called sanitary wastes, became obvious. This was accomplished in one of two ways: (1) discharge of the sanitary wastes into the storm sewers, which then carried both sanitary wastes and stormwater, and were known as *combined sewers,* and (2) construction of a new system of underground pipes for removing the wastewater, which became known as *sanitary sewers.*

Newer cities, and more recently built (post-1900) parts of older cities almost all have separate sewers for sanitary wastes and stormwater. In this chapter, storm sewer design is not covered in detail, since this is discussed in Chapter 9. Emphasis here is on estimating the quantities of domestic and industrial wastewaters, and in the design of the sewerage systems to handle these flows.

ESTIMATING WASTEWATER QUANTITIES

The term *sewage* is used here to mean only domestic wastewater. In addition to sewage, however, sewers also must carry

- industrial wastes
- infiltration
- inflow

The quantity of industrial wastes can usually be established by water use records. Alternatively, the flows can be measured in manholes which serve only a specific industry, using a small flow meter in a manhole. Typically, a parshall flume is used, and the flow rate is calculated as a direct proportion of the flow depth. Industrial flows often vary considerably throughout the day and continuous recording is mandatory.

Infiltration is the flow of groundwater into sanitary sewers. Sewers are often placed under the groundwater table and any cracks in the pipes will allow water to seep in. Infiltration is the least for new, well-constructed sewers, and can go as high as 500 m^3/km-day (200,000 gal/mi-day). Commonly, for older systems, 700 m^3/km-day (300,000 gal/mi-day) is used in estimating infiltration. This flow is of course detrimental since the extra volume of water must go through the sewers and the wastewater treatment plant. It thus makes sense to reduce this as much as possible by maintaining and repairing sewers, and keeping sewerage easements clear of large trees which could send roots into the sewers and cause severe damage.

The third source of flow in sanitary sewers is called *inflow,* and represents stormwater which is collected unintentionally by the sanitary sewers. A common source of inflow is a perforated manhole cover placed in a depression, so that stormwater flows into the manhole. Sewers laid next to creeks and drainageways which rise up higher than the manhole elevation, or where the manhole is broken, are also a major source. Lastly, illegal connections to sanitary sewers, such as roofdrains, can substantially increase the wet weather flow over the dry weather flow. Commonly, the ratio of dry weather to wet weather flow is between 1:1.2 and 1:4.

Domestic wastewater flows vary with season, day of the week and the hour of the day. Figure 6-1 shows a typical daily flow for a residential area.

The three flows of concern when designing sewers are the average flow, the peak or maximum flow, and the extreme minimum. The ratios of average to both the maximum and minimum flows is a function of the total flow, since a higher average daily discharge implies a larger community in which the extremes are evened out. Figure 6-2 is a plot showing commonly experienced ratios of average to the extremes as a function of the average daily discharge.

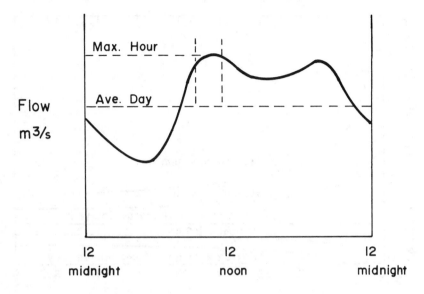

Figure 6-1 Typical dry-weather wastewater flow for a residential area.

SYSTEM LAYOUT

Sewers that collect wastewater from residences and industrial establishments almost always operate as open channels, or gravity flow conduits. Pressure sewers are used in a few places, but these are expensive to maintain and are useful only when there either are severe restrictions on water use, or the terrain is such that gravity flow conduits cannot be efficiently constructed.

A typical system for a residential area is shown in Figure 6-3. Building connections are usually made with clay or plastic pipe, 6 in. in diameter, to the *collecting sewers* which commonly run under the street. Collecting sewers are sized to carry the maximum anticipated peak flows without surcharging (filling up) and are commonly made of clay, asbestos, cement, concrete or cast iron pipe. They discharge in turn into intercepting sewers, known colloquially as *interceptors,* which collect large areas and discharge finally into the wastewater treatment plant.

Collecting and intercepting sewers must be placed at a sufficient grade to allow for adequate velocity during low flows, but not so great as to promote excessively high velocities when the flows are at their maximum. In addition, sewers must have manholes, commonly every 120–180 m (400–600 ft) to facilitate cleaning and repair. Manholes are also necessary

Figure 6-2   Relation of the average daily flow of domestic wastewater to the extremes of flow.

whenever the sewer changes grade (slope), size or direction. Typical manholes are shown in Figure 6-4.

In some cases it becomes either impossible or uneconomical to use gravity flow, and the wastewater must be pumped. A typical packaged pumping station is shown in Figure 6-5, and its use is indicated on the typical system layout in Figure 6-3.

SEWER HYDRAULICS

The design of sewers begins by selecting a reasonable layout and establishing the expected flows within each pipe linking the manholes. For large systems, this often involves the use of economic analysis to determine exactly what routing provides the optimal system. For most smaller systems, this is an unnecessary refinement, and a reasonable system can be drawn by eye.

The average discharge is estimated on the basis of the population

Figure 6-3 Typical wastewater collection system layout [after Clark, J., W. Viessman and M. Hammer *Water Supply and Sewerage* (New York: IEP, 1977)].

served in the drainage area, and the maximum and minimum flows are calculated using Figure 6-2. Once this is done, the design is a search for the right pipe diameter and grade (slope) which will allow the minimum flow to exceed a velocity necessary for the conveyance of solids, while keeping the velocity at maximum flow less than a limit at which undue erosion and structural damage can occur to the pipes. Commonly, the velocity should be held between these limits:

- minimum—0.6 m/sec (~2 ft/sec)
- maximum—3.0 m/sec (~10 ft/sec).

Figure 6-4   Typical manholes used for collecting sewers (courtesy ASCE).

Figure 6-5 Typical pumping station for domestic wastewater (courtesy of Gorman-Rupp).

The velocity in sewers is usually calculated using the Manning equation, derived originally from the Chezy open channel flow equation

$$v = c\sqrt{rs}$$

where   v = velocity of the flow, m/sec
        r = hydraulic radius of the pipe, or the area divided by the wetted perimeter, m
        s = slope of the pipe
        c = the Chezy coefficient.

For circular pipes, the Manning equation is derived by setting

$$c = \frac{k}{n} r^{1/6}$$

where   n = roughness factor
        k = constant

so that

$$v = \frac{k}{n} r^{2/3} s^{1/2}$$

If v is in ft/sec and r in ft, k = 1.486. In metric units with v as m/sec and r as m, k = 1.0. The slope s is, of course, dimensionless, calculated as the fall over distance. The n term is dependent on the pipe material, with n increasing as pipes get rougher. Table 6-1 is a listing of n values for some types of sewer pipe.

Example 6.1
An 8-in. diameter cast iron sewer is to be set at a grade (slope) of 1-m fall per 500-m length. What will be the flow in this sewer when flowing full?
Using common units, and noting that n = 0.013 from Table 6-1,

$$V = \frac{1.486}{0.013} \left[\frac{\pi(8/12)^2}{4} \div (\pi\, 8/12) \right]^{2/3} \left[\frac{1}{500} \right]^{1/2}$$

$$V = 1.54 \text{ ft/sec}$$

and since the area is $[\pi(8/12)^2]/4 = 0.35 \text{ ft}^2$,

$$Q = AV = (0.35)(1.54) = 0.54 \text{ cfs}$$

Table 6-1.   Manning Roughness Coefficient, n

Type of Channel, Closed Conduits	Roughness Coefficient, n
Cast Iron	0.013
Concrete	
straight	0.011
with bends	0.013
unfinished	0.014
Clay, Vitrified	0.012
Corrugated Metal	0.024
Brickwork	0.013
Sanitary Sewers, coated with slime	0.013

For circular conduits flowing full, the Manning formula can be re-arranged as

$$Q = \frac{0.00061}{n} D^{8/3} s^{1/2}$$

where   Q = discharge, cfs
         D = pipe diameter, in.

This calculation can be simplified by using a nomograph which represents a graphical solution of the Manning equation. Figures 6-6 and 6-7 are such nomographs in both common and metric units.

It should be emphasized that the Manning equation is used to calculate the velocities in sewers flowing full, hence the maximum expected flow in the sewer is used. For flows less than the maximum, the velocity is calculated using the *hydraulic elements* graph shown as Figure 6-8. In this chart, the hydraulic radius (area/wetted perimeter) is calculated for various ratios of the depth partially full (d) over the depth full (D), the latter of course being the inside diameter of the pipe. The velocity is calculated using the Manning equation for any depth d, and the discharge is obtained from multiplying the velocity times the area. From experimental evidence, it has been found that the friction factor n also varies slightly with the depth (as noted in Figure 6-8), but this is often ignored in the calculations.

It should be noted that the maximum flow in a given sewer occurs not

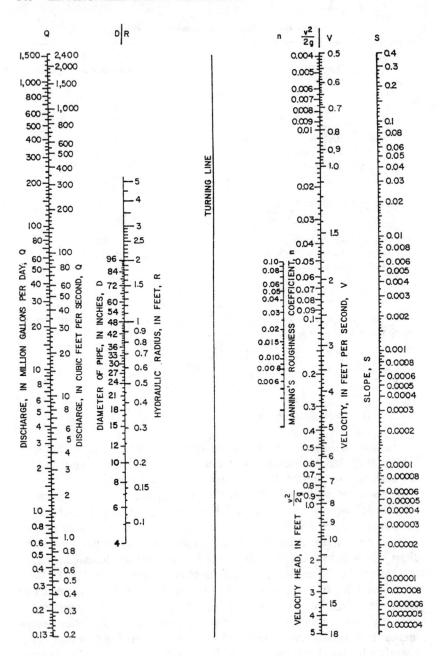

Figure 6-6   Nomograph for the solution of Manning's equation, common units.

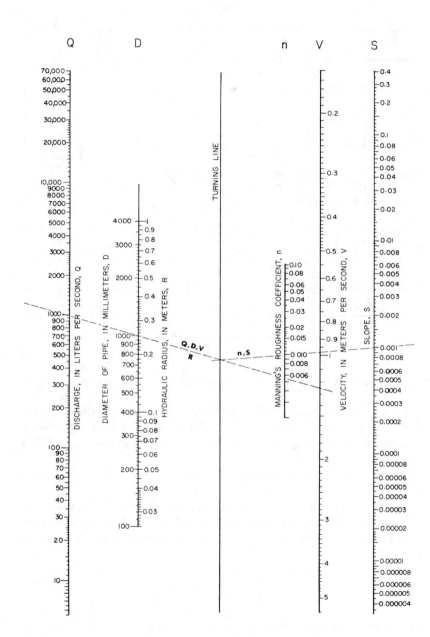

Figure 6-7   Nomograph for the solution of Manning's equation, metric units.

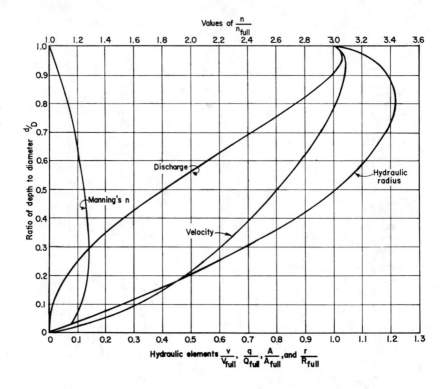

Figure 6-8   Hydraulic elements chart for open channel flow.

when it is full, but when d/D is about 0.95. This is because the additional flow area gained by increasing the d/D from 0.95 to 1.00 is very small compared to the large additional pipe surface area which the flow experiences.

Example 6.2

From Example 6.1, it was found that the flow carried by an 8-in. cast iron pipe flowing full at a grade of 1/500 is 0.54 cfs with a velocity of 1.54 ft/sec. What will be the velocity in the pipe when the flow is only 0.1 cfs?

$$\frac{q}{Q} = \frac{0.1}{0.54} = 0.185$$

From the hydraulic elements chart

$$\frac{d}{D} = 0.33$$

and d = 0.33 (8 in.) = 2.64 in.

$$\frac{v}{V} = 0.64$$

and v = 0.64 (1.54 ft/sec) = 0.98 ft/sec.

In cases where the desired slope between manholes is not sufficient to maintain adequate velocities at minimum flow, additional slope must be provided, or a larger pipe diameter used. On the other hand, if the slope of the pipe is too steep, it must be reduced, and often a drop manhole is used, which is in effect a method of wasting energy (see Figure 6-4). Standard American convention requires that a head loss of 0.1 ft (0.06 m) be allowed for a manhole, and an additional 0.1 ft be allowed for each lateral. The invert elevation of the pipe leaving the manhole is thus at least 0.1 ft lower than the inverted elevation of the pipe coming in.

Example 6.3

The system shown in Figure 6-9 is to be designed given the following flows: maximum flow = 3.2 mgd, minimum flow = 0.2 mgd, and the minimum and maximum allowable velocities are 2 and 12 fps respectively. All manholes should be about 10 ft deep and there is no additional flow between manhole 1 and manhole 4. Design acceptable invert elevations for this system. [Note: The nomograph (Figure 6-6) is used in this solution.]

From 1 to 2: The street slopes at 2/100, so choose the slope of the sewer, s = 0.02. Assume n = 0.013 and try D = 12 in. From the nomograph, connecting n = 0.013 and s = 0.02, hitting the turning line, then connecting the point or the turning line with D = 12 in., read Q = 3/2 mgd and V = 6.1 ft/sec. This is acceptable as the sewer flowing full. To check for minimum velocity, q/Q = 0.2/3.2 = 0.063, and from the hydraulic elements chart (Figure 6-8), read v/V = 0.48, and v = 0.48 (6.1 ft/sec) = 2.9 ft/sec. The downstream invert elevation of manhole 1 is ground elevation minus 10 ft, or 62.0. The upstream invert elevation of manhole 2 is thus 62.0 − 2.0 = 60.0 ft. Allowing 0.1 ft for head loss in the manhole, the downstream invert elevation is 59.9 ft.

From 2 to 3: The slope here will be a problem due to rock. Try a larger pipe, D = 18 in. From the nomograph, V = 2.75 ft/sec for Q = 3.2 mgd. From the hydraulic elements chart, v/V = 0.48 and v = 0.48 (2.75 ft/sec) = 1.32 ft/sec. Too low. So even with a slope of 0.002 (from the nomograph), the velocity at minimum flow is too low. Try s = 0.005 and D = 18 in., reading Q = 4.7 mgd and V = 4.1 ft/sec. From hydraulic elements, v = 1.9 ft/sec, which is close enough. Use D = 18 in., and s = 0.005. Thus

Figure 6-9 Sewer layout for Example 6.3.

the upstream invert of manhole 3 is $59.9 - (0.005)(200) = 58.9$ ft and the downstream is at 58.8, still well above the rock.

From 3 to 4: This street obviously has too much slope. Try using $D = 12$ in., and a $s = (58.8 - 40)/100 = 0.188$ (40 ft is the desired invert elevation of manhole 4). Read $Q = 8.5$ mgd and $V = 19$ ft/sec. But only 3.2 mgd is required at maximum flow, hence $q/Q = 3.2/8.5 = 0.38$ and $v = 0.78 \times 18$ ft/sec $= 14.8$ ft/sec. This is too high. Use a drop manhole, with the invert at, say, elevation 45.0. In this case $s = 0.138$, and $Q = 7.3$ mgd, $V = 14.5$ ft/sec, $q/Q = 3.2/7.3 = 0.44$ and $v = 0.85 \times 14.5$ ft/sec $= 12.3$ ft/sec. Close enough! The upstream invert of manhole 4 is thus at 45.0 ft, and the downstream invert elevation can be at, say, 40.0 ft. There is, of course, little need to check the minimum velocity in this case. The final design is shown in Figure 6-10.

Often it is convenient to organize sewer calculations according to a table such as Figure 6-11. This is a slightly simplified version of a table suggested by the American Society of Civil Engineers and the Water Pollution Control Federation *Manual of Practice on Sanitary Sewer Design*. This volume should be consulted for the details of accepted engineering practice in the design of sewers.

CONCLUSION

The methods discussed in this chapter are crude hand calculations, and few modern consulting engineers who design sewerage systems will use such cumbersome methods. Computer programs which not only solve for the hydraulic parameters but evaluate the economic alternatives (e.g., a deeper cut or bigger pipe) are widely used. Elegant programs, however, are only useful if the design process is well understood.

PROBLEMS

6.1 A vitrified clay pipe sewer is to discharge 4 cfs when laid on a grade of 0.0016 and flowing half full. What is the required diameter?

6.2 An 18-in. sewer, $n = 0.013$, is 15000-ft long and laid on a uniform grade. Difference in elevation of the two ends is 4.8 feet. Find velocity and discharge when sewer is flowing 0.6 full.

6.3 A 48-in. circular sewer is required to discharge 100 cfs when full. What is the required grade, according to the Manning formula, if $n = 0.015$?

Figure 6-10   Final design for Example 6.3.

										LINE NO.
										LOCATION
										FROM / MAN-HOLE NO.
										TO
										LENGTH ft.
										INCREMENT / AREA ACRES
										TOTAL
										INFILTRATION / MAX. DISCHARGE TOTAL
										SEWAGE mgd
										mgd / INFIL-TRATION / SEWAGE
										cfs
										INFILTRATION / MIN. DISCHARGE TOTAL
										SEWAGE mgd
										mgd / INFIL-TRATION / SEWAGE
										cfs
										SLOPE OF SEWER %
										DIAMETER in.
										CAPACITY FULL cfs
										VELOCITY FULL ft/sec
										MIN. VELOCITY ft/sec
										MAX. VELOCITY ft/sec
										MAX. DEPTH ft
										MAX. VEL. HEAD ft
										MAX. ENERGY ft
										MANHOLE LOSS ft
										M.HOLE INVERT DROP ft.
										FALL IN SEWER ft
										UPPER END / SEWER INVERT ELEVATION
										LOWER END
										UPPER END / GROUND SURFACE ELEVATION
										LOWER END

Figure 6-11   Work sheet for the design of sewers.

6.4 A 12-in. concrete sewer, flowing just full, is laid on a grade of 0.00405 ft/ft. Find the velocity and rate of discharge.

6.5 A 12-in. sewer, n = 0.013, is laid on a grade of 3.0 ft/1000 ft. Find the velocity and discharge when the sewer is flowing 0.4 full.

LIST OF SYMBOLS

c = Chezy coefficient
D = diameter of pipe, m (in.)
d = depth of flow in a pipe, m (in.)
n = Manning roughness coefficient
Q = flow rate, full pipe, m³/sec (gal/min)
q = flow rate, partially full pipe, m³/sec (gal/min)
r = hydraulic radius, m (ft)
s = slope
V = velocity, full pipe, m/sec (ft/sec)
v = velocity in a partially full pipe, m/sec (ft/sec)

SUGGESTED READING

Linsley, R. K., and J. B. Franzini. *Elements of Hydraulic Engineering,* (New York: McGraw-Hill Book Company, 1955).

Clark, J. W., W. Viessman, Jr. and M. J. Hammer, *Water Supply and Pollution Control* (New York: IEP, 1977).

Steel, E. W. and T. J. McGhee. *Water Supply and Sewerage* (New York: McGraw-Hill Book Company, 1979).

Simon, A. L. *Practical Hydraulics* (New York: John Wiley & Sons, 1976).

Metcalf & Eddy Inc. *Wastewater Engineering* (New York: McGraw-Hill Book Company, 1979).

Chapter 7
Wastewater Treatment

As civilization developed and cities grew, domestic sewage and industrial wastes were eventually discharged into drainage ditches and sewers, and the entire contents commonly emptied into the nearest watercourse. In the case of major cities, this discharge was often sufficient to destroy even a large body of water. As Samuel Taylor Coleridge observed:

> In Köln, a town of monks and bones
> And pavements fanged with murderous stones
> And rags, and bags, and hideous wenches;
> I counted two and seventy stenches,
> All well defined, and several stinks!
>
> Ye Nymphs that reign o'er sewers and sinks,
> The river Rhine, it is well known,
> Doth wash your city of Köln;
> But tell me Nymphs! What power divine
> Shall hence forth wash the river Rhine?

During the nineteenth century, the River Thames was so grossly polluted that the House of Commons had to have rags soaked in lye stuffed into the cracks in the windows of Parliament to reduce the stench.

Beginning with the pioneering work in the United States and England, sanitary engineering technology eventually developed to the point where it became economically, socially and politically feasible to treat the wastewater so as to reduce its adverse impact on watercourses. In this chapter, this technology is reviewed, beginning with the simplest (and earliest) treatment systems, and concluding with a description of the most advanced systems in use today. The discussion begins, however, by reviewing the characteristics of wastewaters which make their disposal difficult, and showing why wastewater disposal cannot always be done onsite, and sewers and centralized treatment plants become necessary.

149

WASTEWATER CHARACTERISTICS

The discharges in a sanitary sewerage system are comprised of domestic wastewater, industrial discharge and infiltration. Infiltration, of course, is only a problem in that it adds to the total volume of wastewater and seldom is it directly a cause of concern in wastewater disposal. Industrial wastes are of course another matter. These discharges vary widely with the size and type of industry, and the amount of treatment applied by the industry prior to discharge into public sewers. In the United States, the trend has been toward a higher degree of pretreatment (see Chapter 10), prompted in part by regulations limiting discharges and by the imposition of local sewers surcharges. The latter are charges levied by a community to help pay for the extra treatment which an unusual discharge would require. The operation of sewer surcharges is illustrated by the example below.

Example 7.1

A community has a sewer use charge, based on water use, of $0.20/m^3. In addition, for discharges which exceed the limits of BOD = 250 mg/l and suspended solids (SS) = 300 mg/l, the discharger is to pay $0.50/kg of BOD and $1.00/kg SS. A chicken-packing plant uses 2000 m^3 of water per day, and discharges a wastewater with a BOD = 480 mg/l and SS = 1530 mg/l. What is its daily wastewater disposal bill?

The sewer use charge is 2000 m^3/day × $0.20/m^3 = $400/day. The BOD discharged is 480 mg/l × 2000 m^3/day × 1000 l/m^3 = 960 × 10^6 mg/day = 960 kg/day. The allowable BOD is 250 mg/l × 2000 m^3/day × 1000 l/m^3 × 10^{-6} kg/mg = 500 kg/day. Thus the extra BOD is 960 − 500 = 460 kg/day, and the surcharge is 460 kg/day × $0.5/kg = $230/day. Similarly, the allowable SS = 300 × 2000 × 1000 × 10^{-6} = 600 kg/day while the SS discharged is 1530 × 2000 × 1000 × 10^{-6} = 3060 kg/day. The surcharge for SS is thus (3060 − 600) × 1.00 = $2460/day. The total charge to the industry is $400 + $230 + $2460 = $3090/day.

Commonly, the problems with industrial discharges are not BOD and SS, both of which can be readily reduced in a wastewater treatment plant, but chemicals such as toxic metals, radioactive materials, refractory organics, etc. Typically, local communities place tight restrictions on such discharges and thus force the industries to pretreat the wastewater prior to discharging to the public sewers.

The third component of municipal wastewaters, domestic sewage, tends to vary substantially over time and from one community to the next. For example, the BOD and SS for a small community vary as shown in Figure 7-1. During early morning hours, little domestic waste-water is discharged, and a substantial part of the dry weather flow is infiltration, thus resulting in weak sewage.

On the average, however, it is instructive to consider "typical" values for the characteristics of domestic wastewater. Some of these are shown in Table 7-1.

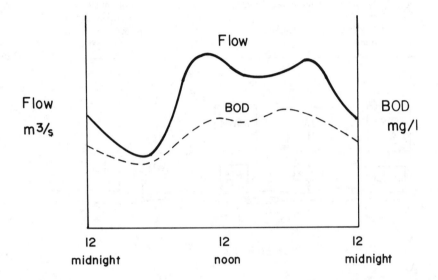

Figure 7-1   Daily variations in flow and BOD in a domestic wastewater from a small community.

ONSITE WASTEWATER DISPOSAL

Environmental engineers have been severely (and sometimes rightly) criticized for having a "sewer syndrome"—they want to collect all waste-water and provide the treatment at a central location. Often this approach does not make much sense.

Consider a situation depicted in Figure 7-2, where two wastewater treat-ment options are shown—a centralized treatment plant, and several smaller plants—all discharging their effluents into the same river. The single large plant obviously must provide extremely good treatment in order to attain acceptable dissolved oxygen levels downstream. On the other hand, the smaller plants could take advantage of the assimilative capacity of the river, and would not necessarily have to provide the same high degree of treatment. The logical extension of this idea is to not have

Table 7-1   Characteristics of a Typical Domestic Wastewater

Parameter	Typical Value for Domestic Sewage
Biochemical Oxygen Demand	250 mg/l
Suspended Solids	220 mg/l
Phosphorus	8 mg/l
Organic and Ammonia Nitrogen	40 mg/l
pH	6.8
Chemical Oxygen Demand	500 mg/l
Total Solids	720 mg/l

SCHEME 1

SCHEME 2

Figure 7-2   Two wastewater treatment options for several small communities.

any treatment plants at all, but dispose of the wastewater onsite, with each house or building having its own treatment system.

The original onsite system of course is the pit privy, glorified in song and fable.* The privy, still used in camps and other temporary residences, consists simply of a deep (perhaps 2 m or 6 ft deep) pit into which human excrement is deposited. When a pit fills up, it is covered and a new pit is dug.

A logical extention of the pit privy idea is a composting toilet which accepts not only human wastes, but food waste as well, and produces a useful compost. With such a system, wastewaters from other sources such as washing and bathing are discharged separately.

By far the greatest number of households with onsite disposal systems use a form of the *septic tank* and *tile field*.

As shown in Figure 7-3, a septic tank consists of a concrete box which removes the solids and promotes partial decomposition. The solid particles settle out and eventually fill the tank, thus necessitating periodic cleaning. The water overflows into a tile drain field which promotes the seepage of water discharged.

A tile field consists of pipe laid in about a 1-m (3-ft)-deep trench, end on end but with short gaps between the pipe. The effluent from the septic tank flows into the tile field pipes and seeps into the ground through these gaps. Alternatively, seepage pits consisting of gravel and sand can be used for promoting the adsorption of the effluent into the ground.

The most important consideration in designing a septic tank and tile field system is the ability of the ground to absorb the effluent. Percolation tests, used to measure the suitability of ground for the fields, are conducted in the following way:

1. Dig a hole about 6-12 in.^{2} and as deep as the proposed tile field trench.
2. Scratch the soil to remove smeared surfaces and provide a more natural soil interface and put some gravel in the bottom of the pit.

---

*Probably the most famous literary work on the theme of the outhouse was by James Whitcomb Riley, who penned "The Passing of the Backhouse." A few lines from this epic:

> "But when the crust was on the snow
>   and the sullen skies were grey,
> In sooth the building was no place where
>   one could wish to stay.
> We did our duties promptly,
>   there one purpose swayed the mind.
> We tarried not nor lingered long
>   on what we left behind,
> The torture of that icy seat
>   would make a Spartan sob..."

Figure 7-3   Septic tank and tile field used for onsite wastewater disposal.

3. Fill the pit with water and let it stand overnight.
4. Next day fill the pit with water to 6 in. above the gravel, and measure the drop in water level in 30 minutes.
5. Calculate the percolation rate as inches per minute.

The U.S. Public Health Service and all county and local departments of health have established guidelines for sizing the tile fields or seepage pits. Typical standards are shown in Table 7-2.

Many areas in the United States have soils which percolate poorly, and septic tank/tile field systems are inappropriate. Fortunately, several options are now available for people wishing to use onsite wastewater disposal, one of which is shown in Figure 7-4.

In urbanized areas, there is insufficient land available to allow for such onsite treatment and percolation. Up until the nineteenth century, this problem was solved by constructing large cesspools or holding basins for the wastewater, which had to be pumped out as they filled up. The public health problems with this system were enormous, as discussed previously (see the Broad Street pump incident, Chapter 2). An obviously better way to move the human waste out of a congested community was to use water as a carrier.

The *water closet,* as it is known in Europe, thus became a standard trapping of our urban society. Actually, the invention of the flushed water closet is mired in controversy. Some credit John Bramah,[1] as being the inventor in 1778, while others present convincing arguments that Sir John Harington[2] was the inventor, in 1596!* The latter argument is strengthened by Sir John's original description of the device, although there is no record of him donating his name to the invention. The first recorded use of that euphemism is found in the regulation at Harvard University, where in 1735 it was decreed that "No Freshman shall go to the Fellows' John."

The wide use of waterborne wastewater disposal however caused another problem. Now the wastes of a community were all concentrated in one place, and a major effort was necessary to clean up the mess. This demand fostered what is known as centralized treatment.

---

[1] Kirby, R. S., et al. *Engineering in History* (New York: McGraw-Hill Book Company, 1956).

[2] Reyburn, W. *Flushed With Pride* (London: McDonald, 1969).

* Sir John, a courtier-poet, installed his invention in his country house at Kelson, near Bath. Queen Elizabeth had one fitted soon afterwards at Richmond Palace. The two books which were written about this innovation bear the strange titles, *A New Discourse on a Stale Subject. Called the Metamorphosis of Ajax, & An Anatomie of the Metamorphased Ajax.* Ajax is a play on words; 'a jakes' is a closet.

Table 7-2   Adsorption Area Requirements
for Private Residences

Percolation Rate (in./min)	Required Area of Adsorption Field, per Bedroom (ft^2)
Greater than 1	70
Between 1 and 0.5	85
Between 0.5 and 0.2	125
Between 0.2 and 0.07	190
Between 0.07 and 0.03	250
Less than 0.03	unsuitable ground

CENTRAL WASTEWATER TREATMENT

The objective of wastewater treatment is to reduce the concentrations of specific pollutants to the level where the discharge of the effluent will not adversely affect the environment. Note that there are two important aspects of this objective. First, wastewater is treated only to reduce the concentrations of selected constituents which would cause harm to the environment or pose a health hazard. Not everything in wastewater is troublesome, and thus is not removed. Secondly, the reduction of these constituents is only to some required level. It is obviously technically possible to produce distilled and deionized H_2O from wastewater, but this is not necessary, and can in fact be detrimental to the watercourse. Fish and other aquatic organisms cannot survive in distilled water.

For any given wastewater in a specific location, the *degree* and *type* of treatment therefore are variables which require engineering decisions.

Often, the degree of treatment is dictated by the assimilative capacity of the recipient. The procedure by which dissolved oxygen sag curves are drawn is reviewed in Chapter 2. As noted there, the amount of oxygen-demanding materials (BOD) discharged determines how far the dissolved oxygen level will be depressed. If this depression (deficit) is too large, some BOD must be removed in the treatment plant. Thus a certain plant on a given watercourse is required to produce a given quality of effluent. Such an *effluent standard* (discussed more fully in Chapter 10) dictates in large part the type of treatment required.

In order to facilitate the discussion of wastewater treatment, a "typical wastewater" (Table 7-1) will be assumed, and it will be further assumed that the effluent from this wastewater treatment must meet the following effluent standards:

Figure 7-4   Alternative onsite disposal system.

$$BOD \leq 15 \, mg$$
$$SS \leq 15 \, mg$$
$$P \leq 1 \, mg$$

Obviously, other criteria might, in given situations, be important. For example, nitrogen is thought to be the limiting nutrient in estuarine waters, and if the discharge was to be a brackish estuary, the total nitrogen would be an important parameter. In our simplified case, however, we are concerned only with these three constituents.

To further facilitate discussion, the treatment system selected to achieve these effluent levels consists of three major components:

- primary treatment—the major objective of which is the removal of solids. Primary treatment systems are always physical processes, as opposed to biological or chemical.
- secondary treatment—which is designed to remove the demand for oxygen. These processes are commonly biological in nature.
- tertiary treatment—a name applied to any number of polishing or cleanup processes, one of which is the removal of nutrients such as phosphorus. These processes can be physical (e.g., filters), biological (e.g., oxidation ponds) or chemical (precipitation of phosphorus).

PRIMARY TREATMENT

The most objectionable aspect of discharging raw sewage into watercourses is the floating material. It is only logical, therefore, that *screens* were the first form of wastewater treatment used by communities, and even today, screens are used as the first step in treatment plants. Typical screens, shown in Figure 7-5, consist of a series of steel bars which might be about 2.5 cm (1 in.) apart. The purpose of a screen in modern treatment plants is the removal of materials which might damage equipment or hinder further treatment. In some older treatment plants screens are cleaned by hand, but mechanical cleaning equipment is used in almost all new plants. The cleaning rakes are automatically activated when the screens get sufficiently clogged to raise the water level in front of the bars.

In many plants, the next treatment step is a *comminutor,* a circular grinder designed to grind the solids coming through the screen into pieces about 0.3 cm (1/8 in.) or smaller. Many designs are in use; one common design is shown in Figure 7-6.

The third treatment step involves the removal of grit or sand. This is necessary because grit can wear out and damage such equipment as

Bars with I-inch spaces

Figure 7-5 Bar screen used in wastewater treatment. The top picture shows a manually cleaned screen, the bottom picture represents a mechanically cleaned screen (photo courtesy Envirex).

Figure 7-6   A comminutor used to grind up large solids.

pumps and flow meters. The most common *grit chamber* is simply a wide place in the channel where the flow is slowed down sufficiently to allow the heavy grit to settle out. Sand is about 2.5 times as heavy as most organic solids and thus settles much faster than the light solids. The objective of a grit chamber is to remove sand and grit without removing the organic material. The latter must be further treated in the plant, but the sand can be dumped as fill without undue odor or other problems.

Following the grit chamber most wastewater treatment plants have a *settling tank* (Figure 7-7) to settle out as much of the solid matter as possible. Accordingly, the retention time* is kept long and turbulence is kept to a minimum. The solids settle to the bottom and are removed through a pipe while the clarified liquid escapes over a V-notch weir, a notched steel plate over which the water flows, promoting equal distribution of liquid discharge all the way around a tank. Settling tanks are also known as *sedimentation tanks* and often as *clarifiers*. The settling tank which follows preliminary treatment such as screening and grit removal is known as a *primary clarifier*. The solids which drop to the bottom of a primary clarifier are removed as *raw sludge,* a name which doesn't do justice to the undesirable nature of this stuff.

Raw sludge is generally odoriferous and full of water, two characteristics which make its disposal difficult. It must be both stabilized to retard further decomposition and dewatered for ease of disposal. In addition to the solids from the primary clarifier, solids from other processes must

---

*Retention time is the total time an average slug of water will spend in the tank. This theoretical time is calculated as the time required to fill up a tank. For example, if the volume is 100 m³, and the flow rate is 2 m³/min, the retention time is 100/2 = 50 min. Some authors use *detention time* synonymously with retention time.

Figure 7-7   Settling tanks (clarifiers). Top drawing shows a rectangular settling tank, and the lower drawing is a circular tank.

similarly be treated and disposed. The treatment and disposal of wastewater solids (sludge) is an important part of wastewater treatment and is discussed further in Chapter 8.

Since settling tanks are both ubiquitous and effective, their operation and design is reviewed in greater detail below.

Settling Tank Design and Operation

As noted above, one of the most efficient means of separating solids from the surrounding liquid is to allow the solids to settle out under the force of gravity. As long as the density of the solids exceeds that of the liquid, it should be possible to achieve solid/liquid separation. Unfortunately, this is not always true, since other forces come into play, as we shall see later.

Particles settle in one of three general ways:

- Class I: as discrete particles, unhindered by the container walls or neighboring particles. This is called *discrete particle settling*.
- Class II: as discrete particles but hindered by their neighbors and changing in size due to particle contact due to close proximity. This is called *flocculent settling*.
- Class III: as a mass of particles, with no interparticle movement so that all particles have the same settling velocity, called *thickening*.

The simplest case is Class I, discrete particle settling. The single particle is an infinite fluid, having attained terminal velocity, and has three forces acting on it: drag, buoyancy and gravity.

If the particle is settling in a fluid at its terminal velocity, these forces balance as

$$F_g = F_D + F_B$$

where
- F_g = force due to gravity = mg
- F_D = drag force = $(C_D \rho v^2 A)/2$
- F_B = buoyancy = $\rho \forall g$
- m = mass of particle = $\rho_s \forall$
- g = gravitational acceleration,
- ρ_s = particle density, kg/m^3
- ρ = fluid density, kg/m^3
- C_D = drag coefficient
- v = velocity of particle, m/sec
- A = projected area of particle, m^2
- $\forall$ = volume of particle, m^3

Substituting, and solving for velocity, we get

$$v = \left[\frac{2 \forall g (\rho_s - \rho)}{C_D \rho A} \right]^{1/2}$$

If we now assume that the particle is a sphere,

$$v = \left[\frac{4}{3} \frac{d g (\rho_s - \rho)}{C_D \rho} \right]^{1/2}$$

which is the well-known Newtonian equation, where d = particle diameter. If, as is the case in most settling in wastewater treatment, the Reynolds number (**R**) is sufficiently low and a laminar boundary layer is maintained, the drag coefficient can be expressed as

$$C_D = \frac{24}{R}$$

where $R = vd\rho/\mu$
μ = fluid viscosity.

If $C_D > 1$, this no longer holds, and the drag coefficient can be approximated as

$$C_D = \frac{24}{R} + \frac{3}{\sqrt{R}} + 0.34$$

For laminar conditions, substituting $C_D = 24/R$ into the Newtonian equation, we get the well-known Stokes equation,

$$v = \frac{d^2 g (\rho_s - \rho)}{18\mu}$$

The velocity of particles can be related to the expected settling tank performance by idealizing a settling tank, as originally proposed by Camp.[3]

As shown in Figure 7-8, a rectangular tank is first divided into four zones: inlet zone, outlet zone, sludge zone and settling zone. The first two zones are designated for the dampening of currents caused by the influent and effluent. The sludge zone is simply storage space for the settled solids. Settling, then, takes place only in the fourth zone.

Several initial assumptions are required in this analysis:

- Uniform flow occurs within the settling zone.
- All particles entering the sludge zone are removed.
- Particles are evenly distributed in the flow as they enter the settling zone.
- All particles entering the effluent zone escape the tank.

Consider now a particle entering the settling zone at the water surface. This particle has a settling velocity of v_o and a horizontal velocity V such that the component of the two defines a trajectory as shown in Figure 7-8. In other words, the particle just is removed (it barely squeaks into the sludge zone). Note that had the particle entered the settling zone at

---

[3]Camp, T. R. "Sedimentation and the Design of Settling Tanks," *Trans. ASCE* 111:895 (1946).

Figure 7-8   Ideal settling tank schematic.

any other height, its trajectory would have always carried it into the sludge zone. Particles having this velocity are termed the "critical particles" in that particles with lower settling velocities are not all removed. For example, the particle having velocity v_s, entering the settling zone at the surface, will end up in the effluent zone and escape. If this same particle, however, entered at some height h, it would have just been removed. Any of these particles that happen to enter the settling zone at height h or lower would thus be removed, and those entering above h would not. Since the particles entering the settling zone are equally distributed, the proportion of those particles with a velocity of v_s that are removed is thus equal to h/H, where H is the height of the settling zone. With reference to Figure 7-8, similar triangles yield

$$\frac{v_o}{H} = \frac{v_s}{h} = \frac{V}{L}$$

The time that the critical particle spends in the settling zone is

$$\bar{t} = \frac{L}{V} = \frac{h}{v_s} = \frac{H}{v_o}$$

hence

$$v_o = \frac{H}{\bar{t}}$$

The time $\bar{t}$ is also equal to the hydraulic retention time, or V/Q, where Q is the flow rate and V is the volume of the settling zone, or $V = AH$, with A = surface area of settling zone. Thus

$$v_o = \frac{H}{\bar{t}} = \frac{H}{AH/Q} = \frac{Q}{A}$$

This equation represents an important design parameter for settling tanks, and is called the *overflow rate*. Note the units:

$$v_o = \frac{m}{s} = \frac{Q}{A} = \frac{m^3/s}{m^2}$$

Commonly, overflow rate is expressed as "gallons/day/ft²," but it actually is a velocity term, and is in fact equal to the velocity of the critical particle. When the design of a clarifier is specified by overflow rate, what is really defined then is the critical particle.

It should also be noted that when any two of the following—overflow rate, retention time, or depth—are defined, the remaining parameter is also fixed.

Example 7.2

A primary clarifier has an overflow rate of 600 gal/day ft^2 and a depth of 6 ft. What is its hydraulic retention time?

$$v_o = 600 \, \frac{\text{gal}}{\text{d-ft}^2} \times \frac{1}{7.48} \, \frac{\text{ft}^3}{\text{gal}} = 80.2 \, \text{ft/d}$$

$$\bar{t} = \frac{H}{v_o} = \frac{6}{80.2} \, \frac{\text{ft}}{\text{ft/d}} = 0.0748 \, \text{d} = 1.8 \, \text{hr}$$

Overflow rate is interesting in that we can obtain a better understanding of settling by looking at individual variables. For example, increasing the flow rate, Q, in a given tank increases the v_o; i.e., the critical velocity increases and thus fewer particles are all removed since fewer particles have a $v_s > v_o$.

We would, of course, like to *decrease* v_o, so more particles can be removed. This is done by either reducing Q, or increasing A. The latter term may be increased by changing the dimensions of the tank so that the depth is shallow, and the length and width are very large. For example, the area can be doubled by taking a 3-m-deep tank, slicing it in half (two 1.5-m slices) and placing them alongside each other. The new shallow tank has the same horizontal velocity, but double the surface area, hence v_o is half of the original value.

Why not then make *very* shallow tanks? Other than the problem of hydraulics, even distribution of flow, and the great expense in concrete and steel,* this seems to be a reasonable conclusion.

We have, however, made a major assumption that isn't really true in wastewater treatment: solids do *not* settle as discrete particles but are indeed influenced by their neighbors, and tend to stick together as they come into contact, thus forming larger particles. This process is commonly called *flocculent settling,* and can be illustrated by setting up a cylinder of dirty water and allowing the particles to settle. If the water is sampled at various elevations, the clarification of the water with time can be observed (Figure 7-9).

If we first introduce particles all with the same settling velocity, we

---

*Remember the max-min problems in calculus? What's the shape of a six-sided prism with the smallest surface area?

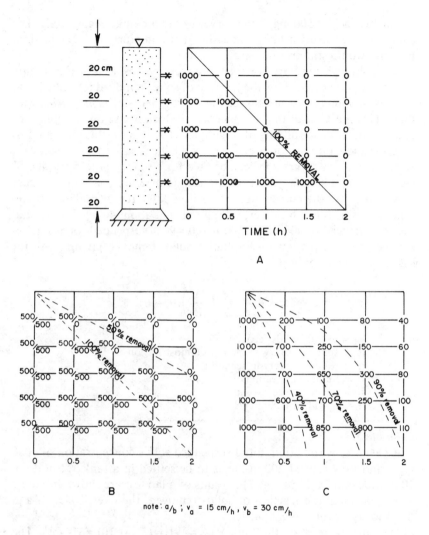

note: a/b ; $v_a = 15$ cm/$_h$, $v_b = 30$ cm/$_h$

Figure 7-9   Settling test using a column of dirty water and periodically sampling at five ports. The initial suspended solids concentration (t = 0) is assumed to be 1000 mg/l.

should find that there is perfect clarification as the highest particle drops to the bottom. If the $v_s = 120$ cm/2 hr, then the curve in Figure 7-9A should result. Everything below the settling curve would be at the original suspended solids concentration (1000 mg/l) and above the line the water would be clean.

If now we had two sizes of particles, $v_a = 60/2 = 30$ cm/hr and $v_b =$

60 cm/hr, each at 500 mg/l, the curves would look like Figure 7-9B. The curves are still straight lines, indicating that a mixture of different sized particles would still produce straight settling lines.

A real-life wastewater, however, produces curves as shown in Figure 7-9C. The curves that show the fraction of solids removed (or reduced) at various sampling ports are curved lines, with their slopes *increasing* with time. This must mean that the increased velocities are due to collisions and the subsequent building of larger particles. It is for this reason (also) that a shallow tank is not as efficient as the ideal tank theory suggests.

Curves such as Figure 7-9C can be used for estimating the removal efficiencies of real slurries. At a retention time of 2 hr, the data in Figure 7-9C show that the entire tank experienced 90% removal. The top section, however, is much clearer, since it only has about 40 mg/l of SS, and thus experienced $[(1000-40)/(1000)]100 = 96\%$ suspended solids removal. The general equation for estimating solids removal, giving credit for the cleaner water on top, is

$$R = P + \sum_{i=1}^{n-1} \left(\frac{h}{H}\right)(P_i - P)$$

where   R = overall solids recovery, %
         P = solids recovery at the bottommost section
         P_i = solids recovery in section i
         n = number of sections
         h = height of each section
         H = height of column (H = nh)

Example 7.3

A chemical waste at an initial suspended solids concentration of 1000 mg/l and flow rate of 200 m³/hr is to be settled in a tank 1.2 m deep, 10 m wide and 31.4 m long. The results of a lab test are shown in Figure 7-9C. Calculate the fraction of solids removed, the overflow rate, and the velocity of the critical particle.

The surface area of the tank is $A = WL = (31.4)(10) = 314$ m². The overflow rate is therefore

$$\text{Overflow rate} = \frac{Q}{A} = \frac{200}{314}$$

$$= 0.614 \text{ m}^3/\text{hr-m}^2$$

The critical velocity is thus $v_o = 0.614$ m/hr which is not germane to the problem since we have flocculent settling.

The hydraulic retention time is

$$\bar{t} = \frac{\Psi}{Q} = \frac{AD}{Q} = \frac{(314)(1.2)}{200} = 1.88 \text{ hr}$$

On Figure 7-9C, the 85% removal line approximately intersects $\bar{t} = 1.88$ hr. Thus 85% of the solids are clearly removed. In addition to this, however, even better removal is indicated at the top of the water column. At the top 20 cm, assume the SS = 40 mg/l, equal to $[(1000 - 40) \times 100]/1000 = 96\%$ removal, or 11% better than the entire column. The second shows $[(1000 - 60) \times 100]/1000 = 94\%$ removal, and so on. The total removed, ignoring the lowermost section, is

$$R = P + \sum_{i=1}^{n-1} \left(\frac{h}{H} \right)(P_i - P)$$

$$R = 85 + \tfrac{1}{6}(11) + \tfrac{1}{6}(9) + \tfrac{1}{6}(5) + \tfrac{1}{6}(4) = 90.9\%$$

Typically, primary clarifier solids capture efficiency for domestic waste can be plotted as a function of the retention time, (which, for a given depth, is also related to the overflow rate) as shown in Figure 7-10.

The third type of settling Class III, *thickening,* where the particles are so tightly packed together that they settle as a single bed of solids, is covered in Chapter 8.

Primary treatment then is mainly a removal of solids, although some BOD is removed as a consequence of the removal of decomposable solids. Typically, the wastewater which was described earlier might now have these characteristics:

	Raw Wastewater	Following Primary Treatment
BOD mg/l	250	175
SS mg/l	220	60
P mg/l	8	7

A substantial fraction of the solids has been removed, as well as some BOD and a little P (as a consequence of the removal of raw sludge).

In a typical wastewater treatment plant, this would now move on to secondary treatment.

SECONDARY TREATMENT

The water leaving the primary clarifier has lost much of the solid organic matter but still contains a high demand for oxygen; i.e., it is

Figure 7-10   Performance of primary clarifiers.

composed of high-energy molecules which will decompose by microbial action, thus creating a biochemical oxygen demand (BOD). This demand for oxygen must be reduced (energy wasted) if the discharge is not to create unacceptable conditions in the watercourse. The objective of secondary treatment is thus to remove BOD while, by contrast, the objective of primary treatment is to remove solids.

Almost all secondary methods use microbial action to reduce the energy level (BOD) of the waste. Although there are many ways the microorganisms can be put to work, the first really successful modern method of secondary treatment was the *trickling filter*.

The trickling filter, shown in Figure 7-11, consists of a filter bed of fist-sized rocks over which the waste is trickled. A very active biological

Figure 7-11   Schematic of a trickling filter.

growth forms on the rocks, and the organisms obtain their food from the waste stream dripping through the bed of rocks. Air is either forced through the rocks or, more commonly, air circulation is obtained automatically by a temperature difference between the air in the bed and ambient temperature. In the older filters the waste is sprayed onto the rocks from fixed nozzles. The newer designs utilize a rotating arm which moves under its own power, like a lawn sprinkler, distributing the waste evenly over the entire bed. Often the flow is recirculated, thus obtaining a higher degree of treatment. The name trickling filter is obviously a misnomer since no filtration takes place.

Around the turn of the century when trickling filtration was already firmly established, some researchers began musing about the wasted space in a filter taken up by the rocks. Could the microorganisms not be allowed to float free and could they not be fed oxygen by bubbling in air? Although this concept was quite attractive, it was not until 1914 that the first workable pilot plant was constructed. It took some time before this process became established as what we now call the *activated sludge system.*

The key to the activated sludge system is the reuse of microorganisms. The system, shown as a block diagram in Figure 7-12, consists of a tank full of waste liquid (from the primary clarifier) and a mass of microorganisms. Air is bubbled into this tank (called the *aeration tank*) to provide the necessary oxygen for the survival of the aerobic organisms. The microorganisms come in contact with the dissolved organics and rapidly adsorb these organics on their surface. In time, the microorganisms decompose this material to CO_2, H_2O, some stable compounds and more microorganisms. The production of new organisms is relatively slow, and most of the aeration tank volume is in fact used for this purpose.

Figure 7-12   Block diagram of the activated sludge system.

Once most of the food has been utilized, the microorganisms are separated from the liquid in a settling tank, sometimes called a *secondary* or *final clarifier.* The liquid escapes over a weir and can be discharged into the recipient. The separation of microorganisms is an important part of the system. In the settling tanks, the microorganisms exist without additional food and become hungry. They are thus activated; hence the term *activated sludge.*

The settled microorganisms, now known as *return activated sludge,* are pumped to the head of the aeration tank where they find more food (organics in the effluent from the primary clarifier) and the process starts all over again. The activated sludge process is a continuous operation, with continuous sludge pumping the clean water discharge.

As mentioned earlier, one of the end products of this process is more microorganisms. If none of the microorganisms are removed, their concentration will soon increase to the point where the system is clogged with solids. It is therefore necessary to waste some of the microorganisms, and this *waste activated sludge* must be processed and disposed of. Its disposal is one of the most difficult aspects of waste treatment.

Activated sludge systems are designed on the basis of loading, or the amount of organic matter (food) added relative to the microorganisms available. This ratio is known as the food-to-microorganisms ratio (F/M) and is a major design parameter. Unfortunately it is difficult to measure either F or M accurately, and engineers have approximated these by BOD and the suspended solids in the aeration tank respectively. The combination of the liquid and microorganisms undergoing aeration is known (for some unknown reason) as *mixed liquor,* and thus the suspended solids are called *mixed liquor suspended solids* (MLSS). The ratio of incoming

BOD to MLSS, the F/M ratio, is also known as the *loading* on the system, calculated as pounds of BOD/day per pound of MLSS in the aeration tank.

If this ratio is low (little food for lots of microorganisms) and the aeration period (retention time in the aeration tank) is long, the microorganisms make maximum use of available food, resulting in a high degree of treatment. Such systems are known as *extended aeration,* and are widely used for isolated sources (e.g., motels, small developments). An added advantage of extended aeration is that the ecology within the aeration tank is quite diverse and little excess biomass is created, resulting in little or no waste activated sludge to be disposed of—a significant saving in operating costs and headaches.

At the other extreme is the "high-rate" system where the aeration periods are very short (thus saving money by building smaller tanks) and the treatment efficiency is lower. The efficiencies and F/M ratios for the three types of activated sludge systems are shown in Table 7-3.

Table 7-3.   Loadings and Efficiencies of Activated
Sludge Systems

Process	Loading $\dfrac{F}{M} = \dfrac{\text{lb BOD/day}}{\text{lb MLSS}}$	Aeration Period (hr)	Efficiency of BOD Removal (%)
Extended Aeration	0.05–0.2	30	95
Conventional	0.2–0.5	6	90
High Rate	1–2	4	85

Example 7.4

The BOD_5 of the liquid from the primary clarifier is 120 mg/l at a flow rate of 0.05 mgd. The aeration tank is $20 \times 10 \times 20$ ft, and the MLSS = 2000 mg/l. Calculate the F/M ratio.

$$\text{lb BOD} = 120\,\text{mg/l} \times 0.05\,\text{mgd} \times 8.34 = 50\,\text{lb/day}$$

(see p. 53 for discussion of how to convert
from flow and concentration to pounds)

$$lb \ MLSS = (20 \times 10 \times 20) \ ft^3 \times 2000 \ mg/l \times 3.83 \ l/gal$$
$$\times \ 7.481 \ gal/ft^3 \times 2.20 \times 10^{-6} \ lb/mg = 229 \ lb$$

$$F/M = \frac{50}{229} = 0.22 \ \frac{lb \ BOD/day}{lb \ MLSS}$$

When the microorganisms first come in contact with the food the process requires a great deal of oxygen. Accordingly, the dissolved oxygen level in the aeration tank drops immediately after the point at which the waste is introduced. If DO levels are measured over the length of a tank, extremely low concentrations are often found at the influent end of the aeration tank. These low levels of DO can be detrimental to the microbial population. Accordingly, two variations of the activated sludge treatment have found some use: *tapered aeration* and *step aeration* (Figure 7-13). The former method consists of blasting additional air

Tapered Aeration   **Step Aeration**

Figure 7-13   Tapered and step aeration schematics.

where needed, while step aeration involves the introduction of the waste at several locations, thus evening out the initial oxygen demand.

The third modification is *contact stabilization,* or *biosorption,* a process in which the sorption and bacterial growth phases are separated by a settling tank. The advantage is that the growth can be achieved at high solids concentrations, thus saving tank space. Many existing activated sludge plants can be converted to biosorption plants when tank volume limits treatment efficiency. Figure 7-14 is a diagram of the biosorption process.

The two principal means of introducing sufficient oxygen into the aeration tank are by bubbling compressed air through porous diffusers or beating air in mechanically. Both diffused air and mechanical aeration are shown in Figure 7-15.

The success of the activated sludge system depends on many factors. Of critical importance is the separation of the microorganisms in the

Figure 7-14   The biosorption modification of the activated sludge process.

final clarifier. The microorganisms in the system are sometimes very difficult to settle out, and the sludge is said to be a *bulking sludge*. Often this condition is characterized by a biomass comprised almost totally of filamentous organisms which form a kind of lattice structure with the filaments and refuse to settle.*

Treatment plant operators should keep a close watch on settling characteristics because a trend toward poor settling can be the forerunner of a badly upset (and hence ineffective) plant. The settleability of activated sludge is most often described by the sludge volume index (SVI), which is determined by measuring the ml of volume occupied by a sludge after settling for 30 minutes in a 1-liter cylinder, and calculated as

$$SVI = \frac{(\text{volume of sludge after 30 min, in ml}) \times 1000}{\text{mg/l of suspended solids}}$$

Example 7.5

A sample of mixed liquor was found to have SS = 4000 mg/l and, after settling for 30 minutes in a 1-liter cylinder, occupied 400 ml. Calculate the SVI.

$$SVI = \frac{400 \times 1000}{4000} = 100$$

---

*You can picture this as filling a glass with cotton balls, then pouring water in it. The cotton is simply not heavy enough to settle to the bottom of the glass.

Diffused Aeration

Figure 7-15   Activated systems with diffused aeration and

Mechanical Aeration

mechanical (surface) aeration (photos courtesy Envirex).

SVI values below 100 are usually considered acceptable, with SVI greater than 200 defined as badly bulking sludges. Some common loadings, as a function of the SVI, for final clarifiers are shown in Figure 7-16.

Figure 7-16   Commonly used solids loading for final clarifiers.

The causes for poor settling (high SVI) are not always known, and hence the solutions are elusive. Wrong or variable F/M ratios, fluctuations in temperature, high concentrations of heavy metals, and deficiencies in nutrients in the incoming wastewater have all been blamed for bulking. Cures include chlorinaton, changes in air supply, and dosing with hydrogen peroxide (H_2O_2) to kill the filamentous microorganisms.

When the sludge does not settle, the return activated sludge becomes thin (low suspended solids concentration) and thus the concentration of microorganisms in the aeration tank drops. This results in a higher F/M ratio (same food input, but fewer microorganisms) and a reduced BOD removal efficiency.*

---

*You can think of the microorganisms as workers in an industrial plant. If the total number of workers is decreased, the production is cut. Similarly, if fewer microorganisms are available, less work is done. This will become evident in the discussion on process dynamics.

Biological Process Dynamics Applied to the Activated Sludge System

The organics in the influent are degraded and oxidized to CO_2 and H_2O. Some of the high-energy organics are used to build new microorganisms. These organics, known in the language of biological process dynamics as *substrate,* are usually measured indirectly as the BOD, although methods such as organic carbon, for example, may be more accurate measures of substrate concentration. The substrate (S) is biodegraded and used by the microorganisms (X), expressed as suspended solids, at a rate dS/dt. The rate of new cell mass (microorganisms) production as a result of the destruction of the substrate is

$$\frac{dX}{dt} = Y \frac{dS}{dt}$$

where Y = the yield, or mass of microorganisms produced per mass of substrate used; commonly expressed as kg SS produced/kg BOD used, or Y = dX/dS.

The expression for substrate utilization commonly employed is the Monod model,[*]

$$\frac{dS}{dt} = \frac{X}{Y} (\mu) = \frac{X}{Y} \left(\frac{\hat{\mu}S}{K_s + S} \right)$$

where μ = is the specific growth rate, in terms of days^{-1}, and equal to $(dX/dt)(1/X)$
$\hat{\mu}$ = maximum growth rate constant, days^{-1}
K_s = saturation constant, mg/l.

This expression is an empirical model, based on experimental work with pure cultures. The two constants, $\hat{\mu}$ and K_s, must be evaluated for each substrate and microorganism culture.

---

[*]The argument for the model validity is beyond the scope of this text. Suffice it to say that the model is empirical, but reasonable. If you are interested in the development of this model, see any number of modern textbooks on wastewater processing.[4-6]

[4]Metcalf & Eddy, *Wastewater Engineering* (New York: McGraw-Hill Book Company, 1979).

[5]Grady, L., and H. C. Lim. *Biological Wastewater Treatment* (New York: Marcel Dekker, 1980).

[6]Benefield, L. D., and C. W. Randall. *Biological Process Design for Wastewater Treatment* (Englewood Cliffs, NJ: Prentice Hall, 1980).

The application of biological process dynamics to the activated sludge process is best illustrated by considering a system shown in Figure 7-17. This is a simple continuous biological reactor, of volume V and an effluent of Q. The reactor is *completely mixed,* meaning that the influent is dispersed within the tank immediately upon introduction, thus there are no concentration gradients in the tank and the quality of the effluent is *exactly* that of the tank contents.

Figure 7-17   A biological reactor without microorganism recycle.

In such a continuous reactor, there are two types of retention times: liquid and solids. The liquid, or *hydraulic retention time* is expressed as

$$\bar{t} = \frac{V}{Q}$$

where $\bar{t}$ = hydraulic retention time, min. The solids retention time is also known as *sludge age,* or the average time a solid (microorganism) particle stays in the system. What wastewater treatment plant operators call sludge age is known also as the *mean cell residence time* in research, and is expressed as

$$\Theta_c = \frac{\text{mass of SS in the system}}{\text{mass rate of solids wasted}}$$

For the system pictured in Figure 7-17, the mean cell residence time is

$$\Theta_c = \frac{VX}{QX}$$

(which of course is equal to the hydraulic retention time). The units of Θ_c commonly are expressed in days.

For the system in Figure 7-17, we now assume that there are no organisms in the influent ($X_o = 0$) and that the reactor has attained steady state (the growth rate of microorganisms is balanced by the loss rate of microorganisms in the effluent). For the sake of simplicity, we also ignore the death of microorganisms.

We can write a mass balance in terms of the microorganisms

$$
\begin{bmatrix} \text{Rate of change} \\ \text{in the} \\ \text{reactor} \end{bmatrix} = \begin{bmatrix} \text{Rate} \\ \text{of} \\ \text{inflow} \end{bmatrix} - \begin{bmatrix} \text{Rate} \\ \text{of} \\ \text{outflow} \end{bmatrix} + \begin{bmatrix} \text{Net rate of} \\ \text{growth of} \\ \text{microorganisms} \end{bmatrix}
$$

$$
\frac{dX}{dt} \mathbb{V} = QX_o - QX + \left(Y \frac{dS}{dt} \right) \mathbb{V}
$$

In a steady-state system, $(dX/dt)\mathbb{V} = 0$, and since we assumed no cells in the inflow, $QX_o = 0$. Substituting the Monod substrate utilization model,

$$
\frac{dS}{dt} = \frac{\hat{\mu} \, X \, S}{Y(K_s + S)}
$$

into the above relationship and introducing the mean cell residence time Θ_c,

$$
\frac{1}{\Theta_c} = \frac{\hat{\mu} \, S}{K_s + S}
$$

and

$$
S = \frac{K_s}{\hat{\mu} \, \Theta_c - 1}
$$

This is an important expression since it implies that the substrate concentration, S, is a function of the kinetic constants (which are beyond our control for a given substrate) and the mean cell residence time. The value of S (and thus the treatment efficiency) is influenced then by the mean cell residence time (or the sludge age as previously defined).

The system pictured in Figure 7-17 is not very efficient, however, since long hydraulic residence times are necessary to prevent the microorganisms from being flushed out. The success of the activated sludge system for wastewater treatment is based on the recycle of the microorganisms. Such a system is shown in Figure 7-18.

We need to make some simplifying assumptions before we can conveniently model this system. First, we again assume that $X_o = 0$, and further that the microorganism separator (the final settling tank or clarifier)

Figure 7-18   A biological reactor with microorganism recycle.

is a perfect device, so that there are no microorganisms in the effluent ($X_e = 0$). We also once again assume steady-state conditions and perfect mixing. The excess microorganisms (waste activated sludge) are removed from the system at a flow rate Q_w and a solids concentration X_r, which is the settler underflow concentration and the concentration of the solids being recycled to the aeration tank. And lastly, we assume that there is no substrate removal in the settling tank and that the settling tank has no volume, so that all of the microorganisms in the system are in the reactor (aeration tank). The volume of the aeration tank is thus the only active volume, and the settling (microorganism separation) is assumed to take place magically in a zero-volume tank, an obviously incorrect assumption. Note that the mean cell residence time in this case is

$$\Theta_c = \frac{\text{microorganisms in the system}}{\text{microorganisms lost or wasted/time}}$$

$$= \frac{X\,V}{Q_w X_r + (Q - Q_w)X_e}$$

and since $X_e = 0$

$$\Theta_c = \frac{X\,V}{Q_w\,X_r}$$

The removal of substrate is often expressed in terms of a *substrate removal velocity* (q) and defined as

$$q = \frac{(\text{mass substrate removed})}{(\text{mass microorganisms under aeration}) \cdot \text{time}}$$

In our notation

$$q = \frac{S_o - S}{X\,\bar{t}}$$

The substrate removal velocity is a rational measure of the substrate removal activity, or the mass of BOD removed in a given time per mass of microorganisms doing the work. This is also called, in some texts, the process-loading factor, with the clear implication that it is a useful operational and design tool.

The substrate removal velocity can be derived by doing a mass balance in terms of substrate or a continuous system with microorganism recycle (Figure 7-18).

$$\begin{bmatrix} \text{Rate of} \\ \text{change of} \\ \text{substrate} \end{bmatrix} = \begin{bmatrix} \text{Rate of} \\ \text{substrate} \\ \text{inflow} \end{bmatrix} - \begin{bmatrix} \text{Rate of} \\ \text{substrate} \\ \text{outflow} \end{bmatrix} - \begin{bmatrix} \text{Rate of} \\ \text{substrate} \\ \text{utilization} \end{bmatrix}$$

$$\frac{dS}{dt} \forall \quad = \quad QS_o \quad - \quad QS \quad - \quad q\,X\,V$$

Note: The rate of substrate utilization is

$$qXV = \left[\frac{\text{mass substrate removed}}{\text{mass microorganisms} \cdot \text{time}} \right] \cdot \left[\frac{\text{microorganisms}}{\text{volume of reactor}} \right]$$

$$\times\ [\text{volume of reactor}] = \frac{\text{mass of substrate removed}}{\text{time}}$$

Solving for q,

$$q = \frac{S_o - S}{X\,\bar{t}}$$

The Monod rate (μ) was defined previously as

$$\frac{dS}{dt} = \frac{\hat{\mu}\,S\,X}{Y(K_s + S)} = \frac{X}{Y}\,(\mu)$$

$$\mu = \frac{dX}{dt}\,\frac{1}{X} = \frac{\text{mass microorganism produced}}{(\text{time}) \cdot (\text{mass of microorganism})}$$

and the yield (Y) as

$$Y = \frac{dX}{dS} = \frac{\text{mass microorganism produced}}{\text{mass substrate removed}}$$

Using now a mass balance in terms of the microorganisms:

$$\begin{bmatrix} \text{Rate of change} \\ \text{in microorganism} \\ \text{concentration} \end{bmatrix} = \begin{bmatrix} \text{Rate of} \\ \text{inflow of} \\ \text{microorganisms} \end{bmatrix} - \begin{bmatrix} \text{Rate of} \\ \text{outflow of} \\ \text{microorganisms} \end{bmatrix} + \begin{bmatrix} \text{Rate of} \\ \text{production of} \\ \text{microorganisms} \end{bmatrix}$$

$$\frac{dX}{dt}\forall \quad = \quad QX_o \quad - Q_wX_r-(Q-Q_w)X_e + \quad \mu\,X\,\forall$$

Assuming again steady state ($[dX/dt] = 0$) and $X_o = X_e = 0$,

$$\mu = \frac{X_r Q_w}{X \, V}$$

Note that this is the reciprocal of previously defined sludge age, or mean cell residence time, so that

$$\Theta_c = \frac{X \, V}{X_r Q_w} = \frac{1}{\mu}$$

The substrate removal velocity can then be expressed as

$$q = \frac{\mu}{Y}$$

$$= \frac{\text{mass microorganisms produced}}{(\text{time}) \cdot (\text{mass microorganism in the reactor})}$$

$$\times \frac{\text{mass substrate removed}}{\text{mass microorganisms produced}}$$

and substituting

$$q = \frac{\hat{\mu} \, S}{Y \, (K_s + S)}$$

Previously we defined q as $(S_o - S)/X\bar{t}$. Equating these two expressions and solving for $S_o - S$, the substrate removal (reduction in BOD),

$$S_o - S = \frac{\hat{\mu} \, S \, X \, \bar{t}}{Y \, (K_s + S)}$$

Since we defined the substrate removal velocity q, as

$$q = \frac{\mu}{Y}$$

the mean cell residence time (sludge age) is

$$\Theta_c = \frac{1}{q \, Y}$$

and the concentration of microorganisms in the reactor (mixed liquor suspended solids) is

$$X = \frac{S_o - S}{\bar{t} \, q}$$

Example 7.6

An activated sludge system operates at a flow rate (Q) of 4000 m³/day, with an incoming BOD (S_o) of 300 mg/l. Through pilot plant work, the kinetic constants for this system were determined to be Y = 0.5 kg SS/kg BOD, K_s = 200 mg/l, μ = 2 day⁻¹. We need to design a treatment system that will produce an effluent BOD of 30 mg/l (90% removal). Determine: (1) what should be (a) the volume of the aeration tank, (b) the mixed liquor suspended solids, and (c) the sludge age, and (2) how much sludge will be wasted daily?

The mixed liquor suspended solids concentration is usually limited by the ability to keep an aeration tank mixed and to transfer sufficient oxygen to the microorganisms. Assume in this case that X = 4000 mg/l. The hydraulic retention time is then obtained from the equation

$$S_o - S = \frac{\hat{\mu} \, S \, X \, \bar{t}}{Y \, (K_s + S)}$$

on p. 185. Rearranged,

$$\bar{t} = \frac{Y \, (S_o - S)(K_s + S)}{\hat{\mu} \, S \, X}$$

$$= \frac{0.5(300 - 30)(200 + 30)}{2(30)(4000)}$$

$$= 0.129 \text{ day} = 3.1 \text{ hr}$$

The volume of the tank is then $\Psi = \bar{t}Q = 4000(0.129) = 516$ m³. The sludge age is

$$\Theta_c = \frac{1}{q \, Y}$$

where

$$q = \frac{S_o - S}{X \, \bar{t}} = \frac{300 - 30}{(4000)(0.129)} = 0.523 \; \frac{\text{kg BOD removed/day}}{\text{kg SS in the reactor}}$$

or, equivalently,

$$q = \frac{\hat{\mu} \, S}{Y \, (K_s + S)} = \frac{2(30)}{(0.5)(200 + 30)} = 0.522 \; day^{-1}$$

and

$$\Theta_c = \frac{1}{q \, Y} = \frac{1}{(0.522)(0.5)} = 3.8 \; days$$

Since

$$\frac{1}{\Theta_c} = \frac{kg \; sludge \; solids \; wasted/day}{kg \; sludge \; in \; aeration \; tank}$$

$$\Theta_c = \frac{X \, V}{X_r \, Q_w}$$

$$X_r \, Q_w = \frac{X \, V}{\Theta_c} = \frac{(4000)(516)(10^3 \; l/m^3)(1/10^6 \; kg/mg)}{3.8}$$

$$= 543 \; kg/day$$

Example 7.7

Using the same data as in the previous example, we now ask what mixed liquor solids concentration is necessary to attain a 95% BOD removal (i.e., $S = 15$ mg).

The substrate removal velocity is

$$q = \frac{\hat{\mu} \, S}{Y \, (K_s + S)} = \frac{2(15)}{0.5(200 + 15)} = 0.28 \; days^{-1}$$

and

$$X = \frac{S_o - S}{\bar{t} \, q} = \frac{300 - 15}{(0.129)(0.28)} = 7890 \; mg/l$$

The sludge age would be

$$\Theta_c = \frac{1}{q \, Y} = \frac{1}{(0.28)(0.5)} = 7.1 \; days$$

Note that more microorganisms are required in the aeration tank if higher removal efficiencies are to be attained.

The SVI (see p. 175) can be used to estimate the return sludge pumping rates. After 30 minutes of settling, the solids in the cylinder are at a SS concentration which would be equal to the expected return sludge solids, or

$$X_r \cong \frac{H}{h}(x)$$

where X_r = the expected return suspended solids concentration, mg/l
X = mixed liquor suspended solids mg/l
H = height of cylinder, m
h = height of settled sludge, m

The mixed liquor solids concentration is of course a combination of the return solids diluted by the influent, or

$$X = \frac{Q_r X_r + QX_o}{Q_r + Q}$$

If we again assume no solids in the influent, $(X_o = 0)$

$$X = \frac{Q_r X_r}{Q_r + Q}$$

The success or failure of an activated sludge system often depends on the performance of the final clarifier. If this settling tank is not able to achieve the required return sludge solids, the MLSS will drop, and of course the treatment efficiency will be reduced.

Final clarifiers act as both settling tanks (flocculent settling, Class II) and thickeners (Class III). The design of these tanks therefore requires that both the overflow rate and solids loading be considered. The latter, more fully explained in the next chapter, is expressed in terms of kg solids/day applied to a surface area of m^2. Figure 7-16 shows some commonly used solids loadings for final clarifiers as a function of the sludge volume index.

Secondary treatment of wastewater then usually consists of a biological step such as activated sludge, which removes a substantial part of the BOD and the remaining solids. Looking once again at the typical wastewater, we now have the following approximate water quality:

	Raw Wastewater	Following Primary Treatment	Following Secondary Treatment
BOD mg/l	250	175	15
SS mg/l	220	60	15
P mg/l	8	7	6

The effluent, in fact, meets our previously established effluent standards for BOD and SS. Only the phosphorus remains high. The removal of inorganic chemicals like phosphorus is accomplished in tertiary (or advanced) wastewater treatment.

TERTIARY TREATMENT

Primary and biological treatments make up the conventional wastewater treatment plant. However, secondary treatment plant effluents still contain a significant amount of various types of pollutants. Suspended solids, in addition to contributing to BOD, can settle out in streams and form unsightly mud banks. The BOD, if discharged into a stream with low flow, can still cause damage to aquatic life by depressing the DO. Neither primary nor secondary treatment is effective in removing phosphorus and other nutrients or toxic substances.

Suspended solids can be effectively removed by a simple device called a *pebble filter* which consists of a box full of pebbles hung over the edge of the secondary clarifier so that all water must travel through the filter bed. The actual mechanism of how the pebbles catch the floc is unknown. What is known is that the effluent through a pebble filter is clear. However, this device will never succeed in the United States because the pebble filter is entirely too simple, with no moving parts. Nobody wants to manufacture it, nor will engineers design such a "primitive" device.

A much more sophisticated, and hence more marketable, gadget is the *microstrainer,* pictured in Figure 7-19. The microstrainer is a large drum covered with a stainless steel sheet with small holes in it. The dirty water is pumped into the drum and the clean water filters through. As the drum rotates, the holes are cleaned with spray.

For BOD removal, by far the most popular advanced treatment method is the polishing pond, often called the *oxidation pond.* This is essentially a hole in the ground, a large pond used to confine the plant effluent before it is discharged. Such ponds are designed to be aerobic, hence light penetration for algal growth is important, and a large surface area is needed. The reactions occurring within an oxidation pond are depicted in Figure 7-20. Oxidation ponds are sometimes used as the only treatment step if the waste flow is small and the pond area is large.

Activated carbon adsorption is another method of BOD removal, and this process has the added advantage that inorganics as well as organics are removed. The mechanism of adsorption on activated carbon is both chemical and physical, with tiny crevices catching and holding colloidal and smaller particles. An activated carbon column is a completely enclosed tube with dirty water pumped up from the bottom and the clear water exiting at the top. As the carbon becomes saturated with various materials, it must be removed from the column and regenerated, or

Figure 7-19   Microstrainer schematic.

cleaned. Removal is often continuous, with clean carbon being added at the top of the column. The cleaning or regeneration is usually done by heating the carbon in the absence of oxygen. A slight loss in efficiency is noted with regeneration, and some virgin carbon must always be added to ensure effective performance.

Figure 7-20   Simplified reactions within an oxidation pond.

*Reverse osmosis** is also finding acceptance as a treatment for various kinds of trace pollutants, organic as well as inorganic. The wastewater is forced through a semipermeable membrane which acts as a superfilter, rejecting dissolved as well as suspended solids.

Nitrogen removal can be accomplished in two ways. The first method makes use of the fact that even after secondary treatment, most of the nitrogen exists as ammonia. Increasing the pH produces the following reaction:

$$NH_4^+ + OH^- \rightarrow NH_3 \uparrow + H_2O$$

Much of the dissolved ammonia gas can then be expelled from the water into the atmosphere. The resulting air pollution problem has not been resolved.

A second method of getting rid of nitrogen is to first treat the waste thoroughly enough to produce nitrate ions. This usually involves longer detention times in secondary treatment, during which bacteria such as *Nitrobacter* and *Nitrosomonas* convert ammonia nitrogen to NO_3^-, a process called *nitrification*. These reactions are

$$2NH_4^+ + 3O_2 \xrightarrow{\textit{Nitrosomonas}} 2NO_2^- + 2H_2O + 4H^+$$

$$2NO_2^- + O_2 \xrightarrow{\textit{Nitrobacter}} 2NO_3^-$$

These reactions are slow and thus require long retention times in the aeration tank, as well as sufficient dissolved oxygen. The kinetics constants for these reactions are low, with very low yields, so that the net sludge production is limited, making washout a constant danger.

Once the ammonia has been converted to nitrate, it can be reduced by a broad range of facultative and anaerobic bacteria such as *Pseudomonas*. This reduction, called *denitrification*, requires a source of carbon, and methanol (CH_3OH) is often used for that purpose.

$$6NO_3^- + 2CH_3OH \rightarrow 6NO_2^- + 2CO_2 \uparrow + 4H_2O$$

$$6NO_2^- + 3CH_3OH \rightarrow 3N_2 \uparrow + 3CO_2 \uparrow + 3H_2O + 6OH^-$$

---

*This term is a misnomer. Reverse osmosis has nothing to do with osmotic pressure, and in the future this process will be known as "ultrafiltration."

Phosphate removal is almost always done chemically. The most popular chemicals used for phosphorus removal are lime $Ca(OH)_2$, and alum, $Al_2(SO_4)_3$. The calcium ion, in the presence of high pH, will combine with phosphate to form a white, insoluble precipitate called calcium hydroxyapatite which is settled out and removed. Insoluble calcium carbonate is also formed and removed and can be recycled by burning in a furnace.

$$CaCO_3 \overset{\Delta}{\rightarrow} CO_2 + CaO$$

Quick lime, CaO, is slaked by adding water,

$$CaO + H_2O \rightarrow Ca(OH)_2$$

thus forming lime which can be reused.

The aluminum ion from alum precipitates out as poorly soluble aluminum phosphate,

$$Al^{+++} + PO_4^{---} \rightarrow AlPO_4 \downarrow$$

and also forms aluminum hydroxides,

$$Al^{+++} + 3OH^- \rightarrow Al(OH)_3 \downarrow$$

which are sticky flocs and help to settle out the phosphates. The most common point of alum dosing is in the final clarifier.

The amount of alum required to achieve a given level of phosphorus removal depends on the amount of phosphorus in the water, as well as other constituents. The sludge produced can be calculated using stoichiometric relationships.

Example 7.8

A wastewater has 6.3 mg/l of P and it is found that an alum dosage of 13 mg/l as Al^{+++} achieves an effluent with a P concentration of 0.9 mg/l.

$$Al^{+++} + PO_4^{---} \rightarrow AlPO_4$$

Molecular weights: 27   95   122
                                (or 31 as P)

or 31 mg of P removal results in the formation of 122 mg of aluminum phosphate sludge.

$$5.4 \text{ mg/l (P removed)} \times \frac{122}{31} = 21.3 \text{ mg/l (AlPO}_4 \text{ sludge produced)}$$

The amount of Al^{+++} used to produce the $AlPO_4$ is

$$5.4 \text{ mg/l (P removed)} \times \frac{27 \text{ mg Al}^{+++}}{31 \text{ mg P}} = 4.7 \text{ mg/l Al}^{+++}$$

The excess aluminum, $13 - 4.7 = 8.3$ mg/l probably went into the formation of aluminum hydroxide. The simplest thing to assume is that

$$Al^{+++} + 3OH^- \rightarrow Al(OH)_3$$

$$27 \quad 3 \times 17 = 51 \quad 78$$

and so the production of $Al(OH)_3$ is

$$8.3 \text{ mg/l Al}^{+++} \left(\frac{78 \text{ mg Al(OH)}_3}{27 \text{ mg Al}^{+++}} \right) = 24 \text{ mg/l (Al(OH)}_3 \text{ sludge)}$$

The total sludge production is then

$$AlPO_4 + Al(OH)_3 = 21.3 + 24 \cong 45 \text{ mg/l}$$

Using such a technique as alum precipitation of phosphorus, (and a bonus of more efficient SS and BOD removal due to the formation of $Al(OH)_3$), we now have attained our effluent goal:

	Raw Wastewater	Following Primary Treatment	Following Secondary Treatment	Following Tertiary Treatment
BOD mg/l	250	175	15	10
SS mg/l	220	60	15	10
P mg/l	8	7	6	0.5

An alternative to high-technology advanced wastewater treatment systems is to spray secondary effluent on land and allow the soil microorganisms to degrade the remaining organics. Such systems, known as *land treatment,* have been employed for many years in Europe, but only recently have been used in North America. They appear to represent a reasonable alternative to complex and expensive systems, especially for smaller communities.

Probably the most promising land treatment method is irrigation. Commonly, from 1000 to 2000 hectares of land are required for every 1 m^3/sec of wastewater flow, depending on the crop and soil. Nutrients such as N and P remaining in the secondary effluent are of course beneficial to the crops.

CONCLUSION

A typical wastewater treatment plant is shown schematically in Figure 7-21. We have discussed primary, secondary, and tertiary treatment. The treatment and disposal of the solids removed from the liquid stream deserves special attention, and the topic is covered in the next chapter.

An aerial view of a typical wastewater treatment plant is shown in Figure 7-22. If such plants are well operated, the effluents are often much less polluted than the stream into which they are discharged.

But not all plants perform that well. The sad fact is that many of the existing wastewater treatment plants are only marginally effective in controlling water pollution, and much of the blame can be placed on plant operation.

The operation of modern wastewater treatment plants is a complex and demanding job.* Unfortunately, operators have historically been considered to be at the bottom of the totem pole, both in terms of pay and community stature, and few qualified people were willing to make plant operation a career. Municipalities were forced to use whatever help was available, which often resulted in poor operation.

Many states now require licensing of treatment plant operators, and their pay and social stature has greatly improved. This is a welcome change, for it makes little sense to entrust unqualified and unreliable workers with the operation of multimillion-dollar facilities.

Wastewater treatment thus requires first the proper design of the plant and second the proper plant operation. One without the other is a waste of money.

PROBLEMS

7.1 The following data were reported on the operation of a wastewater treatment plant:

---

*Although seldom as demanding as the problem faced by an operator finding a pickup truck in the primary clarifier (Figure 7-23).

Figure 7-21   Block diagram of a complete wastewater treatment plant.

	Influent (mg/l)	*Effluent (mg/l)*
BOD$_5$	200	20
SS	220	15
P	10	0.5

(a) What percent removal was experienced for each of these?

(b) What kind of treatment plant would produce such an effluent? Draw a block diagram showing the treatment steps.

7.2 Describe the condition of a primary clarifier one day after the raw sludge pumps broke down.

7.3 One operational problem with trickling filters is "ponding," the excessive growth of slime on the rocks and subsequent clogging of the

Figure 7-22   An aerial view of a secondary wastewater treatment plant (courtesy Envirex).

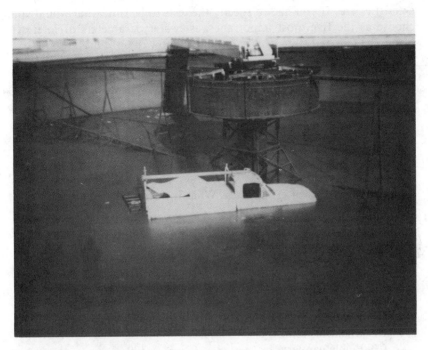

Figure 7-23   An unusual operating problem — a pickup in the primary clarifier (courtesy of Phillip Karr).

spaces so that the water no longer flows through the filter. Suggest some cures for the ponding problem.

7.4 One problem with sanitary sewers is illegal connections. Suppose a family of four, living in a home with a roof area of 70×40 ft, connects the roof drain to the sewer. For a typical rain of 1 in./hr, what percent increase will there be in the flow from their house over the "dry weather" flow (assumed at 50 gal/capita/day)?

7.5 "Good secondary treatment plants remove 90 percent of measured BOD. That does not mean that 90 percent of the total oxygen-demanding wastes are removed, but only the part that is measured by certain laboratory tests." (Environmental Quality, CEQ, 1970). How could 90% of the BOD be removed and not 90% of the oxygen-demanding material?

7.6 Suppose an industry decided to build a wastewater treatment plant and hired an engineer to design it for them. One of the first steps would be to sample the waste and run some analyses to determine what is characteristics are. If you had to specify the tests to be run, what would you choose as the five most important wastewater parameters of interest? Name the five and state why you want to know these values.

7.8 The influent and effluent data for a secondary treatment plant are:

	Influent (mg/l)	Effluent (mg/l)
BOD	200	20
Suspended solids	200	100
Total phosphorus	10	8

Calculate the removal efficiencies. What is wrong with the plant?

7.9 Draw block diagrams of the unit operations necessary to treat the following wastes to effluent levels of $BOD_5 = 20$ mg/l, SS = 20 mg/l, P = 1 mg/l.

Waste	BOD_5 (mg/l)	SS (mg/l)	P (mg/l)
A. Domestic	200	200	10
B. Chemical industry	40,000	0	0
C. Pickle cannery	0	300	1
D. Fertilizer mfg.	300	300	200

7.10 The success of an activated sludge system depends on the settling of the solids in the final settling tank. Suppose the sludge in a system started to "bulk," not settle very well, and the suspended solids concentration of the return activated sludge dropped from 10,000 mg/l to 4000 mg/l.

a. What will this do to the mixed liquor suspended solids?
b. What will this in turn do to the BOD removal? Why?

7.11 If in the above problem the ml of settled sludge (30 min) was 300 ml for both before and during the bulking problem, and the SVI was 100 and 250, what was the MLSS before and after the bulking problem? Does this agree with your answer to problem 7.10?

7.12 If you conducted a percolation test and discovered that in 30 min the water level dropped 5 in., what size percolation field would you need for a two-family house? What size would you need if it dropped 0.5 in.?

7.13 A 1-mgd conventional activated sludge plant has an influent BOD_5 of 200 mg/l. The primary clarifier removes 30% of that BOD. The three aeration tanks are each $20 \times 20 \times 100$ ft. What MLSS are necessary to attain 90% BOD removal in the plant?

7.14 An aeration system with a hydraulic retention time of 2.5 hr receives a flow of 0.2 mgd at a BOD of 150 mg/l. The suspended solids in the aeration tank are 4000 mg/l. The effluent BOD is 20 mg/l and effluent suspended soilds are 30 mg/l. Calculate the F/M ratio (food to microorganism) for this system.

7.15 The MLSS in an aeration tank are 4000 mg/l. The flow from the primary settling tank is 0.2 m^3/sec and the return sludge flow is 0.1 m^3/sec. What must the return sludge suspended solids concentration be to maintain the 4000 mg/l MLSS?

7.16 A wastewater is composed of suspended solids which are all small beads, 0.1 mm diam, with a specific gravity of 2.65.
   a. How long would it take for a settling tank 4 meters deep to clarify the waste?
   b. If a false floor were put into the settling tank so that it was now only 2 meters deep, what fraction of the suspended solids would be removed?
   c. What is the "overflow rate" for this tank?

7.17 A primary clarifier (settling tank) is 80 ft long, 30 ft wide, and 12 ft deep, and receives a flow of 200,000 gpd.
   a. Calculate the overflow rate.
   b. What is the settling velocity of the critical particle (assuming entire tank is used for clarification)?

7.18 A rectangular primary clarifier (settling tank) is 10 m wide, 25 m long and 5 m deep. The flow is 0.5 m^3/sec.
   a. What is the design overflow rate?
   b. Would all particles having a settling velocity of 0.01 cm/min be removed?

7.19 A community receives 0.5 m^3/sec of flow into its wastewater treatment plant. The total surface area of the primary clarifiers is 2700 m^3 and the retention time is 3 hr. Find the overflow rate and the depth of the clarifiers.

7.20 A transoceanic flight on a Boeing 747, with 430 persons on board, takes 7 hours. Estimate the weight of the water necessary to flush the toilets if each flush required 2 gallons. Make any assumptions necessary. What fraction of the total payload (people) would the flush water represent? How could you reduce this weight? (The railroad system, for obvious reasons, is illegal.)

7.21 A family of four wants to build a house on a lot for which the percolation test results show 1.0 mm/min. The county requires a septic tank hydraulic retention time of 24 hr. Find the volume of the tank required, and the area of the tile field. Sketch the system, including all dimensions.

7.22 A consulting engineer has proposed the following design for a primary clarifier: $Q = 0.150$ m^3/sec, influent SS = 310 mg/l, raw primary sludge concentration = 4% solids, SS removal efficiency = 60%, length = 40 m, width = 15 m, depth = 2.2 m, weir length = 75 m. Is this a reasonable design and expected performance?

7.23 A final clarifier is to be able to handle a MLSS of 2800 mg/l at flow of 0.30 m^3/s. Determine the tank depth, diameter solids loading, and weir rate. The design overflow rate is 30 m/day, and the expected SVI is 150.

7.24 A wastewater contains soluble phosphate at a concentration of 4 mg/l. How much ferric chloride would theoretically be necessary to precipitate this nutrient?

7.25 A community with a wastewater flow of 10 mgd is required to meet effluent standards of 30 mg/l for both BOD$_5$ and SS. Pilot plant results with the influent of BOD$_5$ = 250 mg/l estimate the kinetic constants at K$_s$ = 100 mg/l, $\hat{\mu} = 0.25$ day^{-1} and Y = 0.5. It is decided to maintain the MLSS at 2000 mg/l. What is the hydraulic retention time, the sludge age and the required tank volume?

LIST OF SYMBOLS

$$A = \text{area, m}^2$$
BOD = biochemical oxygen demand, mg/l
C_D = drag coefficient
d = diameter, m
F = food (BOD), mg/l
F_B = buoyancy force, N
F_D = drag force, N
F_g = gravitational force, N
g = gravitational acceleration, m/sec^2
H = height of a settling tank or settling cylinder, m
h = height, m
K_s = saturation constant, mg/l
L = length, m

M = microorganisms (SS), mg/l

MLSS = mixed liquor suspended solids, mg/l

m = mass, kg

P = phosphorus, mg/l

Q = flow rate, m^3/sec

Q_w = waste sludge flow rate, m^3/sec

q = substrate removal velocity, sec^{-1}

R = Reynolds number

R = overall recovery of SS in settling tank, %

S = substrate concentration, estimated as BOD, mg/l

S_o = influent substrate concentration, estimated as influent BOD, mg/l

SS = suspended solids, mg/l

t = time, sec

$\bar{t}$ = retention time, sec

v = velocity, m/sec

v_o = settling velocity of a critical particle, m/sec

v_s = settling velocity of any particle, m/sec

X = microorganism concentration, estimated as suspended solids, mg/l

X_e = effluent microorganism concentration, estimated as effluent SS, mg/l

X_r = return sludge microorganism concentration, estimated as return sludge SS, mg/l

X_o = influent microorganism concentration, estimated as influent SS, mg/l

Y = yield, SS produced/BOD utilized

Θ_c = mean cell residence time, or sludge age, days

μ = fluid viscosity, $N \cdot sec/m^2$

μ = growth rate constant, sec^{-1}

$\hat{\mu}$ = maximum growth rate constant, sec^{-1}

ρ = fluid density, kg/m^3

ρ_s = density of a solid, kg/m^3

$\forall$ = volume, m^3

SUGGESTED READING

Ehlers, V. M., and E. W. Steel. *Municipal and Rural Sanitation* (New York: McGraw-Hill Book Company, 1958).

Phelps, E. B. *Public Health Engineering* (New York: John Wiley & Sons, 1948).

Water Pollution Control Federation. *Operation of Wastewater Treatment Plants* (1968).

Fair, G. M., J. C. Geyer and D. A. Okun. *Elements of Water Supply and Wastewater Disposal* (New York: John Wiley & Sons, 1971).

U.S. Public Health Service, *Manual of Septic Tank Practice,* #526 (1967).

Parsons, W. A. (National Lime Association). *Chemical Treatment of Sewage and Industrial Waste* (1965).

Hammer, M. J. *Water and Wastewater Technology* (New York: John Wiley & Sons, 1975).

Sanks, R. L. and T. Asano, Ed. *Land Treatment and Disposal of Municipal and Industrial Wastewater* (Ann Arbor, MI: Ann Arbor Science Publishers, Inc., 1976).

Nemerow, N. L. *Industrial Water Pollution,* (Reading MA: Addison-Wesley Publishing Company, 1978).

Harbold, H. S. *Sanitary Engineering Problems and Calculations for the Professional Engineer* (Ann Arbor, MI: Ann Arbor Science Publishers, Inc., 1979).

Schoeder, E. D. *Water and Wastewater Treatment* (New York: McGraw-Hill Book Company, 1977).

Berthoux, P. M., and D. F. Rudd. *Strategy of Pollution Control* (New York: John Wiley & Sons, 1980).

Clark, J. W., W. Viessman, Jr. and M. J. Hammer. *Water Supply and Pollution Control* (New York: IEP, 1977).

Steel, E. W., and T. J. McGhee. *Water Supply and Sewerage* (New York: McGraw-Hill Book Company, 1979).

Weber, W. J., Jr., Ed. *Physicochemical Processes for Water Quality Control* (New York: John Wiley & Sons, 1972).

Metcalf & Eddy, Inc. *Wastewater Engineering,* (New York: McGraw-Hill Book Company, 1979).

Chapter 8

Sludge Treatment and Disposal

The field of wastewater treatment engineering is littered with unique and imaginative processes for achieving high degrees of waste stabilization at attractive costs. Few of these "wonder plants" have proven themselves in practice, and quite often the problem has been the inattention to the sludge problem. Drawing a flow diagram with a little arrow labeled "SLUDGE TO DISPOSAL" has often been the total extent of the consideration for solids handling, treatment and disposal.

In the past few years the fact that sludge treatment and disposal accounts for over fifty percent of the treatment costs in a typical secondary plant has prompted a renewed interest in this none-too-glamorous, but essential aspect of wastewater treatment.

This chapter is devoted to the problem of sludge treatment and disposal. The sources and quantities of sludge from various types of wastewater treatment systems are examined first, followed by a definition of sludge characteristics. Such solids concentration techniques as thickening and dewatering are discussed next, concluding with consideration for ultimate disposal.

SOURCES OF SLUDGE

The first source of sludge is the suspended solids which enter the treatment plant and are partially removed in the primary settling tank or clarifier. Commonly about 60% of the suspended solids become *raw primary sludge,* which is highly putrescible and very wet (about 96% water).

The removal of BOD is basically a method of wasting energy, and secondary wastewater treatment plants are designed to reduce this high-energy material to low-energy chemicals. This process is typically accomplished by biological means, using microorganisms (the "decomposers" in ecological terms) who use the energy for their own life and procreation. Secondary treatment processes such as the popular activated sludge system are *almost* perfect systems. Their major fault is that the microorganisms convert too little of the high-energy organics to CO_2 and H_2O, and too much of it to new organisms. Thus the system operates with an excess of these microorganisms, or *waste activated sludge*. As defined in the previous chapter, the mass of waste activated sludge produced per mass of BOD removed in secondary treatment is known as the *yield,* expressed as kg SS produced/kg BOD removed.

Phosphorus removal processes also invariably end up with excess solids. If lime is used, the calcium carbonates and calcium hydroxyapatites are formed and must be disposed of. Aluminum sulfate similarly produces solids, in the form of aluminum hydroxides and aluminum phosphates. Even so-called "totally biological processes" for phosphorus removal end up with solids. The use of an oxidation pond or marsh for phosphorus removal is possible only if some organics (algae, water hyacinths, fish, etc.) are periodically harvested.

The quantities of sludges obtained from various wastewater treatment processes can be calculated as shown in Figure 8-1.[1] The symbols are defined as follows:

S_o = influent BOD, kg/hr (5 day 20°C)
X_o = influent suspended solids, kg/hr
h = fraction of BOD not removed in primary clarifier
i = fraction of BOD not removed in aeration or trickling filter
X_e = plant effluent suspended solids, kg/hr
k = fraction of X_o removed in primary clarifier
j = fraction of solids not destroyed in digestion
ΔX = net solids produced by biological action, kg/hr
Y = yield = $\Delta X/\Delta S$, where $\Delta S = hS_o - ihS_o$

Example 8.1

The plant influent has a BOD_5 of 250 mg/l at a flow rate of 1570 m³/hr (10 mgd), and an influent suspended solid of 225 mg/l. The raw sludge produced would be equal to kX_o where k is the fraction of suspended solids removed in the primary clarifier and X_o is the influent suspended solids in kg/hr. If k is (typically) 0.6, and X_o can be calculated as

---

[1]Kormanik, R. A. "Estimating Solids Production for Sludge Handling," *Water & Sew. Works* 118:5 (1972).

Figure 8-1 Schematic of a generalized secondary treatment plant.

$$225 \text{ mg/l} \times 1570 \text{ m}^3/\text{hr} \times 1000 \text{ l/m}^3 \times 10^{-6} \text{ mg/kg} = 353 \text{ kg/hr}$$

The raw sludge produced is then

$$0.6 \times 353 = 212 \text{ kg/hr}$$

Typical values for the constants, for domestic wastewater, are shown below:

$$S_o = \begin{cases} 250 \times 10^{-3} \times Q = \text{kg/h if Q is m}^3/\text{hr} \\ 250 \times 8.34 \times Q = \text{lb/day if Q is mgd} \end{cases}$$

$$X_o = \begin{cases} 220 \times 10^{-3} \times Q = \text{m}^3/\text{h if Q is m}^3/\text{hr} \\ 220 \times 8.34 \times Q = \text{lb/day if Q is mgd} \end{cases}$$

$$k = 0.6$$
$$h = 0.7$$

$$X_e = \begin{cases} 20 \times 10^{-3} \times Q = \text{m}^3/\text{h if Q is m}^3/\text{hr} \\ 20 \times 8.34 \times Q = \text{lb/day if Q is mgd} \end{cases}$$

$$j \simeq 0.5 \text{ for anaerobic}$$
$$j \simeq 0.8 \text{ for aerobic}$$
$$i = 0.1 \text{ for well operated activated sludge}$$
$$i = 0.2 \text{ for trickling filters}$$
$$Y = 0.5 \text{ for activated sludge}$$
$$Y = 0.2 \text{ for trickling filters}$$

Quantities of chemical sludges formed due to the precipitation of phosphorus, for example, must be added to the total sludge produced.

CHARACTERISTICS OF THE SLUDGES

The characteristics of sludges of interest depend entirely on what is to be done with a sludge. For example, if the sludge is to be thickened by gravity, its settling and compaction characteristics are important. On the other hand, if the sludge is to be digested anaerobically, the concentrations of volatile solids, heavy metals, etc., are important.

Another fact of immense importance in the design of sludge handling and disposal operations is the variability of the sludges. In fact this variability can be stated in terms of three "laws":

1. No two wastewater sludges are alike in all respects.
2. Sludge characteristics change with time.
3. There is no "average sludge."

The first statement reflects the fact that no two wastewaters are alike, and that if the variable of treatment is added, the sludges produced will have significantly different characteristics.

The second statement is often overlooked by designers. For example, the settling characteristics of chemical sludges from the treatment of plating wastes [e.g., $PB(OH)_2$, $Zn(OH)_2$ or $Cr(OH)_3$] vary with time simply because of uncontrolled pH changes. Biological sludges are of course continually changing, with the greatest change occurring when the sludge changes from aerobic to anaerobic (or vice versa). It is thus quite difficult to design sludge handling equipment, since the sludge may change in some significant characteristic in only a few hours.

The third "law" is constantly violated. Tables showing "average values" for "average sludges" are useful for illustrative and comparative purposes only, and should not be used for design.

With that caveat, we now proceed to discuss some characteristics of "average sludges," tabulated in Table 8-1.

The first characteristic, solids concentration, is perhaps the most important variable, defining the volume of sludge to be handled, and determining whether the sludge behaves as a liquid or a solid. The importance of volatile solids is, of course, in the disposability of the sludge. With high volatiles, a sludge would be difficult to dispose of into the environment. As the volatiles are degraded, gases and odors are produced, and thus a high volatile solids concentration would restrict the methods of disposal.

The rheological characteristics of sludge are of interest in that this is one of only a few truly basic parameters describing the physical nature of a sludge. Two-phase mixtures like sludges, however, are almost without exception non-Newtonian and thixotropic. Sludges tend to act like pseudoplastics, with an apparent yield stress and a plastic viscosity. The rheological behavior of a pseudoplastic fluid is defined by a rheogram shown as Figure 8-2. The term thixotropic relates to the time dependence of the rheological properties.

Sludges tend to act more like plastic fluids as the solids concentration increases. True plastic fluids can be described by the equation

$$\tau = \tau_y + \eta \, \frac{du}{dy}$$

where
τ = shear stress
τ_y = yield stress
η = plastic viscosity
du/dy = rate of shear, or the slope of the velocity (u)-depth (y) profile

Although sludges are seldom true plastics, the above equation can be used as an approximation of the rheogram.

Table 8-1.   Sludge Characteristics

	Physical				Chemical		
Type of Sludge	Solid Concentration (mg/l)	Volatile Solids (%)	Yield Strength (dyn/cm²)	Plastic Viscosity (g/cm s)	Nitrogen (% as N)	Phosphorus (% as P₂O₂)	Potassium (% as K₂O)
Water	–	–	0	0.01	–	–	–
Raw Primary	60,000	60	40	0.3	2.5	1.5	0.4
Mixed Digested	80,000	40	15	0.9	4.0	1.4	0.2
Waste Activated	15,000	70	0.1	0.06	4.0	3.0	0.5
Alum, ppt	20,000	40	–	–	2.0	2.0	–
Lime, ppt	200,000	18	–	–	2.0	3.0	–

Figure 8-2 Typical rheogram for sludges.

The yield stress can vary from above 40 dyn/cm² for 6% raw sludge, to only 0.07 dyn/cm² for a thickened activated sludge. The large differences suggest that rheological parameters could well be used for scale-up purposes. Unfortunately, few researchers have bothered to measure the rheological characteristics (such analyses are not even included in *Standard Methods*) and many gaps exist in the available data.

The chemical composition is important for several reasons. First, the fertilizer value of the sludge is dependent on the availability of N, P and K, as well as trace elements. A more important measurement, however, is the concentration of heavy metals and other toxins which would make the sludge toxic to the environment. The ranges of heavy metal concentrations are very large (e.g., cadmium can range from almost zero to over 1000 mg/kg). Since a major source of such toxins is in industrial discharges, a single poorly operated industrial firm may contribute enough toxins to make the sludge worthless as a fertilizer. Most engineers agree that although it would be most practical to treat the sludge at the plant in order to achieve the removal of toxic components, there are no effective methods available for removing heavy metals, pesticides and other potential toxins from the sludge, and that the control must be over the influent (that is, tight sewer ordinances).

In addition to the physical and chemical characteristics, the biological parameters of sludge can also be important. The volatile solids parameter is in fact often interpreted as a biological characteristic, the assumption being the VSS is a gross measure of viable biomass. Another important parameter, especially in regard to ultimate disposal, is the concentration of pathogens, both bacteriological and viral. The primary clarifier seems to act like a viral and bacteriological concentrator, with a substantial fraction of these microorganisms existing in the sludge instead of the liquid effluent.

SLUDGE TREATMENT

A great deal of money could be saved, and troubles averted, if sludge could be disposed of as it is drawn off the main process train. Unfortunately, the sludges have three characteristics which make such a simple solution unlikely: they are aesthetically displeasing, they are potentially harmful, and they have too much water.

The first two problems are often solved by stabilization, such as *anaerobic* or *aerobic digestion*. The third problem requires the removal of water by either thickening or dewatering. Accordingly, the next three sections cover the topics of stabilization, thickening and dewatering, followed finally by considerations of ultimate disposal.

Before we embark on a discussion on sludge treatment, however, we should be reminded that the processes should be used in wastewater treatment only when necessary. It makes little sense to spend time and money to treat sludges if their ultimate disposal does not require it. The "Get the water out!" syndrome often found in engineering offices is often associated with "Gotta digest it!" disease. These processes all cost money, and should be considered only if direct disposal in undigested liquid form is not practical.

Sludge Stabilization

The objective of sludge stabilization is to reduce the problems associated with two of the detrimental characteristics listed above: sludge odor and putrescibility, and the presence of pathogenic organisms.

There are three primary means of sludge stabilization

- lime
- aerobic digestion
- anaerobic digestion.

Lime stabilization is achieved by adding lime (either as hydrated lime, $Ca(OH)_2$ or as quicklime, CaO) to the sludge and thus raising the pH to about 11 or above. This significantly reduces the odor and helps in the destruction of pathogens. The major disadvantage of lime stabilization is that it is temporary. With time (days) the pH drops and the sludge once again becomes putrescible.

Aerobic stabilization is merely a logical extension of the activated sludge system. Waste activated sludge is placed in dedicated aeration tanks for a very long time, and the concentrated solids allowed to progress well into the endogenous respiration phase, in which food is obtained only by the destruction of other viable organisms. This results in a net reduction in total and volatile solids. Aerobically digested sludges are, however, more difficult to dewater than anaerobic sludges.

The third commonly employed method of sludge stabilization is anaerobic digestion. The biochemistry of anaerobic decomposition of organics is illustrated in Figure 8-3. Note that this is a staged process, with the solution of organics by extracellular enzymes being followed by the production of organic acids by a large and hearty group of anaerobic microorganisms known, appropriately enough, as the *acid formers*. The organic acids are in turn degraded further by a group of strict anaerobes called *methane formers*. These microorganisms are the prima donnas of wastewater treatment, getting upset at the least change in their environment. The success of anaerobic treatment thus boils down to the creation of a suitable condition for the methane formers. Since they are strict anaerobes, they are unable to function in the presence of oxygen and very sensitive to environmental conditions such as temperature, pH and toxins. If a digester goes "sour," the methane formers have been inhibited in some way. The acid formers, however, keep chugging away, making more organic acids. This has the effect of further lowering the pH and making conditions even worse for the methane formers. A sick digester is therefore difficult to cure without massive doses of lime or other antacids.

Most treatment plants have two kinds of digesters—primary and secondary (Figure 8-4). The primary digester is covered, heated and mixed to increase the reaction rate. The temperature of the sludge is usually about 35°C (95°F). Secondary digesters are not mixed or heated and are used for storage of gas and for concentrating the sludge by settling. As the solids settle, the liquid supernatant is pumped back to the main plant for further treatment. The cover of the secondary digester often floats up and down, depending on the amount of gas stored. The gas is high enough in methane to be used as a fuel, and is in fact usually used to heat the primary digester.

Figure 8-3   Generalized biochemical reactions to anaerobic digestion.

Anaerobic digesters are commonly designed on the basis of solids loading. Experience has shown that domestic wastewaters contain about 120 g (0.27 lb) suspended solids per capita. This can be translated, knowing the population served, into total suspended solids to be handled. To this, of course, must be added the production of solids in secondary treatment. Once the solids production is calculated, the digester volume is estimated by assuming a reasonable hydraulic detention time for the sludge (generally 30 days) and using a loading factor such as 4 kg/m³·day (0.27 lb/ft³·day). This loading factor is decreased if a higher reduction of volatile solids is desired.

Example 8.2

Raw primary and waste activated sludge at 4% solids is to be anaerobically digested at a loading of 3 kg/m³·day at a hydraulic detention time of 30 days. The total sludge produced in the plant is 1500 kg/day. Calculate the required volume of the primary digester.

Figure 8-4   Anaerobic sludge digesters (photo courtesy Dorr Oliver Inc.).

Each day's production of sludge thus requires

$$\frac{1500 \text{ kg/day}}{3 \text{ kg/m}^3 \cdot \text{day}} = 500 \text{ m}^3$$

The detention time is 30 days, so the total volume required is $30 \times 500 = 15{,}000 \text{ m}^3$. The total volume of sludge pumped to the digester is

$$\frac{1500 \text{ kg/day}}{0.04} = 37,500 \text{ kg/day}$$

and since one liter of sludge weighs about 1 kg, the volume of sludge is 37,500 l/day or 37.5 m^3/day.

The production of gas from digestion varies with the temperature, solids loading, solids volatility and other factors. Typically, about 0.6 m^3 of gas per kg volatile solids added (10 ft^3/lb) has been observed. This gas is about 60% methane and burns readily, usually being used to heat the digester and answer additional energy needs within a plant. It has been found that an active group of methane formers operates at 35°C (95°F) in common practice, and this process has become known as *mesophilic digestion*. As the temperature is increased, to about 45°C (115°F), another group of methane formers predominates, and this process is tagged *thermophilic digestion*. Although the latter process is faster and produces more gas, it is also more difficult and expensive to maintain such elevated temperatures.

Finally, a word of caution should be expressed about the problem of mixing a primary digester. The assumption is invariably made that the tank is totally mixed, either by mechanical means or by bubbling gases through it. Unfortunately, digester mixing is quite difficult, and some recent studies have shown that on the average, only about 20% of the tank volume is well mixed!

All three stabilization processes reduce the concentration of pathogenic organisms, but to varying degrees. Lime stabilization achieves a high degree of sterilization, due to the high pH. Further, if quicklime (CaO) is used, the reaction is exothermic and the elevated temperatures assist in the destruction of pathogens. Aerobic digestion at ambient temperatures is not very effective in the destruction of pathogens.

Anaerobic digesters have been well studied from the standpoint of pathogen viability since the elevated temperatures should result in substantial sterilization. As early as 1958, however, it was found that *Salmonella typhosa* organisms and many other pathogens can survive digestion. Polio viruses similarly survive with little reduction in virulence. An anaerobic digester cannot, therefore, be considered a method of sterilization.

Sludge Thickening

Sludge thickening is a process in which the solids concentration is increased and the total sludge volume is correspondingly decreased, but the sludge still behaves like a liquid instead of a solid. Typically, for mixed digested sludges, the point at which sludge begins to have the properties of a solid is between 15 and 20% solids. Thickening also implies that the process is gravitational, using the difference between particle and fluid densities to achieve greater compacting of solids.

The advantages of sludge thickening in reducing the volume of sludge to be handled are substantial. With reference to Figure 8-5 a sludge with 1% solids thickened to 5% results in an 80% volume reduction. A 20% solids concentration, which might be achieved by mechanical dewatering (discussed in the next section) would result in a 95% reduction in volume. The savings in treatment, handling and disposal costs accrued can be substantial.

Two types of nonmechanical thickening operations are presently in use: the gravity thickener and the flotation thickener. These are not very good names, since the latter also uses gravity to separate the solids from the liquid. For the sake of simplicity, however, we will continue to use the two descriptive terms.

A typical gravity thickener is shown in Figure 8-6. The influent, or feed, enters in the middle, and the water moves to the outside, eventually leaving as the clear effluent over the weirs. The sludge solids settle as a blanket and are removed out the bottom.

The thickening characteristics of a sludge have for many years been described by the sludge volume index (SVI). The parameter is defined as the volume occupied by one gram of sludge after settling for 30 minutes. The sludge is commonly settled in a 1-liter cylinder and the volume of sludge (in ml) is measured after 30 minutes. The SVI is calculated as

Figure 8-5   Volume reduction due to sludge thickening.

Figure 8-6 Gravity thickener (photo courtesy Dorr Oliver Inc.).

$$SVI = \frac{\text{ml of sludge after 30 min} \times 1000}{\text{sludge SS mg/l}}$$

Treatment plant operators usually consider sludges with a SVI less than 100 as well settling sludges and those with a SVI greater than 200 as potential problems. These numbers are of some use in estimating the settleability of activated sludge, and the SVI is without question a valuable tool in running a secondary treatment plant. But the SVI also has some drawbacks and potential problems.

One of these problems is that the SVI is not necessarily independent of solids concentration. Substantial increases in SS can change the SVI severalfold. Further, for high suspended solids, the SVI has a maximum above which it can never occur. Take for example a lime sludge with a suspended solids concentration of 40,000 mg/l, and suppose that it doesn't settle at all (the sludge volume after 30 min is 1000 ml). Calculating the SVI gives us

$$SVI = \frac{1000 \times 1000}{40,000} = 25$$

How can a sludge that does not settle at all have a SVI = 25, when we originally noted that a SVI of less than 100 indicated a good settling sludge?

The point is that the maximum SVI for a sludge with a suspended solids of 40,000 is 25. Similar maximum values can be calculated and plotted as shown in Figure 8-7.

A better way to describe the settling characteristics of a sludge is a "flux plot." This graph is developed by first running a series of settling tests at various solids concentrations and recording the sludge-water interface height with time. The velocity of settling is calculated as the slope of the initial straight line portion of the curve and plotted against the solids concentration. Multiplying a velocity by its corresponding solids concentration yields solids flux, which is plotted against the concentration (as shown in Figure 8-8). The flux curve is a "signature" of the sludge thickening characteristics. Further, it can be used to design thickeners. Note that at any concentration C_i, the interface settling velocity is v_i and thus the solids flux at the concentration is $C_i v_i (kg/m^3 \times m/hr = kg/m^2 \cdot hr)$.

The design involves choosing a desired thickened sludge solids concentration, C_u (estimated in laboratory tests) and drawing a line from this value tangent to the underside of the flux curve. The intercept at the ordinate is the limiting flux, G_L, or that solids flux which controls the

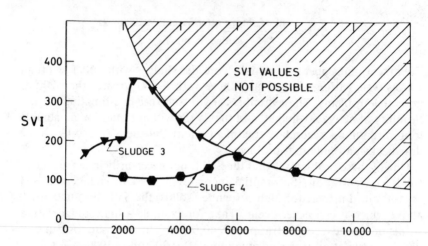

Figure 8-7   Maximum values of SVI.

thickening operation. In a continuous thickener, it is impossible to pass more solids than that (at the stated concentration) through a unit area.

The units of flux are kg solids/m^2·hr (or other comparable units), and are defined as

$$G = \frac{Q_o C_o}{A}$$

where   G = solids flux
Q$_o$ = inflow rate (m^3/hr)
C$_o$ = inflow solids concentration (mg/l)
A = surface area of thickener (m^3)

If G is the limiting flux, the necessary thickener area can be calculated as

$$A_L = \frac{Q_o C_o}{G_L}$$

where the subscript L specifies the limiting area and flux.

This graphical procedure can be used to develop an optimal process. If the calculated thickener area is too big, a new (less ambitious) underflow solids concentration can be selected and a new required area calculated.

Example 8.5

If, on Figure 8-8, the C$_u$ was chosen as 25,000 mg/l and the G$_L$ was then read off the graph as 3 kg/m^2·hr, what is the required area for a

Figure 8-8   Calculation of limiting solids flux in gravitational thickening.

thickener if the feed is 60 m³/hr of sludge with 1% solids? Suppose the C_u was next chosen as 40,000 mg/l and the G_L read off as 1.8 kg/m²·day. What is the required area in the second case?

The limiting area in the first case is

$$A_L = \frac{Q_o C_o}{G_L} = \frac{60 \text{ m}^3/\text{h} \times 0.01 \times 1000 \text{ kg/m}^3}{3 \text{ kg/m}^2 \cdot \text{hr}} = 200 \text{ m}^2$$

and in the second case

$$A_L = \frac{60 \times 0.01 \times 1000}{1.8} = 333 \text{ m}^2$$

Note that the area requirement increases as the desired underflow concentration increases.

In the absence of laboratory data, thickeners are designed on the basis of solids loading, which is another way of expressing the limiting flux. Gravity thickener design loadings for some specific sludges are shown in Table 8-2.

A flotation thickener, shown in Figure 8-9, operates by forcing air under pressure to dissolve in the return flow, and releasing the pressure as the return is mixed with the feed. As the air comes out of solution, the tiny bubbles attach themselves to the solids and carry them upward, to be scraped off.

Table 8-2. Design Loadings for Gravity Thickeners[a]

Sludge	Design Loading (kg solids/m² · hr)
Raw Primary	5.2
Waste Activated	1.2
Raw Primary + Waste Activated	2.4
Trickling Filter Sludge	1.8

[a]Note: $0.204 \times kg/m^2 \cdot hr = lb/ft^2 \cdot hr.$

Figure 8-9   Flotation thickener.

Sludge Dewatering

As defined above, dewatering differs from thickening in that the sludge should behave as a solid after it has been dewatered. Dewatering is seldom used as an intermediate process, unless the sludge is to be incinerated. Most wastewater plants use dewatering as a final method of volume reduction prior to ultimate disposal.

In the United States five dewatering techniques have been most popular: sand beds, vacuum filters, pressure filters, belt filters and centrifuges. Each of these is discussed below.

Sand beds have been in use for a great many years and are still the most cost-effective means of dewatering when land is available and labor is not exorbitant. The beds consists of tile drains in gravel, covered by about 26 cm (10 in.) of sand. The sludge to be dewatered is poured on the beds at about 15 cm (6 in.) deep. Two mechanisms combine to separate the water from the solids: seepage and evaporation. Seepage into the sand and through the tile drains, although important in the total volume of water extracted, lasts for only a few days. The sand pores are quickly clogged, and all drainage into the sand ceases. The mechanism of evaporation takes over, and this process is actually responsible for the conversion of liquid sludge to solid. In some northern areas sand beds are enclosed under greenhouses to promote evaporation as well as prevent rain from falling into the beds.

For mixed digested sludge, the usual design is to allow for 3 months drying time. Some engineers suggest that this period be extended to allow a sand bed to rest for a month after the sludge has been removed. This seems to be an effective means of increasing the drainage efficiency once the sand beds are again flooded.

Raw sludge will not drain well on sand beds and will usually have an obnoxious odor. Hence raw sludges are seldom dried on beds. Raw secondary sludges have a habit of either seeping through the sand or clogging the pores so quickly that no effective drainage takes place. Aerobically digested sludge can be dried on sand, but usually with some difficulty.

If dewatering by sand beds is considered impractical, mechanical dewatering techniques must be employed. The first successful mechanical dewatering device employed in sludge treatment was the *vacuum filter*. As shown in Figure 8-10, the vacuum filter consists of a perforated drum covered with a fabric. A vacuum is drawn in the drum and the covered drum dipped into sludge. The water moves through the filter cloth, leaving the solids behind eventually to be scraped off.

The effectiveness of a vacuum filter in dewatering a specific sludge is measured most frequently by the *specific resistance to filtration* test. The resistance of a sludge to filtration can be stated as

$$r = \frac{2PA^2b}{\mu W}$$

where　P = vacuum pressure, N/m^2
　　　　A = area of the filter, m^2
　　　　μ = filtrate viscosity, $N \cdot sec/m^2$
　　　　w = cake deposited per volume of filtrate (for dry cakes this can be approximated as the feed solids concentration) kg/m^3
　　　　b = slope of the filtrate volume vs time/filtrate volume curve.

Using the units above, specific resistance is in terms of m/kg.

The factor b can be determined from a simple test with a Büchner funnel, as shown in Figure 8-11. The sludge is poured onto the filter, the

Figure 8-10.   Vacuum filter.

vacuum exerted, and the volume of the filtrate recorded against time. The plot of these data yields a straight line with a slope b.

Again, although there are no "average sludges," it may be instructive to list some data from experience and compare the filterability of different sludges. Such a listing is shown in Table 8-3.

The Büchner funnel test for specific resistance is rather clumsy and

Figure 8-11   Büchner funnel test for determining specific resistance to filtration.

Table 8-3.   Specific Resistance of Typical Sludges

	Specific Resistance[a] (m/kg)
Raw Primary	$10\text{--}30 \times 10^{14}$
Mixed Digested	$3\text{--}30 \times 10^{14}$
Waste Activated	$5\text{--}20 \times 10^{14}$
Lime and Biological Sludge	$1\text{--}5 \times 10^{14}$
Lime Slurry	$5\text{--}10 \times 10^{13}$
Alum	$2\text{--}10 \times 10^{13}$

[a] As a rule of thumb, a sludge will not filter well if the specific resistance exceeds 1×10^{12} m/kg.

time consuming. An indirect method for estimating how well a sludge will filter uses the *capillary suction time* (CST) apparatus, developed in Great Britain. The CST device, as illustrated in Figure 8-12, allows the water to seep out of a sludge (in effect a filtration process) and onto a blotter. The speed at which the water is taken up by the blotter is measured by timing of the travel over a set distance. This time, in seconds, can be correlated to specific resistance. A short time would indicate a highly filterable sludge, whereas a long time would portend problems in filtration. Some typical CST values are shown in Table 8-4.

Figure 8-12   Capillary suction time apparatus.

Table 8-4.   Typical CST Values

Sludge	CST (sec)
Raw Primary Sludge	80
Activated Sludge	10
Mixed Digested Sludge	50
Alum Sludge	30

The specific resistance test, although the closest thing we have to a basic parameter for filtration, should not be used for design. The *filter leaf* test is by far superior, simply because it simulates the workings of a vacuum filter so well. Figure 8-13 is a diagram showing the filter leaf apparatus. The filter cloth (a sample of the proposed fabric) is wrapped around a hollowed-out disc, the disc (filter leaf) is inserted into the sludge for about 20 seconds (the length of time a typical vacuum filter might be in contact with the sludge), raised out of the sludge for maybe 40 seconds (simulating the drying cycle) and the vacuum cut off. The cake developed is a very good approximation of how a continuous filter would operate.

Figure 8-13 Filter leaf apparatus for vacuum filter design.

The design parameter calculated from the filter leaf test is the *filter yield,* in terms of kg of dry solids/m² of filter area/hr. If this value is known, the required filter size can be determined by selecting the hours of operation and establishing how much sludge needs to be removed from the treatment plant.

Typical filter yields for some representative sludges with polymer conditioning are shown in Table 8-5.

The *pressure filter,* shown in Figure 8-14, uses positive pressure to force the water through a filter cloth. Typically, the pressure filters are built as plate-and-frame filters, where the sludge solids are captured in the spaces between the plates and frames, which are then pulled apart to allow for sludge cleanout.

Table 8-5. Filter Yields for Typical Sludges

	Filter Yield[a] (kg/m² · hr)
Raw Primary (with polymer)	40
Mixed Anaerobically Digested (with polymer)	20
Waste Activated	(will not filter)
Lime and Bilogical Sludge	15

[a] kg/m² · hr × 0.204 = lb/ft² · hr.

Figure 8-14 Pressure filters (photo courtesy Envirex).

The *belt filter,* shown in Figure 8-15, operates as both a pressure filter and by a gravity drainage. As the sludge is introduced onto the moving belt, the free water drips through the belt but the solids are retained. The belt then moves into the dewatering zone where the sludge is squeezed between two belts. These machines are quite effective in dewatering

Figure 8-15   Belt filter.

many different kinds of sludges and are being widely installed in small wastewater treatment plants.

Centrifugation became popular in wastewater treatment only after organic polymers were available for sludge conditioning. Although the centrifuge will work on any sludge (unlike the vacuum filter which will not pick up some sludges, resulting in zero filter yield), without good conditioning, most sludges can not be centrifuged with greater than 60 or 70% solids recovery.

The centrifuge most widely used is the solid bowl decanter, which consists of a bullet-shaped body rotating on its long axis. The sludge is placed into the bowl, the solids settle out under about 500 to 1000 gravities (centrifugally applied) and are scraped out of the bowl by a screw conveyor (Figure 8-16).

Although laboratory tests are of some value in estimating centrifuge applicability, tests using continuous models are considerably better and highly recommended whenever possible.

A centrifuge must be able to first settle the solids, and then to move them out of the bowl. Accordingly, two parameters have been suggested for scale-up between two geometrically similar machines.

Settling characteristics are measured by the *Sigma equation*.[2] Without going into the derivation of this parameter, it is simply assumed that if two machines (1 and 2) are to have equal effects on the settling within the bowl, the relationship

$$\frac{Q_1}{\Sigma_1} = \frac{Q_2}{\Sigma_2}$$

must hold. The symbol Q represents the liquid flow rate into the machine, and Σ is a parameter comprised of machine (not sludge!) characteristics. For solid bowl centrifuges, Sigma is calculated as

$$\Sigma = \frac{\omega}{g \, \ell n \, (r_2/r_1)}$$

where ω = rotational velocity of bowl, rad/sec
         g = gravitational constant, m/sec^2
         r_2 = radius from centerline to inside bowl wall, m
         r_1 = radius from centerline to surface of sludge, m

Thus if machine 1, at a certain Σ_1 as calculated from the machine parameters, at the flow rate Q_1 produced a satisfactory result, it is expected that a second (geometrically similar) machine with a (larger) Σ_2 will achieve equal dewatering performance at a flow rate of Q_2.

This analysis does not, however, consider the movement of the solids out of the bowl; a very necessary component of centrifugation. The

---

[2]Ambler, C. M. "The Evaluation of Centrifuge Performance" *Chem. Eng. Prog.* 48:3 (1952).

L —   Bowl length   S —   Scroll pitch
B —   Beach length α—   Beach angle
D —   Bowl diameter ω_B—   Bowl speed
Z —   Pool depth ω_S—   Scroll speed

Figure 8-16   Solid bowl centrifuge (photo courtesy Indersoll Rand and I. Krüger).

solids movement can be calculated based on the *Beta equation,*[3] where again, for two machines,

$$\frac{Q_1}{\beta_1} = \frac{Q_2}{\beta_2}$$

where Q now is the terms of solids (lb/hr or other such dimension) and where $\beta = (\Delta\omega)SN\pi Dz$. These terms are defined as

$\Delta\omega$ = difference in rotational velocity between the bowl and conveyor, rad/sec
S = scroll pitch (distance between blades), m
N = number of leads
D = bowl diameter, m
z = depth of sludge in the bowl, m

The scale-up procedure involves the calculation of Q_2 for liquid as well as solids throughput, and the lowest value would govern the centrifuge capacity.

Example 8.6

A solid bowl centrifuge (Machine 1) was found to perform well if fed at 0.5 m^3/hr of 1% solids sludge. It was decided to scale-up to a larger machine (Machine 2) and it is necessary to determine the flow rate at which this geometrically similar machine would perform equally well. The machine variables are as follow:

	Machine 1	Machine 2
Bowl diameter (cm)	20	40
Pool depth (cm)	2	4
Bowl speed (rpm)	4000	3200
Bowl length (cm)	30	72
Conveyor speed (rpm)	3950	3150
Pitch (cm)	4	8
Number of leads	1	1

Looking first at the settling (Sigma) scale-up: The volume of sludge in the pool can be approximated as

$$V = 2\pi \left(\frac{r_1 + r_2}{2} \right) (r_2 - r_1)L$$

---

[3] Vesilind, P. A. *Treatment and Disposal of Wastewater Sludges* (Ann Arbor, MI: Ann Arbor Science Publishers, Inc., 1979).

where the variables are defined in Figure 8-16. The volumes of the two machines are

$$V_1 = 2(3.14) \left(\frac{8 + 10}{2} \right) (10 - 8)(30) = 3400 \text{ cm}^3$$

$$V_2 = 2(3.14) \left(\frac{16 + 20}{2} \right) (20 - 16)(72) = 32,600 \text{ cm}^3$$

Since

$$\omega = (\text{rpm}) \left(\frac{1}{60} \frac{\text{min}}{\text{sec}} \right) \left(2\pi \frac{\text{rad}}{\text{rev}} \right)$$

$$\omega_1 = 4000 \left(\frac{1}{60} \right) (2)(3.14) = 420 \text{ rad/sec}$$

$$\omega_2 = 3200 \left(\frac{1}{60} \right) (2)(3.14) = 335 \text{ rad/sec}$$

$$\Sigma_1 = \frac{(420)^2}{980} \left(\frac{3400}{\ln(10/8)} \right) = 2.76 \times 10^6$$

$$\Sigma_2 = \frac{(335)^2}{980} \left(\frac{32,600}{\ln(20/16)} \right) = 16.8 \times 10^6$$

Thus

$$Q_2 = \frac{\Sigma_2}{\Sigma_1} Q_1 = \frac{16.8}{2.78} (0.5) = 3.2 \text{ m}^3/\text{hr}$$

Next looking at the solids loading (Beta), the solids flow rate is

$$Q_1 = 0.5 \frac{\text{m}^3}{\text{h}} \times 0.01 \times 1000 \frac{\text{kg}}{\text{m}^3} = 5 \text{ kg/hr}$$

$$\beta = (\Delta\omega)(S)(N)(\pi)(D)(z)$$

$$\beta_1 = (4000 - 3950)(4)(1)(3.14)(20)(2) = 25,120$$

$$\beta_2 = (3200 - 3150)(8)(1)(3.14)(40)(4) = 200,960$$

$$Q_2 = \frac{200,960}{25,120} (5) = 40 \text{ kg/hr}$$

corresponding to a liquid flow rate of

$$\frac{40}{0.01} \times 10^{-3} = 4 \text{ m}^3/\text{hr}$$

The liquid loading thus governs, and the big machine (Machine 2) cannot be fed at a flow rate greater than 3.2 m²/hr.

ULTIMATE DISPOSAL

The options for ultimate disposal of sludge are limited to air, water and land. Strict controls on air pollution complicate incineration, although this certainly is an option. Disposal of sludges in deep water (such as oceans) is decreasing due to adverse or unknown detrimental effects on aquatic ecology. Land disposal can be either dumping in a landfill, or spreading the sludge out over land and allowing natural bio-degradation to assimilate the sludge back into the environment. Because of environmental and cost considerations, only incineration and land disposal are presently commonly encountered.

Incineration is actually not a method of disposal at all, but rather a sludge treatment step in which the organics are converted to H_2O and CO_2 and the inorganics drop out as a nonputrescible residue.

Two types of incinerators have found use in sludge treatment: multiple hearth and fluid bed. The multiple hearth incinerator, as the name implies, has several hearths stacked vertically, with rabble arms pushing the sludge progressively downward through the hottest layers, and finally into the ash pit (Figure 8-17).

The fluid bed incinerator is full of hot sand, suspended by air injection, and the sludge is incinerated within the moving sand. Due to the violent motion within the fluid bed, scraper arms are unnecessary. The sand acts as a "thermal flywheel," allowing intermittent operation.

The second method of disposal—land spreading—is a science in its infancy. It has become popular only in the last few years and design and operating data are only now being developed.

The ability of land to absorb sludge and to assimilate it depends on such variables as soil type, vegetation, rainfall, slope, etc. In addition, the important variable of the sludge itself will influence the capacity of a soil to assimilate sludge.

Generally, sandy soils with lush vegetation, low rainfall and gentle slopes have proven most successful. Mixed digested sludges have been spread from tank trucks, and activated sludges have been sprayed from both fixed and moving nozzles. The application rate has been variable but 100 dry tons/acre/yr is not an unreasonable estimate. Most unsuccessful land application systems can be traced to overloading the soil. Given enough time (and absence of toxic materials) any soil will assimilate sprayed liquid sludge.

Transporting liquid sludge is often expensive, and volume reduction by dewatering is necessary. The solid sludge can then be deposited on land and disked in. A higher rate (tons/acre/yr) can be achieved by trenching where 1-m² (3-ft²) trenches are dug with a backhoe, the sludge deposited and covered. The sludge seems to assimilate rapidly, with undue leaching of nitrates or toxins.

In the last few years a method of chemically bonding the sludge solids so that the mixture "sets" in a few days has found use in industries which

Figure 8-17   Multiple hearth incinerator (courtesy Nichols Engineering and Research Corp.).

have especially critical sludge problems. Although *chemical fixation* is expensive, it is often the only alternative for besieged industrial plants. The leaching from the solid seems to be minimal.

The toxicity can be interpreted in several ways: toxicity to vegetation, toxicity to the animals (including people) who eat the vegetation, and the poisoning of groundwater supplies. Most domestic sludges do not contain sufficient toxins such as heavy metals to cause harm to vegetation. The total body burden of heavy metals is of some concern, however. It is possible to precipitate out the metals during sludge treatment, but the most effective means of controlling such toxicity seems to be to prevent these metals from entering the sewerage system. Strong enforced sewer ordinances are necessary and can be cost-effective. The regulatory aspects of sludge disposal are discussed further in Chapter 10.

CONCLUSION

Sludge disposal still represents a major headache for many municipalities. And quick relief does not seem to be in sight. After all, sludge represents the true residues of our civilization, and its composition reflects our style of living, our technological development and our ethical concerns. "Pouring things down the drain" that might cause damage or health problems in the ultimate disposal of the sludge is too often done without thought, or malice. We return to the question of moral responsibilities in the chapter on environmental ethics.

PROBLEMS

8.1 Using reasonable values, estimate the sludge production in a 10-mgd wastewater treatment plant. Express the sludges both as volumes per time and as dry solids per time.

8.2 A 1-liter cylinder is used to measure the settleability of 0.5% suspended solids sludge. After 30 min, the settled sludge solids occupy 600 ml. Calculate the SVI.

8.3 What measures of "stability" would you need if a sludge from a wastewater treatment plant was to be
    a. placed on the White House lawn
    b. dumped into a trout stream
    c. sprayed on a playground
    d. spread on a vegetable garden.

8.4 A sludge is thickened from 2000 mg/l to 17,000 mg/l. What is the reduction in volume, in percent?

8.5 A laboratory batch thickening test produced the following results:

Solids concentration (%)	Settling velocity (cm/min)
0.6	0.83
1.0	0.25
1.4	0.11
1.8	0.067
2.2	0.041

Calculate the required thickener area if the desired underflow concentration (C_u) is to be 3% solids and the feed to the thickener is 0.5 m³/min at a solids concentration of 2000 mg/l.

8.6 A Büchner funnel test for specific resistance to filtration, conducted at a vacuum pressure of 6×10^4 N/m², with an area of 400 cm², and a solids concentration of 60,000 mg/l, yielded the following data:

Time (min)	Filtrate Volume (ml)
0	0
1	10
2	20
3	29
4	37
5	43

Calculate the specific resistance to filtration. Is this sludge amenable to dewatering by vacuum filtration?

8.7 Two geometrically similar centrifuges have the following machine characteristics:

	Machine 1	Machine 2
Bowl diameter, cm	25	35
Pool depth, $(r_2 - r_1)$ cm	3	4
Bowl length, cm	30	80

If the flow rates are 0.6 and 4.0 m³/hr and the first machine operates at 1000 rpm, at what speed should machine 2 be operated to achieve similar performance?

LIST OF SYMBOLS

A = area of thickener, m^2
A_L = limiting area in a thickener, m^2
b = slope of filtrate volume vs time curve
C_i = solids concentration at any level i
C_u = underflow solids concentration mg/l
C_o = influent solids concentration, mg/l
CST = capillary suction time
D = diameter, m
du/dy = rate of shear, or slope of the velocity (u) - depth (y) profile
G = flux in a thickener, $kg/m^2 \cdot sec$
G_L = limiting flux in a thickener, $kg/m^2 \cdot sec$
h = fraction of BOD not removed in the primary clarifier
i = fraction of BOD not removed in the biological treatment step
j = fraction of solids not destroyed in digestion
k = fraction of influent SS removed in the primary clarifier
L = length of the bowl cylinder, m
N = number of leads on a scroll
P = vacuum pressure, N/m^2
Q_o = influent flow rate, m^3/sec
r = radius, m
r = specific resistance to filtration, m/kg
S = scroll pitch, m
S_o = influent BOD, kg/hr
SVI = sludge volume index
v = interface velocity at solids concentration C_i
w = cake deposited per volume of filtrate, kg/m^3
X_e = effluent SS, mg/l
x_o = influent SS, kg/hr
Y = yield, $\Delta X / \Delta S$
z = depth of sludge in a bowl, m
β = beta factor
ΔS = net BOD utilized in secondary treatment, kg/hr
ΔX = net solids produced by the biological step, kg/hr
$\Delta \omega$ = difference between bowl and conveyor rotational velocity rad/sec
η = plastic viscosity, $N \cdot sec/m^2$
μ = viscosity, $N \cdot sec/m^2$ (poise)
Σ = sigma factor
τ = shear stress, N/m^2

τ_y = yield stress, N/m^2
ω = rotational velocity, rad/sec
V = filtrate volume, m^3

SUGGESTED READING

Bitton, G., et al., Eds. *Sludge—Health Risks of Land Application* (Ann Arbor, MI: Ann Arbor Science Publishers, Inc., 1980).

Culp, G. L. *Handbook of Sludge Handling Processes* (New York: Garland STPM Press, 1979).

Eckenfelder, W. W. Jr., and C. J. Santhanam. *Sludge Treatment* (New York: Marcel Dekker Inc., 1981).

EPA. *Multimedium Management of Municipal Sludge.* Report to the U.S. EPA, Commission on Natural Resources, National Research Council, National Academy of Sciences, Washington, D.C. 1978.

Vesilind, P. A. *Treatment and Disposal of Wastewater Sludges* (Ann Arbor, MI: Ann Arbor Science Publishers, Inc., 1979).

Chapter 9

Nonpoint Source
Water Pollution

As rain falls and strikes the ground, a complex runoff process begins, and nonpoint source water pollution is the unavoidable result. Even before people entered the picture, the rains came, raindrops picked up soil particles, muddy streams formed, and major water courses became clogged with sediment. Witness the formation of the Mississippi River delta which has been forming for tens of thousands of years. We can safely surmise that, even before man, rivers were "polluted" by this natural series of events; sediment clogged fish gills and fish probably even died. This natural runoff is classified as "background" nonpoint source runoff, and not generally labeled as "pollution."

Now view the world as it has been since the dawn of mankind—a busy place where man's activities continue to influence our environment. For years, these activities have included farming, harvesting trees, constructing buildings and roadways, mining, and disposing of liquid and solid wastes. Each activity has led to disruptions in the surface of the earth's soil and/or involves the application of chemicals to the soil. This increased transport of soil particles, i.e., increased sediment loads to watercourses, and the application of chemicals to the soil, are generally labeled as pollution.

This "pollution" is the focus of Chapter 9, where we address:

- the runoff process
- loading functions for sediment, a critical pollutant
- control technologies applicable to nonpoint source pollution

Table 9-1 gives a shopping list of nonpoint source categories, including sources ranging from agricultural practices to air pollution fallout to "natural" background. The focus of this chapter is the seven major activities of concern: agriculture, urban stormwater, construction, silviculture, residuals management and onsite sewage disposal. Table 9-1 also summarizes the relative significance of the concentrations of pollutants which are attributable to the list of activities.

THE RUNOFF PROCESS

The complex runoff process includes both the detachment and transport of soil particles and chemical pollutants. For the remainder of this chapter, we simply umbrella all people-induced soil erosion and chemical applications under the term "nonpoint source pollutant." Chemicals can be bound to soil particles and/or be soluble in rainwater; in either case, water movement is the prime mode of transport for solid and chemical pollutants. The *characteristics of the rain* indicate the ability of the rainwater to splash and detach the pollutants. This rain energy is defined by droplet size, velocity of fall, and the intensity characteristics of the particular storm.

Soil characteristics impact both the detachment and transport processes. Pollutant detachment is a function of an ill-defined motion of soil stability. Size, shape, composition and strength of soil aggregates and soil clods all act to determine how readily the pollutants are detached from the soil to begin their movement to streams and lakes within a region. Pollutant transport is influenced by the permeability of the soil to the water. Soil permeability, the ease by which water passes through the soil, helps determine the infiltration capabilities and drainage characteristics of the surface receiving the rainfall. Pollutant transport is also a function of soil porosity, which affects storage and movement of water, and soil surface roughness, which tends to create a potential for temporary and long-term detention of the pollutants. (The porosity of selected materials is listed in Table 4-1).

Slope factors also help define the transport component of the nonpoint source problem. The slope gradient as well as slope length influence the flow and velocity of runoff, which in turn influence the quantity of pollutants that are moved from the soil to the water course.

Land cover conditions also impact the detachment and transport of pollutants. Vegetative cover helps to:

• provide protection from the impact of raindrops, thus reducing detachment
• make the soil aggregates less susceptible to detachment by protecting soil from evaporation and thus keeping the soil moist

- furnish roots, stems and dead leaves which help slow overland flow and hold pollutant particles in place.

Table 9-2 shows the different types of erosion which may contribute to nonpoint source pollution.

Table 9-2.   Types of Erosion Which May Contribute
to Nonpoint Source Pollution

Type of Erosion	Description
Raindrop	From direct impact of falling raindrops Detached soil and chemicals can be transported downhill
Sheet	From raindrop splash and runoff in wide, thin layers of surface water Wall of water transports pollutants downhill
Gully/Rill	From concentrated rivulets of water cutting 5- to 10-cm rills in soil Rills grow into gullies
Stream/Channel	From confluence of rills and gullies Increased volumes and velocities cause erosion along streams and riverbeds

Only a portion of the pollution detached and transported from upland regions in a watershed is actually carried all of the way to a stream or a lake. In many cases, significant portions of the materials are deposited at the base of slopes or on flood plains. The portion of the pollution detached, transported and actually delivered from its source to the receiving waterway is defined as the *delivery ratio*.

Numerous factors influence the pollutant delivery ratio.[1,2] Where chemical pollutants are involved, the whole spectrum of factors which determine reaction rates act to limit the delivery ratio: temperature, times of transport, presence of other chemicals, and presence of sunlight to name just a few. Whenever sediment and/or chemical pollutants become a problem, the list of physical factors becomes quite long, and includes:

[1] USDA Soil Conservation Service, *National Engineering Handbook,* Section 3, "Sedimentation," U.S. Government Printing Office, Washington, DC (1978).

[2] USDA Agriculture Research Service, "Present and Prospective Technology for Predicting Sediment Yield and Sources," Proceedings of the Sediment Yield Workshop, USDA Sedimentation Laboratory, Oxford, Mississippi (November 1972).

Table 9-1.   Relative Significance of Concentrations of Pollutants[a]

NPS Category	Suspended Solids/Sediment	BOD	Nutrients	Toxic Metals	Pesticides	Pathogens	Salinity/TDS[c]	Acids	Heat
Urban Storm Runoff[b]	M	L-M	L	H	L	H	M	N	N
Construction[b]	H	N	L	N-L	N	N	N	N	N
Highway Deicing	N	N	N	N	N	N	H	N	N
In-Stream Hydrologic Modifications	H	N	N	N	N	N	N-M	N	M
Non-Coal Mining[b]	H	N	N	M-H	N	N	M-H	H	N
Agriculture[b]									
– Nonirrigated Crop Production	H	M	H	N-L	H	N-L	N	N	N
– Irrigated Crop Production	L	L-M	H	N-L	M-H	N	H	N	N
– Pasture and Range	L-M	L-M	H	N	N	N-L	N-L	N	N
– Animal Production	M	H	M	N-L	N-L	L-H	N-L	N	N

Forestry[b]									
– Growing	N	N	L	N	L	N-L*	N	N	N
– Harvesting	M-H	L-M	L-M	N	N	N	N	N	M
Residuals Management[b]	N-L	L-H	L-M	L-H	L	L-H	N-H	N-H	N
On-Site Sewage Disposal[b]	L	M	H	L-M	L	H	N	N	N
In-Stream Sludge Accumulation/Resuspension	H	H	M-H	L-H	M	L	N	N	N
Direct Precipitation	N	N	N-M	L	L	N-L	N	N-M	N
Air Pollution Fallout	M	L	L-M	L-H	L-M	N-L	N	L-M	N
"Natural" Background	L-H	L-H	M	N-M	N	N-L	N-H	N-H	N-M

[a] N = Negligible; L = Low; M = Moderate; H = High.
[b] Considered in this chapter as a major contributor of nonpoint source pollution.
[c] TDS = total dissolved solids.

1. *Magnitude of sediment sources.* Whenever the quantity of sediment available for transport is greater than the capability of the runoff transport system, disposition will occur and the delivery ratio will be decreased.

2. *Proximity of pollutant sources to receiving waterways.* Pollutants entrapped in runoff often move only short distances and, due to factors such as surface roughness and slope, may be deposited far from the lake or stream. Areas close to a receiving waterway, or areas where channel-type erosion takes place, may be characterized with a relatively high delivery ratio.

3. *Velocity and volume of water.* The characteristics of the pollutant transport system, particularly the velocity and volume of water from a given storm, impact the delivery ratio. A small storm may not supply enough water to carry a load of pollution to a lake or stream, resulting in a zero or very low delivery ratio. A large, lengthy rainfall may have the opposite effect and transport a very large portion of the pollution that is detached from its source to the receiving waterway.

By understanding rainfall characteristics, soil properties, slope factors, and vegetative covers, the loads of different nonpoint source pollutants to lakes and rivers can be predicted and possibly controlled.

THE ART AND SCIENCE OF DEVELOPING AND APPLYING LOADING FUNCTIONS FOR SEDIMENT DETACHMENT AND TRANSPORT

Pollutant loading is defined as the quantity of pollution that is detached and transported into surface watercourses. For engineering studies, mathematical models exist which predict these loadings as a function of soil characteristics, slope and vegetative cover, including models for estimating:[3]

- sediment from:
    sheet and rill erosion
    urban runoff
    feedlots
- nitrogen
- phosphorus

---

[3]U.S. EPA Office of Research and Development (OR&D), *Loading Functions for Assessment of Water Pollution from Nonpoint Sources,* National Technical Information Service, Springfield, Virginia (1976).

- organic matter
- pesticides
- insoluble pesticides
- salinity
- acid mine drainage
- heavy metals and radiation.

In theory, a distinction is made between solutional transport and sediment transport, because one is essentially chemical and the other is mechanical. A second distinction which is also useful to note is the differences between processes which move large quantities of soil (mass movement) and those processes which move soil as individual dispersed particles in a fluid (fluid transport). Under mass movement conditions, transport of pollution is largely related to a component of the sediment's own weight. Under fluid transport, the motion of the water is the major controlling factor. The fluid mechanics which ties this theory to the engineering world is outside the scope of this presentation; the reader is encouraged to pursue additional reading on the subject.

In practice, the modeling of nonpoint source pollution becomes a mixture of art and science. Complicated mathematical expressions often boil down to a guess here or a gut feeling there by an engineer trying to map stormwater runoff.

In this section, we present sample calculations for using one empirical method of predicting sediment loads from sheet and rill runoff on agricultural land. Other mathematical models exist for sediment as well as other pollutants, and this representation has some limitations. However, throughout the discussion, we point to these limitations and draw on this limited methodology as one example for estimating sediment loadings.

This method for estimating the sediment from sheet and rill erosion is a loading function based on the often-criticized *universal soil loss equation* (USLE):

$$Y_{(S)_E} = \sum_{i=1}^{n} A_i (R \cdot K \cdot L \cdot S \cdot C \cdot P \cdot S_d)_i$$

where $Y_{(S)_E}$ = sediment loading from surface erosion (ton/yr)
    n = number of subareas identified for the region, with a subarea defined for each portion of the watershed having similar characteristics in terms of slope, vegetative cover, permeability
    A_i = acreage of subarea (acre)
    R = rainfall factor, expressing the erosion potential of average annual rainfall in the locality. R is the summation of the individual storm products of the kinetic energy of rainfall

(hundreds of ft-tons/acre) and the maximum 30-min rainfall
intensity (in./hr) for all significant storms.

K = soil erodibility factor (ton/acre/R unit)
L = slope-length factor (dimensionless ratio)
S = slope-steepness factor (dimensionless ratio)
C = cover factor (dimensionless ratio)
P = erosion control practice factor (dimensionless ratio)
S_d = sediment delivery ratio (dimensionless ratio) obtained from
USDA Soil Conservation Service field office in each region
of the nation

The USLE was developed primarily for agriculture by the U.S. Department of Agriculture (USDA), and the Soil Conservation Service. It is
used in every state in the nation by federal, state, regional and local
agencies.

The terms on the right-hand side of the equation are best defined for
pasture and cropland, and are not necessarily well defined for silviculture, construction, mining and other nonagricultural land uses.
Although the model may not predict absolute values for sediment yields,
it is useful for projecting relative changes in sediment yields if different
nonpoint source control technologies are designed and constructed.

The *rainfall factor* expresses the erosion potential of precipitation in a
given locality. This index is a measure of the summation of the individual
storm products of the kinetic energy of rainfall E and the maximum
30-min rainfall intensity I, for all significant storms occurring within
the time period under consideration. The product $R = E \times I$, which has
been calculated at the macro level for all regions of the nation, reflects
the combined potential of rainfall impact and runoff turbulence to
detach and transport soil particles (see Figure 9-1).

The *soil erodibility factor* is a quantitative measure of the rate at which
a soil will erode. K is expressed as the soil loss in tons per acre unit of R,
standardized for a plot of land with 9% slope, 72.6 ft long, under continuous cultivation. This factor has been developed for most soils in
the nation and the data are available from Soil Conservation Services
offices. Figure 9-2 offers one method of determination.

The *topographic factor,* L·S, combines both the slope-length and
slope-steepness factors, and reflects the fact that soil loss is affected by
both length of flow and steepness of slope. These two factors together
account for the capability of runoff to detach and transport soil material. Slope-length is defined as the distance from the origin of overland
flow to either the point where the slope decreases to the extent that disposition begins, or to the point where runoff enters a well defined channel
that is part of the drainage network, whichever distance is less. Slope-

NOTE: In the shaded Northwestern states, runoff from snowmelts contributes significantly to erosion. Therefore, R is segmented to R_f (indicated on map above) plus R_s. The annual R_s is obtained by multiplying the average annual total snowfall (in inches of water depth) by the constant 1.5.

Figure 9-1   Average annual values of rainfall factor, R.

Figure 9-2 Soil erodibility nomograph for calculating values of K.

length can be determined reasonably accurately by site inspection, by measurements from aerial photographs, or as a last resort, by topographic maps. Slope-steepness data may be obtained from similar sources of information, or the Soil Conservation Service. Figure 9-3 presents estimates for this topographic factor.

The *cover factor* C represents the ratio of soil quantity eroded from land that has a vegetative cover to the soil quantity eroded from clean-

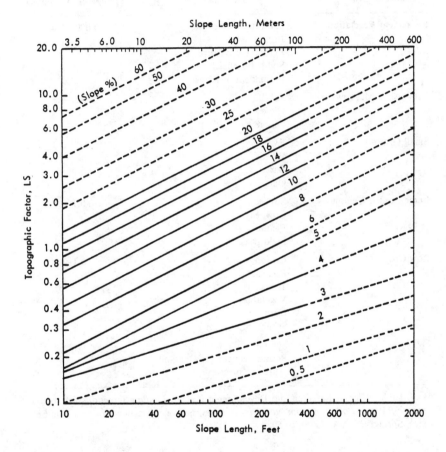

NOTE: The dashed lines are estimates for slope dimensions beyond the range of lengths and steepness for which data are available.

Figure 9-3   Slope effect for areas east of the Mississippi River.

tilled soil under identical slope and rainfall conditions. Local Soil Conservation Services offices keep records of recommended values of C for given localities, or can be estimated from Table 9-3.

Table 9-3.   Relative Protection of Ground Cover Against
Erosion: Values for C Factor[4]

Land-Use Groups	Examples	Range of "C" Values
Permanent Vegetation	Protected woodland Prairie Permanent pasture Sodded orchard Permanent meadow	0.0001–0.45
Established Meadows	Alfalfa Clover Fescue	0.0004–0.3
Small Grains	Rye Wheat Barley Oats	0.07–0.5
Large-Seeded Legumes	Soybeans Cowpeas Peanuts Field peas	0.1–0.65
Row Crops	Cotton Potatoes Tobacco Vegetables Corn Sorghum	0.1–0.70
Fallow	Summer fallow Period between plowing and growth of crop	1.0

[4]U.S. EPA Office of Research and Development (ORD); *Loading Functions for Assessment of Water Pollution from Nonpoint Sources,* National Technical Information Service, Springfield, VA (1976), p. 59.

The *sediment recovery ratio* S_d is defined as the fraction of total erosion which is delivered to a stream. The historical method of determining an average delivery ratio is by comparing the magnitude of sediment yield at a given point in a watershed with the total amount of erosion. Estimates of delivery ratios are also available from the Soil Conservation

Service. If S_d is omitted from the USLE, the projected sediment loading is the tonnage of soil that is eroded from a given area, but not necessarily the tonnage of soil that reaches a waterway. Some fraction of what is eroded is deposited prior to entering the lake or stream; this variable is estimated for each region of the nation by the Soil Conservation Service.

This soil loss equation has many limitations. For one, it predicts soil loss from sheet and rill erosion, but does not predict losses from gullies, streambanks, landslides or wind. This model was developed and calibrated primarily for croplands, and has been applied mostly in the humid regions east of the Rocky Mountains if not mostly east of the Mississippi River. The model generally is not applicable for western croplands, and its utility in predicting sediment loading from silvicultural activities, construction sites and mining locations is questionable.

Nevertheless, this model is useful because it can predict erosion rates by storm event, season and annual averages, and a nationwide data collection effort has resulted in an adequate approximation of the independent factors in the model.

Example 9.1

An 830-acre watershed is located 5 miles south of Indianapolis in central Indiana. Given the information presented below, compute the projected sediment from sheet and rill erosion in terms of average daily loading.

Cropland: 180 acres

- Corn
- Conventional tillage, average annual yield   40 to 45 bushels/acre
- Cornstalks remain in fields after harvest
- Contour strip-cropped planting method
- Soil is Fayette silt loam
- Slope is 6%
- Slope length is 250 ft

Pasture: 220 acres

- No canopy of trees or brush
- Cover at surface is grass and grass-like plants
- 80% ground cover
- Soil is Fayette silt loam
- Slope is 6%
- Slope length is 200 ft

Woodland: 430 acres

- 50% of area covered with tree canopy
- 80% of area covered with leaf and limb litter
- Undergrowth is managed and therefore minimal
- Slope is 12%
- Slope length is 150 ft

Calculations for loading from cropland:

$$Y(s)_{annual\ cropland} = R \cdot K \cdot L \cdot S \cdot C \cdot P \cdot S_d$$

R = 200 from Figure 9-1

K = 0.37 from Figure 9-2

$L \cdot S$ = 1.0% from Figure 9-3

C = 0.49 from Table 9-3 and USDA Soil Conservation Service field office in Indianapolis

P = 0.25 from Table 9-4

S_d = 0.60 from USDA Soil Conservation field office in Indianapolis

$$Y(s)_{annual\ cropland} = 200 \cdot 0.37 \cdot 1.08 \cdot 0.49 \cdot 0.25 \cdot 0.60$$

$$= 5.87\ ton/acre/yr$$

$$Y(s)_{avg.\ daily\ cropland} = 5.87\ ton/acre/yr \div 365\ days$$

$$= 0.016\ ton/acre/day$$

$$= 32\ lb/acre/day$$

Calculations for loading from pasture:

$$Y(s)_{annual\ pasture} = R \cdot K \cdot L \cdot S \cdot C \cdot P \cdot S_d$$

$$R = 200$$

$$K = 0.37$$

$$L \cdot S = 0.95$$

$$C = 0.013$$

Table 9-4.   "P" Values for Erosion Control Practices
on Cropland[5]

		Type of Control Practice			
Slope	Up-and Downhill	Cross-Slope Farming Without Strips	Contour Farming	Cross-Slope Farming with Strips	Contour Strip-Cropping
2.0–7	1.0	0.75	0.50	0.37	0.25
7.1–12	1.0	0.80	0.60	0.45	0.30
12.1–18	1.0	0.90	0.80	0.60	0.40
18.1–24	1.0	0.95	0.90	0.67	0.45

[5]U.S. EPA Office of Research and Development (ORD); *Loading Functions for Assessment of Water Pollution from Nonpoint Sources,* National Technical Information Service, Springfield, VA (1976), p. 64.

$$P = 1.00$$

$$S_d = 0.60$$

$$Y(s)_{\text{annual pasture}} = 200 \cdot 0.37 \cdot 0.95 \cdot 0.013 \cdot 1.00 \cdot 0.6$$

$$= 0.548 \text{ ton/acre/yr}$$

$$= 1,100 \text{ lb/acre/yr}$$

$$Y(s)_{\text{avg. daily pasture}} = 0.548 \text{ ton/acre/yr} \div 365 \text{ days}$$

$$= 0.0015 \text{ ton/acre/day}$$

$$= 3 \text{ lb/acre/day}$$

Calculations for loading from woodland:

$$Y(s)_{\text{annual woodland}} = R \cdot K \cdot L \cdot S \cdot C \cdot P \cdot S_d$$

$$R = 200$$

$$K = 0.32$$

$$L \cdot S = 2.75$$

$$C = 0.003$$

$$P = 1.00$$

$$S_d = 0.60$$

$$Y(s)_{annual\ woodland} = 200 \cdot 0.32 \cdot 2.75 \cdot 0.003 \cdot 1.00 \cdot 0.60$$

$$= 0.3168\ ton/acre/yr$$

$$Y(s)_{avg.\ daily\ woodland} = 0.3168\ ton/acre/yr \div 365\ days$$

$$= 0.0009\ ton/acre/day$$

$$= 1.8\ lb/acre\ day$$

Calculations for cross loading from watershed:

Cropland:	180 acres · 0.016 ton/acre/day =	2.88 ton/day
Pasture:	220 acres · 0.0015 ton/acre/day =	0.33 ton/day
Woodland:	430 acres · 0.0009 ton/acre/day =	0.39 ton/day
Total	$Y(s)_{avg\ total}$ =	3.60 ton/day

Finally, this loading function is designed to predict long-term average loadings. Sediment loadings for a specific year may vary drastically from this long-term average because of differences in the number, size and timing of erosive rainstorms.

Because of the large uncertainties in this model, it is advisable that all predictions be utilized only as reasonable estimates and not labeled as absolute values. Sensitivity analysis, the varying of one independent variable at a time to observe the change in predicted loadings, is advisable in all cases. The model is also useful in areas outside agriculture for projecting relative impacts of such nonpoint source control measures as the addition of vegetative covers and slope modifications. Data have been collected by the Soil Conservation Service since the early 1930s, and models have been developed and revised to the present form of USLE. This model is used extensively by engineers in the Soil Conservation Service, and understanding its strengths and weaknesses is critical to environmental engineers.

CONTROL TECHNOLOGIES APPLICABLE TO NONPOINT SOURCE POLLUTION

The importance of controlling nonpoint source pollution and the obstacles to achieving that goal are becoming more and more evident. Control over municipal and industrial point sources of pollution historically has received considerable federal and corporate attention through the nationwide construction grants and permit programs. Yet, public and private investment to significantly reduce point source pollution may be ill spent in cases where water quality is governed instead by nonpoint source discharges.

Agriculture

Many control measures exist for reducing nonpoint source pollution from agricultural areas. These options range from the management of surface cover and tillage to mechanical conservation measures. Alternative systems have been developed to reduce the erosion potential from tilled (inverted or plowed) land.

Terracing, which breaks the slope of a field into shorter segments, is often applied in fields where contouring and other tillage systems do not offer adequate soil stabilization. Terraces consist of a combination of ridges and channels constructed across the slope which collect surface water from above. Water is either held until it is absorbed or diverted in a controlled manner.

Diversions are large, individually designed terraces constructed across the slope of a field to intercept and divert runoff to a stable outlet. Usually constructed above croplands or above such critical erosion areas as gullies, they act to reduce the volume of runoff entering the problem area.

Other options available to control agricultural nonpoint source pollution include grassed waterways (the use of year-round grasses and areas of surface water movement), and cover plants (grasses, trees and shrubs planted in critical areas to control severe erosion problems near stream banks).

Silviculture

Silviculture is defined as that part of forest management that deals with the process of utilizing forest crops. Surface water pollution results mainly from three silvicultural activities: the building and use of transportation systems, the harvesting of timber, and intermediate practices such as thinning and spraying for the control of fire, insects and disease.

From a water quality standpoint, the most critical decisions are made during the planning phase of the crop utilization process. Transportation networks, the most significant nonpoint source of water pollution from silviculture in the Northwest and other areas of the United States, can be designed to minimize pollution runoff. Design considerations include optimizing the number of roadways and their layout, and minimizing the number of times they cross natural water courses. The three surfaces associated with a given segment of roadway offer distinct control options: the cut-slope, the roadway surface itself, and the fill-slope. Geotechnical investigations should precede the construction of any road or staging area. In areas with a slope of greater than 60%, it is generally prescribed that roads be built only as a last resort; such log removal techniques as skylining, helicoptering and ballooning should be investigated. The techniques by which logs are cut and transported influence the amount of nonpoint source pollution.

A component of the runoff problem can be controlled if transportation and harvesting techniques are planned to avoid or minimize the impact on environmentally sensitive flood plains, and buffer strips along waterways should be protected. In addition, chemical applications should follow best management practices. Erosion from staging areas can be controlled by diverting runoff around the site. Bars and ditches can be 70% effective in reducing sediment loading to receiving waterways.[6]

Construction

The erosion of soil can cause construction problems onsite as well as water quality problems offsite. Additionally, the loss of soil is often regarded as the loss of a valuable natural resource. Home buyers expect a landscaped yard, and lost topsoil is often costly for the contractor to replace. The builder of homes, highways and other construction views soil erosion as a process that must be controlled to maximize economic gain.

The planning phase of construction activities considers controlled clearing of the proposed construction site, as the area to be disturbed during the construction and site restoration phases is held to a minimum. Environmentally sensitive areas must be designated, and if any clearing is required in such areas, it should be limited as much as possible. Such critical areas include: steep slopes, unaggregated soils (sands, etc.), natural

---

[6]"Silvicultural Activities and Non-point Pollution Abatement: A Cost-Effectiveness Analysis Procedure," Environmental Research Laboratory, U.S. EPA, Athens, GA, undated, pp. 63–74.

sediment ponds, natural waterways (including intermittent streams), and flood plains.

The planning phase should also consider a comprehensive erosion control system. The components of such a system include planned access as well as techniques for use in the operational and site restoration phases of the construction activity.

During construction, several pollution-abatement techniques appear to be effective. Velocity regulation methods attempt to reduce the rate at which water moves over the construction site. Velocity reduction minimizes particle uptake by the water and can lead to particle disposition in instances where the reduction is sufficient. The result is a decrease in erosion. There are alternative methods of achieving velocity reduction, and all involve the application of some material to the exposed soil: filter inlets, jute mesh, seeding, fertilizing and mulching can reduce sediment runoff. The best method, and its corresponding cost, must be determined on a site by site basis.

Stormwater deflection methods attempt to reduce the amount of water passing over the construction site by diverting it before it is properly installed. However, the dike does not limit or control runoff generated by rain falling directly on the site, and its effectiveness is thus limited. Stormwater channeling methods attempt to reduce the effect of water passing over the construction site by controlling its movement through the site. Chutes, flumes and flexible downdrains are effective in certain areas, but their costs are quite high. Their outfall must be handled to ensure that a secondary source of pollution is not created.

Restoration of the site after construction is necessary if water pollution is to be controlled. Regrading costs are necessary expenditures at most construction sites if an effective revegetation process is to be undertaken. Regrading is only effective if it is followed by seeding, fertilizing and mulching. Such revegetation costs depend on the type of mulch used, and the type of mulch applicable on a given site is highly dependent on the slope of the site. More expensive practices generally are required on steeper slopes. Wood fiber mulch applied by a hydroseeder is very effective on relatively flat landscapes. The more expensive and more permanent excelsior mats and jute netting are also available.

Mining

Mining is discussed in this section under two major headings: surface mining and subsurface mining. The major pollutants from both types of operations are sediments, toxic substances and acids.

There are several forms of surface mining, including strip, open pit and dredging. Strip mining requires removing a large amount of overlying material (overburden) to expose the desired ore. The distinction between strip mining and open pit mining is that open pit is used in areas where there is relatively little overburden. In such areas, most of the material being removed is the desired mineral, while most of the material being removed during strip mining is overburden and waste. Most strip mining is done to obtain coal, while open pit mining is performed to mine a range of minerals. Dredging is used to recover minerals from underwater mines, with gravel accounting for the majority of dredging production. Hydraulic mining is carried out by directing a high-velocity stream of water at a disperse mineral deposit. It is used almost exclusively for gold.

These forms of mining result in varying amounts of nonpoint source water pollution. Siltation is a major problem, and chemical pollution may occur when surface mining results in accelerating pollution-forming chemical reactions. Water pollution control is generally achieved by changing the conditions responsible for the pollution. Combinations of several techniques are usually required to achieve adequate water pollution abatement. As an example, regrading should be accompanied by revegetation and possibly by water diversion.

A given area of surface mining may simultaneously exhibit characteristics of three phases of operation: planning of a portion of the dig area may be taking place, mining operations may be taking place in a second portion of the dig area, and restoration may be taking place in a third portion of the dig area.

The planning phase of surface mining should consider a technique referred to as controlled mineral extraction. This procedure is based on the fact that pollution-forming materials are not usually distributed evenly throughout a mineral seam. Water quality data, hydrologic considerations, and core bore sampling often indicate where high pollution potentials exist in any prospective mining area. Planning mineral extraction around such problem spots can result in the abatement of nonpoint source water pollution from surface mines. The cost of sampling and the opportunity cost associated with not mining a critical area must be developed on an individual mine basis.

The planning phase must also consider a comprehensive erosion control system. Elements in such a system include access roads and transportation considerations. Roadways are a major source of pollution in active mine sites. Additionally, the erosion control system must consider water pollution from mine tailings.

During the operational phase of surface mining, several water pollution abatement strategies appear to be useful. The utility of each, however, is site specific, depending on each mine's geography and overburden characteristics. Overburden segregation has been employed many times in the United States. When properly utilized with regrading and revegetation, it is believed to be one of the most effective methods for controlling water pollution from surface mines. The cost of overburden segregation is site specific. In the contemporary reclamation method, only the overburden and waste from the first dig is placed on land adjacent to the mine. The overburden and waste from successive digs is placed in the mined portion of the area. The technique appears to be a good method to control water pollution. It has the distinct advantage to the mineral industry in that most of the overburden is handled only once, and a cost savings is available. Further, regrading and revegetation costs are often reduced. Water diversion is a technique that focuses on collecting surface water before it enters the mine area and diverting it around the mine site. The technique is effective in reducing nonpoint source water pollution, and it can be applied to any mine surface or mine waste pile. The cost of the technique may prove to be excessive for large mining areas.

During the site restoration phase of a surface mining operation there are basically two types of water pollution control techniques: regrading and revegetation. Regrading is undertaken both to prevent significant erosion and to restore the site to a usable form. In general, regrading will significantly reduce surface water pollution from surface mining. To be totally effective, regrading must be accompanied by revegetation. Revegetation includes not only seeding and planting an abandoned mine site, but also preparing the soil, burying harmful substances, and maintaining the vegetation cover through several growing seasons to ensure that the vegetation is permanent. If this method is properly undertaken and coupled with regrading, it can eliminate essentially all pollution resulting from a surface mining activity. However, many questions remain unanswered concerning the economics of both regrading and revegetation.

Sub-surface mines are identified as nonpoint sources of water pollution because the water that runs out of mine openings is often extremely acidic (with a pH as low as 2) and high in metals which may be toxic to aquatic and human life. The source of the water in the mines is usually groundwater which seeps through the porous structure, boreholes, and fractures in the strata adjacent to the mine. The pollutants carried in the water are a result of ores which have been oxidized due to exposure to air. Therefore pollution controls for subsurface mines are aimed at restricting waterflows from the mines and/or maintaining anaerobic atmosphere in the mine.

Several actions can be implemented during the planning phase of a project to reduce the potential for pollution. The actual location of the mine can be selected, and areas with high pollution potential within the site can be avoided. Next, an overall erosion control program can be designed accounting for surface disturbance activities such as milling and mine tailing disposal. The mine itself can be designed to minimize the production of pollutants through planning contemporaneous backfilling and roof fraction control.

Controls applicable during the actual operation of the mine include many different technologies. The most commonly applied method is the plugging of boreholes and the grouting of fracture zones. Other methods call for the diversion of groundwater flows, and restriction of free oxygen into the mine, the reduction of rainwater infiltration in the above watershed, and other diverse ideas. The costs of these activities are not estimable for a typical mine and will vary significantly from site to site. Along with these controls, the structural controls which required preplanning to be most effective, such as backfilling and the use of roof supports, are implemented during this phase.

After the production phase is completed, site restoration activities must be initiated. For subsurface mines this involves sealing the openings to the mines and at times inducing cave-ins to fill the void spaces. The cost of these activities is dependent on the number of openings to the mine and the strata around the deposit.

Urban Stormwater Runoff

The methods for controlling urban stormwater runoff range from nonstructural urban housekeeping practices, such as litter control regulations, to structural collection and treatment systems such as settling tanks and possibly even secondary treatment. No single control can be used in all situations or locations. Factors affecting the choice of controls for a given site include:

- the type of sewerage system (separate or combined)
- the status of development in the area (planned, developing or established urban area)
- the land use (residential, commercial or industrial).

Controls are grouped into three categories: planning controls, pollutant accumulation controls, and collection and treatment controls. These control categories correspond to those applicable to planned, developing and developed urban areas.

Several planning controls can be undertaken to reduce the number of pollution sources. Street litter can be reduced by passage and enforcement of antilittering laws. Air pollution, which becomes a source of water pollution when it settles on urban surfaces (particularly streets and rooftops) can be reduced via effective air pollution abatement planning. Transportation residues such as oil, gas and grease from cars, and particulates from deteriorating road surfaces can be reduced via transportation planning, selection of road surfaces which are less susceptible to deterioration, and automobile inspection programs. Preventative actions can be taken as part of land use planning strategies to reduce potential runoff pollution, such as avoiding development in environmentally sensitive areas, or in areas where urban runoff is an existing problem. Floodplain zoning, one type of land use regulation, often creates a buffer strip which is effective in reducing urban runoff pollution by filtering solids from overland flows and by stabilizing the soils of the floodplain.

Several control techniques prevent pollutants from accumulating on urban surfaces such as streets and parking lots. By preventing such buildup, the total pollutant loadings and the concentration of pollutants in the first flush of an area are reduced. The high concentration of pollutants in this first flush is the cause of many negative water quality impacts associated with urban runoff. The most common methods of street cleaning include street sweeping, street vacuuming and street flushing. Street sweeping is the oldest technique and is used in most urban areas. It is also the least expensive of the three. Street sweeping will reduce soil loadings in the runoff, but fails to pick up the finer particulates, which often are the more significant source of pollution (biodegradables, toxic substances, nutrients, etc.). Street vacuuming is more efficient in collecting the small particulates, but is more expensive.

Catch basin cleaning refers to the periodic removal of refuse and other solids from catch basins. Significant reductions in biodegradables, nutrients and other pollutants can result from regular cleaning. The basins can be cleaned by hand or by vacuum.

Urban stormwater runoff pollution can also be controlled after it enters the stormwater drainage system. Detention systems reduce runoff pollutant loadings by retarding the rate of runoff and by encouraging the settling of suspended solids. These systems range from low technology controls such as rooftop storage to intermediate technology controls such as small detention tanks interspersed in the collection network. In general the size and number of units are directly proportional to the effectiveness of the system. Detention basins act like settling tanks and can be expected to remove 30% of the biodegradables and 50% of the suspended solids in the stormwater.

In storage and treatment systems the first flush of an area is retained in the collection network, in a storage unit, or in a flow equalization basin. The stormwater is then treated at a nearby wastewater treatment facility when the sanitary flow volume and the design capacity of the facility allow. Several innovative storage methods exist, ranging from storing the stormwater in the drainage network, to routing the flow using a computerized network of dams and regulators (in Seattle), to the digging of a subterranean storage tunnel (in Chicago). The effectiveness of these systems depends on the quantity of pollutants captured, and sizing of the storage units and the extent of treatment received at the local wastewater facility. The storage capacity is the most critical factor. A debate exists on what size storm should be used as the design storm. The larger the design storm, the more costly the system, and (generally) the more effective the system. The discussion focuses on whether treatment should be planned for a 1-month, a 6-month, or even a less common storm.

CONCLUSION

Nonpoint sources contribute major pollutant loadings to the waterways throughout the nation. Control techniques are readily available but vary considerably in both cost and effectiveness, and are typically implemented through a generally confused institutional framework. This institutional setting sometimes poses an obstacle to nonpoint source abatement.

In addition, costs of control for abatement of nonpoint source pollution vary widely and are site-specific. Planning controls (preventative measures), however, are relatively inexpensive when compared to operational and site restoration controls (remedial measures), and thus may be the most efficient way to control nonpoint pollution.

PROBLEMS

9.1 Use Table 9-1 and rank the concentrations of nonpoint source pollutants coming from construction sites. How would you rank these same pollutants in terms of harm to the environment? Compare and discuss both rankings.

9.2 Contact the SCS agent in your area and calculate the soil loss from 500 acres of cropland using the universal soil loss equation. Assume:

- conventional tillage
- cornstalks remain in field after harvest

- farmer uses contour stripped planting
- soil is Fayette silt loam
- slope is $\simeq 7\%$
- slope length $\simeq 200$ ft

9.3 Onsite wastewater treatment and disposal systems are also criticized for creating nonpoint source water pollution. Describe the pollutants that are factors in backyards utilizing such systems, and identify alternative technologies to control the problems.

9.4 Discuss the wastewater treatment technologies in Chapter 7 that are particularly applicable to help control pollution from urban stormwater runoff.

9.5 Calculate the cost for controlling runoff from the construction of a 5-mile highway link across rolling countryside near your home town. Assume hay bales are sufficient along the construction site if they are coupled with burlap barriers located at key locations. No major collection and treatment works are required. Document your assumptions, including labor and materials charges, as well as time commitments.

LIST OF SYMBOLS

A_i = acreage of subarea i (acres)
C = cover factor (dimensionless ratio)
K = soil erodibility factor (ton/acre/R unit)
L = slope-length factor (dimensionless ratio)
n = number of subareas identified in a region
P = erosion control practice factor (dimensionless ratio)
R = rainfall factor
S = slope-steepness factor (dimensionless ratio)
S_d = sediment delivery ratio (dimensionless ratio)
SCS = USDA Soil Conservation Service
USDA = U.S. Department of Agriculture
USLE = Universal Soil Loss Equation
$Y_{(s)E}$ = sediment yield from surface erosion (ton/yr)

SUGGESTED READING

Berner, R. A. *Principles of Sedimentary Geochemistry* (New York: McGraw-Hill Book Company, 1971).
Bureau of Water Management, *Michigan Soil Erosion and Sedimenta-*

tion Control Guidebook, Michigan Department of Water Resources, Lansing, MI (1975).

Healy, R. G. *Land Use and the States, Resources for the Future* (Baltimore, MD: Johns Hopkins University Press, 1977).

Linsley, R. R., M. A. Kohler and J. L. H. Paulhus. *Hydrology for Engineers* (New York: McGraw-Hill Book Company, 1958).

Office of Research and Development, *Loading Functions for Assessment of Water Pollution From Nonpoint Sources,* U.S. Environmental Protection Agency, Washington, DC (1976).

Office of Water Planning and Standards, *Nonpoint Source Control Guidance: Construction Activities,* U.S. Environmental Protection Agency, Washington, DC (1976).

Overcash, M. R., and J. M. Davidson. *Environmental Impact of Nonpoint Source Pollution* (Ann Arbor, MI: Ann Arbor Science Publishers, Inc., 1980).

Statham, I. *Earth Surface Sediment Transport* (Oxford, England: Clarendon Press, 1977).

Thurow, C. *Performance Controls for Sensitive Lands,* Planning Advisory Service, Number 307 and 308, American Society of Planning Officials, Chicago (1975).

Chapter 10

Water Pollution Law

A complex system of laws requires industries and towns to treat their wastewater flows prior to discharge to receiving waterways. In this system, *common law* and *statutory law* are intertwined to form the regulatory basis for pollution control.

The American legal tradition is based on common law, a body of law vastly different from statutory law as written by Congress and state governments across the nation. This common law is the aggregate body of decisions made in courtrooms as judges decide individual cases. An individual or group of individuals damaged by water pollution or any other wrong (the plaintiffs) historically could seek relief in the courtroom in the form of an injunction to stop the polluter (the defendant) and/or in the form of payment for damages. Because of the possibilty that the polluter would have to pay damages, the pollution of waterways can be reduced.

Court rulings in these cases were, and are, based on *precedents*. The underlying theory of precedents is that if a similar case or cases were brought before any court in the past, then the present-day judge would violate the rules of fair play if the present-day case were not decided in the same manner and for the same party that the precedent cases dictated. Similar cases, defined to be so by the judge, theoretically have similar endings. If no precedent exists, then the plaintiff essentially rolls the dice in hopes of convincing the court to make a favorable precedent-setting decision.

Statutory law, on the other hand, is a set of rules mandated by a representative governing body, be it the Congress of the United States or the state legislatures. Such legislation supplements or changes the effect of existing common law in areas where Congress or state legislatures perceive shortcomings. For example, environmental quality in general and public health in particular were continually harmed under the common laws as they related to dirty water. Common law courts were taking years to reflect changed societal conditions because the courts were bound to precedents set during times in the nation's history when clean water was plentiful and essentially free. Finally, Congress decided to take the initiative with a series of laws aimed at abating water pollution and cleaning the surface waters of the nation.

In this chapter, we discuss the evolution of water law from the common law courtrooms through the legislative chambers of Congress to the administrative offices of the U.S. Environmental Protection Agency (EPA) and state agencies.

COMMON LAW

There are two major theories of common law as it applies to water. One theory, labeled the *riparian doctrine,* says that conflicts between plaintiffs and defendants must be decided by the ownership of the land underlying or adjoining a body of surface water. The second theory, known as the *prior appropriations doctrine,* takes a different focus and simply states that water use is rationed on a first-come first-serve basis, regardless of land ownership. Note that in both of these doctrines, the focus is on water quantity, on deciding how to apportion a finite body of clean surface water. Common law is generally unclear about water quality consideration.

The principle underlying the riparian doctrine is that water is owned by the owner of the land underlying or adjoining the stream, and that the owner generally is entitled to use the water as long as the quantity is not depleted nor the quality degraded. Riparian land is land bordering a surface waterway.

The doctrine has a somewhat confused history. It was originally introduced to the New World by the French and adopted by several colonies. The English court system eventually adopted the theory as common law in several court cases, and it thus eventually became the official law of the New World. In colonial courtrooms, the riparian doctrine was a workable concept. The land owner was entitled to use the water for domestic purposes, such as washing and watering stock, but common law held that it could not be sold to nonriparian parties, simply because the water in the stream or river would be diminished in quantity and the downstream user would then not have access to the total flow. Even at the present time in sparsely populated farming areas where water is plentiful, this system is still applicable.

In urban and more densely populated areas, courts generally found that the riparian doctrine could not be applied in its pure form. Accordingly, several variations or ground rules were developed in the courts: the *principle of reasonable use* and the *concept of prescriptive rights.*

The *principle of reasonable use* holds that a riparian owner is entitled to make reasonable use of the water, taking into account the needs of other riparians. Reasonable use is defined on a case by case basis by the courts. Obviously, this opens tremendous loopholes, which have been used in numerous litigations. Possibly the most famous example is the case of *New York City vs. the States of Pennsylvania and New Jersey.* In the 1920s, New York City began to pipe drinking water from the upper reaches of the Delaware River, and as the city grew, the demand increased until the people downstream from these impoundments found that their rivers had disappeared. Many resort owners simply went out of business. After prolonged court battles it was finally determined that since the city did own the land around the impounded streams, and the use of this water was "reasonable," the city could continue to use the water. There was some monetary compensation for the downstream riparian owners, but in retrospect the failure of common law is quite evident.

The *concept of prescriptive rights* evolved to the point where, if a riparian owner doesn't use the water and an upstream user "openly and notoriously" abuses the water quantity or quality, then the upstream user is entitled to continue this practice. This concept holds that, through lack of use, the downstream riparian has forfeited the water rights.

This concept was established in a famous case in 1886: *Pennsylvania Coal Co. vs. Sanderson.* Anthracite coal mines north of Scranton, Pennsylvania, at the headwaters of the Lackawanna River, were polluting the river and eventually made it unfit for aquatic life and human consumption. Mrs. Sanderson, a riparian landowner, built a house near the river before the polluted water quality conditions became noticeable. Her intent was to live there indefinitely, but the water quality soon deteriorated, eliminating her opportunities to benefit from the resource. She took the mining company into a courtroom—and lost. The court basically held that the use of the river as a sewer was "reasonable," since the company had been in operation before Mrs. Sanderson had built her house and that since water pollution was a necessary result of coal mining, the coal company could continue its open and notorious practice. This illustrates another example of where common water law started to break down in its ability to serve the people.

Although historically important, the riparian doctrine is declining in use. It is, after all, applicable to sparsely populated areas with no severe water supply and water quality problems. Most of its applicability is limited to areas east of the Mississippi River where there is sufficient rainfall to enable the system to work.

The other important water law concept is the appropriations doctrine, which states that water users "first in time" are necessarily "first in line." In other words, if a user put surface water to some "beneficial use" before the next person, the first user is guaranteed that quantity of water for as long as the use demands. Land ownership and user location, upstream or downstream, are irrelevant.

The doctrine began in the mid-1800s as gold miners in the western part of the country sought to stake their claims to water in the same manner they would stake their claims to a mining area. The concept works well in such areas where water is in short supply, and has been adopted by most of the western states. It is a doctrine which recognizes the scarcity of water and guarantees the availability of this resource if the miner, farmer or industrialist decides to invest in the development of the land. Additionally, this right to water may be sold like a parcel of land. The purchaser need not be a riparian owner, and the only restriction generally is that the water must be put to beneficial use. The owner can lose the right to the water only if it is not used.

Since the flow of most streams is highly variable, it is possible to own a water right and a dry stream bed simultaneously. This conflict is resolved under the appropriation doctrine by prior claim. For example, if a user has first claim of 1 million gallons per day (mgd), a second claimant has 3 mgd and the third has 2 mgd, as long as the river flows at 6 mgd everyone is happy. If the flow drops to 4 mgd, the third claimant is completely out of business. If the third claimant happens to be upstream from 1 and 2, the 4 mgd of water must be permitted to flow past that claimant's water intake without the removal of even a single drop.

As is the case with the riparian doctrine, the appropriations doctrine says very little about water quality. Under the modified appropriations doctrine, the upstream user who is senior in time generally may pollute. If the downstream user is senior, the court has in the past directed payments for losses. If the cost of cleanup is greater than the downstream benefits, however, courts have often found it "reasonable" to allow the pollution to continue. Under the appropriations doctrine, a downstream owner who is not actually using the water has no claim whatever.

The common law theories of public and private nuisance have been found to have applicability in certain cases. Nuisance, however, has found more application in air pollution control, and is discussed in some detail in Chapter 21.

STATUTORY LAW

Citing the shortcoming in common law and continued water pollution problems, Congress and state governments have passed a series of laws designed to clean the surface waters across the nation. Although most states had some laws regulating water quality, it wasn't until 1965 that a concerted push was made to curb water pollution. In that year the U.S. Congress passed the Water Quality Act which, among other provisions, required each state to submit a list of water quality standards and to classify all streams by these standards.

Most states adopted a system similar to the *ambient water quality stream classifications* shown in Table 10-1. Streams were classified according to their anticipated maximum beneficial use. This allowed some states to classify certain streams low-quality waterways and others as virgin trout streams. The method of stream classification theoretically forces the states to limit industrial and municipal discharges and prevents a stream from decaying further. As progress in pollution control is made, the classifications of various streams can be improved. Lowering a stream classification is generally not allowed by state and federal regulatory agencies.

In order to attain the desired water quality, restrictions on wastewater discharges are necessary. Such restrictions, known as *effluent standards,* have been used by various levels of government for many years. For example, an effluent standard for all pulp and paper mills may require that the discharge not exceed 50 mg/l BOD. The total loading (in pounds of BOD per unit time) of the pollution or its effect on a specific stream is thus not considered. Using perfectly ''reasonable'' effluent standards, it is still possible that the effluent from a large mill, although it meets the

Table 10-1.   Typical Stream Classification System

Stream Classification	Best Use	Dissolved Oxygen (mg/l)	Coliforms (per 100 ml)
A	Drinking water (all uses)	>5	<50
B	Water contact sports, fishing, etc. (all but drinking)	>4	<500
C	Noncontact sports, no swimming. Fish and wildlife propagation.	>4	<5000
D	Agricultural,industrial. No fishing, swimming, boating or drinking.	>3	—

effluent standards, completely destroys a stream. On the other hand, a small mill on a large river, which may in fact be able to discharge even untreated effluent without producing any appreciable detrimental effect on the water quality, must meet the same effluent standards.

This dilemma can be resolved by developing a system where minimum effluent standards are first set for all discharges, and then these standards are modified based on the actual effect the discharge would have on the receiving watercourse. For example, the large mill above may have a BOD standard of 50 mg/l, but because of the severe detrimental impact it has on the receiving water quality, thus may be reduced to 5 mg/l BOD. This concept requires that each discharge be considered on an individual basis, a process spelled out in the 1972 Federal Water Pollution Control Act, which also established a nationwide policy of zero discharge by 1985. In plain terms, Congress mandated the EPA to ensure that all waste be removed prior to discharge to a receiving waterway. The control mechanism to achieve a reduction in pollution was the EPA's prohibition of any discharge of pollutants into any public waterway unless authorized by a permit. The permit system, known as the National Pollutant Discharge Elimination System (NPDES), is administered by the EPA, with direct permitting power transferred to states able to convince the EPA that their administering agency has the authority and expertise to conduct the program. In Wisconsin, for example, the state government has been granted the authority to administer the Wisconsin Pollutant Discharge Elimination System, or WPDES.

In situations where an industry wishes to discharge into a municipal sewage treatment system, the industry must agree to contractual arrangements developed with the local governments to ensure compliance with federal industrial pretreatment requirements. For selected industries, pretreatment guidelines are being developed that require facilities to treat their wastewater flows prior to discharge to municipal sewer systems. These pretreatment rules are discussed in detail later in this chapter.

The 1977 Amendments to the Clean Water Act have, in recognition of limits of technology and management of wastewater treatment systems to achieve a zero discharge, proposed that eventually all discharges be treated using "best conventional pollutant control technology," even though this would not be 100% removal. In addition, the EPA is responsible now for setting effluent limits to a list of about 100 toxic pollutants, which must be controlled using "best available control technology."

An industrial facility has two choices in the disposal of wastewater:

- discharge to a watercourse—in which case a NPDES permit is required, and the discharge will have to be continually monitored

• discharge to a public sewer.

The latter method seems highly desirable, since the only restrictions are local sewer ordinances. As a result of this, many industries now discharge to public sewers and transfer the treatment problems to the publicly owned treatment works (POTW).

Some industrial discharges can, however, cause severe treatment problems in the POTWs, and tighter restrictions on what industries can and cannot discharge into public sewers have become necessary. This has evolved into what is now known as the *pretreatment* program.

Pretreatment Guidelines

The pretreatment regulations have been developed by the EPA.[1] Under these general regulations, any municipal facility or combination of facilities operated by the same authority with a total design flow greater than 5 mgd and receiving pollutants from industrial users is required to establish a pretreatment program. The EPA regional administrator may require that a municipal facility with a design flow of 5 mgd or less develop a pretreatment program if it is found that the nature or the volume of the industrial effluent disrupts the treatment process, causes violations of effluent limitations, or results in the contamination of municipal sludge.

In addition to these general pretreatment regulations, the EPA is developing specific regulations for the 33 major industries listed in Table 10-2. The regulations for each industry are designed to limit the concentrations of certain pollutants which may be introduced into sewerage systems by the respective industries. The standards require limitations on the discharge of pollutants that are toxic to human beings as well as to aquatic organisms, such as cadmium, lead, chromium, copper, nickel, zinc and cyanide.

Table 10-3 summarizes pretreatment data for six large cities, and presents estimates of the industrial contribution of cadmium to municipal sewerage systems. These data also display the effectiveness of existing pretreatment guidelines in reducing cadmium. One significant conclusion that can be drawn from Table 10-3 is that we actually know very little about industrial discharges in large cities. This lack of understanding is probably equally true for smaller communities.

---

[1] "General Pretreatment Regulations for Existing and New Sources of Pollution," *Federal Register* Part IV (June 28, 1978), pp. 27736–27773.

Table 10-2.   Pretreatment Industries

Timber Processing	Laundries
Leather Tanning	Soaps and Detergents
Steam Electric	Machinery
Petroleum Refining	Copper Working
Iron and Steel	Aluminum
Paving and Roofing	Plastics
Nonferrous	Batteries
Paint and Ink	Coated Coils
Printing	Enamel Products
Coal Mining	Photographic Supplies
Ore Mining	Foundries
Organics	Adhesives
Inorganics	Explosives
Plastics and Synthetics	Gum and Wood
Textiles	Pharmaceuticals
Pulp and Paper	Pesticides
Rubber	Electroplating

Table 10-3.   Pretreatment of Six Sample Cities

	Industrial Contribution: % of Cd in Influent	Effects of Existing Pretreatment Guidelines	
		% Reduction of Cd in Influent	% Reduction of Cd in Sludge
Los Angeles County	86	30	—
Buffalo	100	—	50
Philadelphia	—	80	—
Chicago	38	21	—
Milwaukee	74	—	—
New York City	53	—	—

Drinking Water Standards

Drinking water standards are equally if not more important to public health than stream standards. These standards have a long history. In 1914, faced with the questionable quality of potable water in the towns along their routes, the railroad industry asked the U.S. Public Health Service (USPHS) to suggest standards which would describe a good drinking water. As a result of this problem, the first USPHS Drinking Water Standards were born. There was no law passed to require that all towns abide by these standards, but it was established that interstate transportation would not be allowed to stop at towns which could not provide water of adequate quality. Over the years most water supplies in the United States have not been closely regulated, and the high-quality water provided by municipal systems has been as much the result of the professional pride of the water industry personnel as any governmental restrictions.

Because of a growing concern with the quality of some of the urban water supplies and reports that not all waters are as pure and safe as people have always assumed, the federal government passed the Safe Drinking Water Act in late 1974. This law authorizes the EPA to set minimum national drinking water standards. The EPA has published some of these standards, which are quite similar to the USPHS Water Standards. Some of these representative numbers are shown in Table 10-4. Potable water standards used to describe these contaminants can be divided into three categories: physical, bacteriological and chemical.

Physical standards include color, turbidity and odor, all of which are not dangerous in themselves but could, if present in excessive amounts, drive people to drink other, perhaps less safe, water.

Bacteriological standards are in terms of coliforms, the indicators of pollution by wastes from warm-blooded animals. Tests for pathogens are almost never attempted. The present EPA standard calls for a concentration of coliforms of less than 1 per 100 ml water. This standard is a classical example of how the principle of expediency is used to set standards.

Table 10-4.   Selected EPA Drinking Water Standards

Physical
   Turbidity   5 units
   Color   15 units
   Odor   3 (threshold odor)
Bacteriological
   Coliforms 1 coliform/100 ml

	Suggested (mg/l)	Maximum (mg/l)
Chemical		
Arsenic	0.01	0.05
Chloride	250	—
Copper	1	—
Cyanide	0.01	0.2
Iron	0.3	—
Phenols	0.001	—
Sulfate	250	—
Zinc	5	—

Before modern water treatment plants were commonplace, the bacteriological standard stood at 10 coliforms/100 ml. In 1946, this was changed to the present level of 1/100 ml. In reality, with modern methods we can attain about 0.01 coliforms/100 ml. It is, however, not expedient to do this because the extra measure of public health attained by lowering the permissible coliform level would not be worth the price we would have to pay.

Chemical standards include a long list of chemical contaminants beginning with arsenic and ending with zinc. Two classifications exist, the first being a suggested limit, the latter a maximum allowable limit. Arsenic, for example, has a suggested limit of 0.01 mg/l. This concentration has, from experience, been shown to be a safe level even when ingested over an extended period. The maximum allowable arsenic level is 0.05 mg/l, which is still under the toxic threshold but close enough to create public health concern. On the other hand, some chemicals such as chlorides have no maximum allowable limits since at concentrations above the suggested limits the water becomes unfit to drink on the basis of taste or odor.

CONCLUSION

Over the years, the battles for clean water have moved from the courtroom through the congressional chambers to the administrative offices of EPA and state departments of natural resources. Permitting systems have replaced inconsistent, one-case-at-a-time judicial proceedings as ambient water quality standards and effluent standards are sought. Tough decisions lie ahead as current water programs are administered, particularly the NPDES permits for polluters discharging to waterways and the pretreatment guidelines for polluters discharging to municipal sewer systems. Even tougher decisions must be faced in the future as regulations are developed for the control of toxic substances.

PROBLEMS

10.1 In your home town, describe the NPDES reporting requirements for the local wastewater treatment facility. What data are required, how often are summary forms completed, and what agency reviews the data on the forms?

10.2 Health departments often require that chlorine be added to water as it enters municipal distribution systems. Discuss the benefits and risks associated with these rules, and describe alternative ways to ensure potable water at the household tap.

10.3 Many industrial processes are water-intensive. That is, in order to produce a product that will sell in the marketplace, many gallons of water must flow into the factory. Develop a sample listing of such industries and discuss the legal and administrative problems generally associated with securing this water for new factories. Compare and contrast these problems with respect to the generally wet eastern states and dry western states.

10.4 Federal regulations are designed to achieve "zero-discharge" of pollutants from point sources located along surface waterways. Land application of liquid waste is an option often proposed in many sections of the nation. Discuss the advantages and disadvantages of such systems particularly in terms of heavy metal pollutants, and outline possible restrictions on land where such wastes have been applied.

10.5 Assume you work for the EPA and are assigned to propose a standard for the allowable levels of arsenic for household drinking water. What data would you collect, where would you go to get the data (literature and/or laboratory), and in what professions would you seek experts to help guide you?

LIST OF SYMBOLS

BOD	=	biochemical oxygen demand
EPA	=	U.S. Environmental Protection Agency
NPDES	=	National Pollutant Discharge Elimination System
POTW	=	publicly owned (wastewater) treatment works
USPHS	=	U.S. Public Health Service
WPDES	=	Wisconsin Pollutant Discharge Elimination System

SUGGESTED READING

Chanlett, E. *Environmental Protection* (New York: McGraw-Hill Book Company, 1973).

Council on Environmental Quality, *Environmental Quality,* Washington, DC (Annual Reports).

Landau, N. J., and P. D. Rheingold. *The Environmental Law Handbook,* (New York: Ballantine Books, 1971).

MacDonald, J. B., and J. H. Beuscher. *Water Rights* (Madison, WI: American Printing and Publishing, 1973).

Waste Management and Control (Washington, DC: National Academy of Science, 1966).

Office of the President, "Floodplain Management, Executive Order 11988," *42 Federal Register 101,* U.S. Government Printing Office, (May 25, 1977).

Phelps E. B. *Public Health Engineering* (New York: John Wiley & Sons, 1948).

Safe Drinking Water Act: Public Law 93–523, U.S. Congress, U.S. Government Printing Office (1974).

Water Pollution Control Act and Amendments: Public Law 92–500, U.S. Congress, U.S. Government Printing Office (1977).

Chapter 11

Solid Waste

In this chapter, solid wastes other than hazardous materials and nuclear wastes are considered. Such solid wastes are often called *municipal solid waste* (MSW) and consist of all the solid and semisolid materials discarded by a community. The fraction of MSW produced by a household is called *refuse*.

Refuse until fairly recently was mostly food waste, but new materials such as plastics, new packages for products such as beer cans, and new products such as garbage grinders have all changed the composition of municipal solid waste. Industry creates about 2000 new products each year, all of which eventually find their way into municipal refuse and contribute to individual disposal problems.

The components of refuse are *garbage,* and food wastes; *rubbish,* which includes glass, tin cans and paper; and *ashes,* still a problem where coal is used for heating homes. Lastly, *trash* refers to such larger items as tree limbs, old appliances, etc., which are not normally deposited into garbage cans.

The relationship between solid wastes and human disease is intuitively obvious, but difficult to prove. For example, if a flea, sustained by a rat which in turn is sustained by an open dump, transmits murine typhus to a human, the absolute proof of the pathway is to find *the* rat and *the* flea, an obviously impossible task. Nevertheless, we are certain that improper solid waste disposal is a true health hazard, for we know that at least 22 human diseases are associated with solid wastes.

The two most important vectors* of human disease in regard to solid wastes are rats and flies. The fly is a prolific breeder (70,000 flies can be

---

*Vectors are means by which disease organisms are transmitted. Water, air and food can all be vectors.

produced in 1 cubic foot of garbage) and a carrier of many diseases, e.g., bacillary dysentery. Rats not only destroy property and infect by direct bite, but are also dangerous as carriers of insects which can also act as vectors. For example, the plagues of the Middle Ages were directly associated with the rat populations.

As serious as the public health aspects of solid wastes are, health is seldom of primary concern when municipalities decide on a method of disposal. The overriding criterion is still money! What is the cheapest way to "get rid of" this stuff?

In this chapter, the quantities and composition of this "stuff" are discussed first, followed by a brief introduction to disposal options and the specific problem of litter. In the following chapter, disposal is discussed further and Chapter 13 is devoted to the problems and promises of recovering energy and materials from refuse.

QUANTITIES AND CHARACTERISTICS OF MUNICIPAL SOLID WASTE

The quantities of MSW generated in a community can be estimated by one of three techniques:

- input analysis
- secondary data analysis
- output analysis.

In the first case, the MSW is estimated based on the products people use. For example, if a community purchases 100,000 steel beer cans per week, it can be expected that MSW (including litter pickup) will include about 100,000 beer cans per week. Unfortunately, this technique is very difficult to use in any but very isolated small communities.

Secondary data can be used to estimate solid waste production by some empirical relationship. For example, one study[1] concluded that solid waste generation could be predicted as

$$W = 0.0179(S) - 0.00376(F) - 0.00322(D) + 0.0071(P) - 0.0002(I) + 44.7$$

where   W = waste generated, ton (2000 lb)
         W = number of stops made by the truck

---

[1]Shell, R. L., and D. S. Shute. "A Study of the Problem Predicting Future Volumes of Wastes," *Solid Waste Management* (March 1972).

F = number of families served
D = number of single-family dwellings
P = population
I = adjusted gross income per dwelling unit ($)

Such models are rarely sufficiently accurate, however, and solid waste generation is almost always measured directly using an output analysis.

In weighing the trucks arriving at a disposal site, it is important to keep in mind that refuse generation varies with time—day of the week and week of the year. The weather conditions also affect the weight of the refuse, since the moisture content may vary from 15 to 30%. If the solid waste disposal site has truck scales, refuse weights can be easily calculated. Without such scales, however, wheel scales must be used to weigh incoming and exiting trucks. In such surveys, seldom is it possible to weigh all trucks and thus a sample must be used. This introduces another error. Statistical methods for estimating the confidence level in such a sampling survey must be used.[2]

Characteristics

Among the characteristics of refuse which are important for developing engineered solutions to refuse management are

- moisture
- particle size
- chemical makeup
- density
- composition

The moisture concentration of MSW can vary between 15 and 30% water, with 20% being normal. The moisture is measured by drying a sample at 77°C (170°F) for 24 hours and calculating it as

$$M = \frac{w - d}{w} \times 100$$

where   M = moisture content, percent
        w = initial (wet) weight of sample
        d = final (dry) weight of sample

[2]Klee, A. *Quantitative Decision-Making,* Vol. III of *Design and Management for Resource Recovery,* P. A. Vesilind, Ed. (Ann Arbor, MI: Ann Arbor Science Publishers, Inc., 1981).

The particle size distribution is especially important in refuse processing for resource recovery, and this is discussed further in Chapter 13.

The chemical composition of typical refuse is shown in Table 11-1. The use of both the proximate and ultimate analysis in the combustion of MSW and its various fractions is discussed further in Chapters 12 and 13.

The density of refuse varies depending on where it is. Table 11-2 shows some typical densities of MSW.

Refuse composition is possibly the most important characteristic affecting its disposal or the recovery of materials and energy from refuse. Composition can vary significantly from one community to the next, and with time in any given community.

Table 11-1.   Proximate and Ultimate Chemical Analysis of MSW[3]

Proximate Analysis	
Moisture (%)	15–35
Volatile Matter (%)	50–60
Fixed Carbon (%)	3–9
Noncombustibles (%)	15–25
Higher Heat Value (Btu/lb)	3000–6000
Ultimate Analysis	
Moisture (%)	15–35
Carbon (%)	15–30
Hydrogen (%)	2–5
Oxygen (%)	12–24
Nitrogen (%)	0.2–1.0
Sulfur (%)	0.02–0.1

[3]"Incinerator Guidelines," U.S. Department of Health, Education and Welfare (1969).

Table 11-2.   Refuse Densities

	kg/m³	(lb/yd³)
Loose refuse	60–120	(100–200)
Refuse from a collection vehicle, after dumping	200–240	(350–400)
Refuse in a collection vehicle	300–400	(500–700)
Refuse in a landfill	300–540	(500–900)
Baled refuse	470–700	(800–1200)

Refuse composition is expressed either in terms of "as-generated" or "as disposed" since during the disposal process moisture transfer takes place, thus changing the weights of the various fractions. Table 11-3 shows typical breakdowns of average refuse components for the United States. It should be reemphasized that such numbers are useful only as guidelines and that each community has unique characteristics which influence its solid waste production and composition.

COLLECTION

In the United States and most other countries, solid waste is collected by trucks. In some instances, these are open-bed trucks which carry trash or bagged refuse. The usual vehicle however is the packer, a truck that uses hydraulic rams to compact the refuse to reduce its volume and thus is able to carry larger loads (Figure 11-1). Commercial and industrial collections are facilitated by the use of containers which are either emptied

Table 11-3.   Average Composition of MSW
in the United States (1973)

Category	As-Generated (millions of tons)		(%)		As-Disposed (millions of tons)	(%)
Paper	37.2		29.0		44.9	34.9
Glass	13.3		10.4		13.5	10.5
Metal	12.1		9.6		12.6	9.8
Ferrous		10.8		8.6		
Aluminum		0.9		0.7		
Other Nonferrous		0.4		0.3		
Plastics	4.4		3.4		4.9	3.8
Rubber and Leather	2.6		3.3		3.4	2.6
Textiles	2.1		1.6		2.2	1.7
Wood	4.9		3.8		4.9	3.8
Food Waste	22.8		17.8		19.1	14.9
Yard Waste	26.0		20.2		20.0	16.3
Miscellaneous	1.9		1.5		2.0	1.6
Total	128.2		100.0		128.5	100.0

Figure 11-1   Packer truck used for residential refuse collection.

into the truck using a hydraulic mechanism, or where the entire container is carried by the truck to the disposal site (Figure 11-2).

On the average, of the total cost of solid waste management, a full 80% is spent on collection. The common method of collection is by packer truck with three workers: one driver and two loaders. These workers fill the truck and then drive it to the disposal area. The entire operation is a study in inefficiency. The time spent by the loaders traveling to and from the dump, for example, is pure waste. Accordingly, many new devices and methods have been proposed to cut collection cost. Some that are already in use are discussed below.

Garbage grinders reduce the amount of garbage in refuse. If all homes had garbage grinders, the frequency of collection could be cut in half, since the twice-a-week collection is most communities is necessary only because of the rapid decomposition of the garbage component. Obviously, garbage grinders put an extra load on the wastewater treatment plant, but sewage is relatively dilute and ground garbage can easily be accommodated both in the sewers and in treatment plants.

Pneumatic pipes have been installed in some small communities, mostly in Sweden and Japan. The refuse is ground at the residence and sucked through underground lines. One system in the United States is in Walt Disney World in Florida. Collection stations scattered throughout

Figure 11-2 Containerized collection system (courtesy Dempster Systems).

the park receive the refuse, and pneumatic pipes deliver the waste to a central processing plant (Figure 11-3). There are no garbage trucks in the Magic Kingdom.

It is not at all unreasonable to expect that pneumatic pipes will be the collection method of the future.

Figure 11-3   Solid waste collection system at Disney World (courtesy AVAC Inc.).

Compactors in the kitchen, the first new appliances introduced to the household market in many years, have distinct possibilities of reducing collection costs, but only if everyone has one. It is obvious that if all residences in a community had garbage grinders and refuse compactors, the cost of solid waste disposal (and thus local taxes) could be reduced considerably. Stationary compactors for commercial establishments and apartment houses have already had significant influence on collection practices.

Transfer stations are applicable to almost all larger communities. A typical system (Figure 11-4) involves several stations scattered around a city to which ordinary collection trucks bring the refuse. The drive to the nearest station is fairly short for each truck, so the workers spend more time collecting and less traveling. At the transfer station, bulldozers cram this refuse into large cans, which in turn take the material to the ultimate disposal. In some towns the refuse is baled into convenient desk-sized blocks before disposal in a landfill.

Figure 11-4 Transfer station method of solid waste collection.

Green cans on wheels are now widely used for the transfer of refuse to the truck. Shown in Figure 11-5, the green cans are pushed curbside by the householders and emptied by means of a hydraulic lift. Not only does this system save money but it has a dramatic effect on the incidence of injuries to solid waste collection personnel, who have by far the highest lost-time accident rate of any municipal or industrial workers.

Route optimization can result in significant savings to a city. Several computer programs are available for selecting the least-cost routes and collection frequencies. Such optimization techniques have resulted in increased effectiveness and lower cost of refuse collection.

The problem of route optimization was actually first addressed in 1736 by the famous Prussian mathematician, Leonard Euler. He was asked to design a parade route for Königsberg such that the parade would not cross any bridge over the River Pregel more than once (Figure 11-6). Not only did Euler show that this was not possible, but generalized the prob-

Figure 11-5   The "green can" system of solid waste collection.

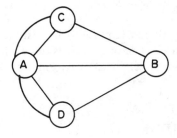

Figure 11-6   The seven bridges of Königsberg; the routing problem presented to Euler.

lem by observing that *nodes* must have an even number of *links* to make such an *Euler's tour* possible. Clearly the bridge problem had 4 nodes (two shores C and D, and two islands A and B) and 7 links (the roads), with an odd number of links to each node.

The objective of truck routing is to create an Euler's tour, and eliminate *deadheading,* traveling twice down the same street. Although sophisticated computer programs are available for routing trucks so as to achieve the most efficient tour, it is often just as simple to develop a route by common sense or *heuristic* means. Some common sense rules for routing trucks are as follows:[4,5]

1. Routes should not overlap, but should be compact and not fragmented.

---

[4]Liebman, J. C., J. W. Male and M. Wathne. "Minimum Cost in Residential Refuse Vehicle Routes," *J. Environ. Eng. Div., ASCE* 101(EE3):339–412 (1975).

[5]Shuster, K. A., and D. A. Schur. "Heuristic Routing for Solid Wastes Collection Vehicles," U.S. EPA OSWMP SW–113 (1974).

2. The starting point should be as close to the truck garage as possible.
3. Heavily traveled streets should be avoided during rush hours.
4. One-way streets that cannot be traversed in one line should be looped from the upper end of the street.
5. Dead-end streets should be collected when on the right side of the street.
6. On hills, collection should proceed downhill so that the truck can coast.
7. Clockwise turns around blocks should be used whenever possible.
8. Long, straight paths should be routed before looping clockwise.
9. For certain block patterns, standard paths, as shown in Figure 11-7, should be used.
10. U-turns can be avoided by never leaving one two-way street as the only access and exit to the node.

Figure 11-7 shows three examples of heuristic routing. The first two show a route where each side of a street is to be collected separately, the third example shows a system where both sides of the street are collected at once.

DISPOSAL OPTIONS

Historically, the disposal of municipal solid waste did not represent a problem for municipalities, following the invention of city dumps by the Romans. When cities began to grow closer together, and the amount of refuse generated mushroomed, the disposal problem became serious. At first, the solution advocated by engineers was the incinerator (or "reducer," as it was originally called). In time, however, it became clear that these incinerators caused substantial air pollution, did not in all cases burn the combustibles sufficiently, and were expensive to operate. The sanitary landfill then became the accepted method of solid waste disposal. It was promoted as a reasonably inexpensive and environmentally sound alternative and became the accepted method of disposal. Unfortunately, not all landfills were successful, or inexpensive, and there was well-placed concern with the idea of throwing away materials which might be useful. As a result, the idea of processing wastes so as to reclaim materials and energy was born. The options for resource recovery are discussed further in Chapter 13.

LITTER

One of the most ubiquitous environmental insults is litter. It is also the easiest to control—at least in theory.

Figure 11-7   Heuristic routing examples.

Litter is not only unsightly, it is also unhealthy (as a breeding ground for rats and other rodents) and damaging to wildlife. Deer and fish, attracted to aluminum can pop-tops, ingest them and die in agony. Plastic sandwich bags are mistaken for jellyfish by tortoises, and death results. Litter is, in short, a most uncivilized by-product of our civilization, and considerable efforts have recently been directed toward curbing this insult.

Public awareness campaigns have been ongoing for many years. The most recent efforts toward volunteer participation have been funded by bottle manufacturers and bottlers, hoping to avert legal restrictions on their products. Unfortunately, people still litter and alternative solutions to the problem are being sought.

One report has recommended that from the purely economic standpoint the best way to reduce litter is to hire more street cleaners and road crews to collect the litter. Common sense has happily prevailed and this solution has not been suggested as the final solution to the litter problem.

The availability of trash cans seems to be an important factor influencing littering. A person will walk or ride only a short distance out of the way to deposit waste into a litter can. Increasing the availability and the frequency of maintenance of litter deposits can have a marked effect on litter. Incidentally, experiments have also determined that the type, construction and color of the litter deposit can all influence the frequency of volunteer participation.

A much more drastic assault on the litter problem is restrictive legislation. The most notorious legislative act is the "Oregon Bottle Law" which prohibits the use of pop-top cans and discourages nonreturnable glass beverage bottles. The effect of the law on litter on Oregon has been either negligible or fantastic—depending on whose press releases one believes. It has been reported that the percent of beverage containers collected as litter has dropped by 92%, and most of the 8% comes from other states, a statistic indicating success. Oregonians do not mind the mild inconveniences and seem to be quite pleased with the results. Similar legislation has been passed or is pending in other states.

The bottling people argue that it is people, not bottles, that cause litter, and they are right in principle. The trick is to solve the problems by changing the habits of the people without resorting to laws. Thus far all such efforts have been unsuccessful.

CONCLUSION

The solid waste problem has three facets: source, collection and disposal. We have little chance of coping with this problem unless we begin to attack all three areas. The methods of collection and disposal are discussed in the next chapter. The source of solid waste is perhaps the most difficult of the three to tackle. New concepts in packaging, use of natural resources, and evaluation of planned obsolescence are necessary if progress is to be made. In this respect, however, we are swimming against the economic current. The economies in all political systems are oiled with money, and the worths of objects are in terms of cost in money. It is not unreasonable to speculate that this method may some day need to be changed. We must, in short, create a "new economy" with regard to productivity, obsolescence and waste before we can begin to be at peace with our ecosystem and have any hopes of long-term survival.

PROBLEMS

11.1 Walk along a stretch of road and collect the litter into two bags, one for beverage containers only, and one for everything else. Calculate the
  a. number of items per mile
  b. number of beverage containers per mile
  c. weight of the litter per mile
  d. weight of the beverage containers per mile
  e. % of beverage containers by count
  f. % beverage containers by weight
If you were working for the bottle manufacturers, how would you report your data, as e or f? Why?

11.2 How would it be possible to tax the withdrawal of natural resources? What effect would this have on the economy?

11.3 What effect will the following have on the composition of municipal solid waste: (a) garbage grinders, (b) home compactors, (c) no more returnable beverage containers, and (d) a newspaper strike?

11.4 What effect would the Oregon Bottle Law have on your consumption practices? How would you change your life style?

11.5 Drive along a measured stretch of highway and count the pieces of litter visible from the car. (It's best to do this with a friend riding shotgun who does the counting.) Then walk along the same stretch and pick up the litter, counting the pieces and weighing the full bags. What percent of litter (by weight and by piece) is visible from a car?

11.6 On a map of your campus (or any other convenient map) develop an efficient route for refuse collection, assuming that every blockface must be collected.

11.7 Using a study hall or social lounge as a laboratory, study the prevalence of litter by counting the items in the receptacles vs the items improperly disposed of. Each day (weekdays only), vary the conditions as follows:

Day 1: Normal conditions (baseline)
Day 2: Remove all receptacles except one
Day 3: Add additional receptacles (more than normal)
If possible, run several more experiments with different numbers of receptacles. Plot the percent of properly disposed of material vs number of receptacles. Discuss the implications.

11.8 Using heuristic routing, develop an efficient route for the map shown in Figure 11-8, if (a) both sides of the street are to be collected together, (b) one side of a street is to be collected at a time.

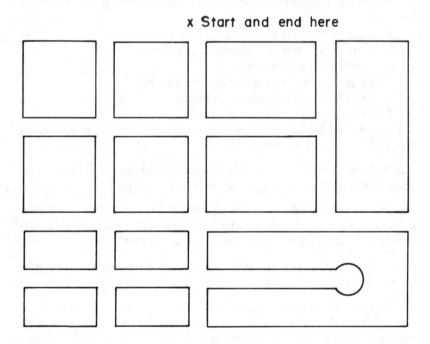

x Start and end here

Figure 11-8   Route for Problem 11.8.

SUGGESTED READING

Carruth, D. E., and A. J. Klee. *Analysis of Solid Waste Composition,* U.S. Department of Health, Education and Welfare (1969).
Goldstein, J. *Garbage as You Like It* (Emmaus, PA: Rodale Books, 1970).
Kirov, N. Y., Ed. *Waste Management, Control, Recovery and Reuse* (Ann Arbor, MI: Ann Arbor Science Publishers, Inc., 1976).

National Center for Resource Recovery, *Municipal Solid Waste Collection* (Lexington, MA: D. C. Heath and Company, 1973).

Small, W. E. *Third Pollution* (New York: Frederick A. Praeger, Inc., 1970).

"Solid Waste/Disease Relationships," U.S. Environmental Protection Agency, SW–1c (1967).

Tchobanoglous, G., H. Theisen and R. Eliassen. *Solid Wastes* (New York: McGraw-Hill Book Company, 1977).

Vesilind, P. A., and A. E. Rimer. *Unit Operations in Resource Recovery Engineering,* (Englewood Cliffs, NJ: Prentice Hall, 1981).

Wang, L. K., and N. C. Pereira (Ed.) *Handbook of Environmental Engineering,* (Clifton, NJ: Humana Press, 1980).

Chapter 12

Solid Waste Disposal

The disposal of solid wastes is defined as placement of the waste so it no longer impacts society. This is achieved either by assimilating the residue so it can no longer be identified in the environment (e.g., fly ash from an incinerator) or by hiding the wastes well enough so they cannot be readily found.

Solid waste can also be processed so that some of its components can be recovered, a procedure popularly known as recycling. Before disposal or recycling, however, the waste must be collected. All of these—collection, disposal and/or recovery—form a part of the total solid waste management system. The collection operation is discussed in the previous chapter, and the next chapter is devoted to the recovery of energy and materials from refuse. This chapter covers the disposal of solid wastes.

Refuse can be disposed of either as is, or after suitable processing. This processing may be thermal or physical, and is performed only for the purpose of converting refuse to a more readily disposable form, and not as a method of energy or materials recovery.

DISPOSAL OF UNPROCESSED REFUSE

The only two realistic options for disposal are in the oceans (or other large bodies of water) and on land. In the United States, the former is presently forbidden by federal law, and it is becoming similarly illegal in most other developed nations. Little else need thus be said of ocean disposal, except perhaps that its use was a less than glorious chapter in the annals of public health and environmental engineering.

The place for solid waste disposal on land is called a *dump* in the United States and a *tip* in Great Britain (as in "tipping"). The dump is by far the least expensive means of solid waste disposal, and thus was the

295

original method of choice for almost all inland communities. The operation of a dump is simple, and involves nothing more than making sure that the trucks empty out at the proper spot. Volume is often reduced by setting the refuse on fire, which prolongs dump life.

The problems with rodents, odor, air pollution and insects at the dump, however, can become serious public health and aesthetic problems, and an alternative method of refuse disposal then becomes necessary. For larger communities, this alternative often was the incinerator. Smaller towns, however, could not afford such a capital investment and sought lower cost options.

The term *sanitary landfill* was first used for the method of disposal employed in the burial of waste ammunition and other material after World War II. The concept of refuse burial had, however, been used by several communities in the midwest, and had proven highly successful.

The sanitary landfill differs markedly from open dumps in that the latter are simply places to dump wastes, but sanitary landfills are engineered operations, designed and operated according to acceptable standards.

The basic principle of a landfill operation is to deposit the refuse, compact it with bulldozers, and cover the material with at least 15 cm (6 in.) of dirt at the conclusion of each day's operation, and a final cover of 60 cm (2 ft) when the area is full (Figure 12-1). The 2-ft depth is necessary to prevent rodents from burrowing into the refuse.

Figure 12-1   The sanitary landfill.

The selection of a landfill site is a sticky problem. The engineering aspects include: (1) drainage—rapid runoff will lessen mosquito problems, but proximity to streams or well supplies might result in water pollution; (2) wind—it is preferable that the landfill be downwind from the community; (3) distance from collection; (4) size—a small site with limited capacity is generally not acceptable since the trouble of finding a new site is considerable; and (5) ultimate use—can the area be utilized for public or private use after the operation is complete?

Perhaps even more important than the engineering problems are the social and psychological problems. No one wants a sanitary landfill in the back yard.

Surprisingly enough, however, there have been many cases where property values have actually been enhanced by a landfill or, more correctly, by what was done with landfill site after the operation was completed. Golf courses, playgrounds and tennis courts can be rewards for tolerating a landfill operation for a few years. If the operation is conducted according to accepted practice, there should be little adverse environmental impact from landfills. This is, as one might suspect, a difficult thing to explain to the community, especially since most "sanitary landfills" have in the past been glorified dumps.

The landfill operation is actually a biological method of waste treatment. Municipal refuse deposited as a fill is anything but inert. In the absence of oxygen, anaerobic decomposition steadily degrades the organic material to more stable forms. But this process is very slow. After 25 years the decomposition can still be going strong.

The liquid produced during the decomposition process, as well as the water that has seeped through the groundcover and worked its way out of the refuse, is known as leachate. This liquid, although small in volume, is extremely high in pollutional capacity. Table 12-1 shows some typical values of leachate composition.

Table 12-1.   Typical Sanitary Landfill Leachate Composition

Component	Typical Value
BOD_5	20,000 mg/l
COD	30,000 mg/l
Ammonia Nitrogen	500 mg/l
Chloride	2,000 mg/l
Total Iron	500 mg/l
Zinc	50 mg/l
Lead	2 mg/l
pH	6.0

The effect of leachate on groundwater can be severe. In a number of cases, the leachate has polluted wells around a landfill to the point where they ceased to be a source of potable water. One example of such pollution is shown in Figure 12.2, which depicts the problems encountered with a landfill on Long Island. The town of Islip's Sayville Landfill was started in a sand and gravel pit in 1933, and is still in operation. The waste at this disposal site, which extends from about 20 feet above grade to the water table about 30 feet below grade, covers 17 acres. Initially, the site was an open dump and received all types of wastes. Presently it receives mostly incinerator residue, and some individually hauled residential wastes. The site is underlain by mostly coarse sand with streaks of gravelly sand.

The leachate plume at Islip's Sayville Landfill extends more than 5000 feet downgradient of the site, 170 feet in depth, and up to 1300 feet in width. About 0.22 square mile and one billion gallons of groundwater have been contaminated. Three residential wells near the disposal site and in the leachate plume were contaminated and had to be abandoned. A laundry, sink fixtures, pipes and a water heater were among the items damaged as a result of the wells' contamination.

The amount of leachate produced by a landfill is difficult to predict. The only available method is to use a water balance—that is, the water entering a landfill has to equal the water flowing out of a landfill (i.e., the leachate).

The total water entering the top soil layer is

$$C = P(1 - R) - S - E$$

where C = total percolation into the top soil layer, mm
         P = precipitation, mm
         R = runoff coefficient
         E = evapotranspiration, mm
         S = storage, mm

The percolation for three typical landfills is shown in Table 12-2.

Using these figures, it is possible to predict when landfills produce leachate. Obviously, Los Angeles' landfills will never produce leachate, it will take 15 years to leach through a 7.5-m (25-ft)-deep landfill in Orlando, and it will take only 11 years for a 20-m (65-ft)-deep landfill in Cincinnati to produce leachate.

A second by-product of a landfill is gas. Since landfills are anaerobic biological reactions, they produce mostly methane and carbon dioxide (see page 210 for a review of anaerobic decomposition).

Landfills go through four distinct stages. As illustrated in Figure 12-3,

Table 12-2.   Percolation in Three Landfills[1]

Location	Precipitation, P (mm)	Runoff Coefficient, R	Evapotranspiration, E (mm)	Percolation, C (mm)
Cincinnati, OH	1025	0.15	658	213
Orlando, FL	1342	0.07	1172	70
Los Angeles, CA	378	0.12	334	0

the first stage is aerobic and may last for a few days to several months, during which time aerobic organisms are active and affect the decomposition. As the organisms use up all available oxygen, however, the landfill enters the second stage where anaerobic decomposition begins, but where methane-forming organisms have not yet taken hold, and the acid formers cause a buildup of CO_2. This stage may also vary with environmental conditions. The third stage is the anaerobic methane production buildup stage, during which the percent of CH_4 progressively increases, along with an increase in landfill temperature to about 55°C (130°F). The last steady-state condition occurs when the fractions of CO_2 and CH_4 are about equal and microbial activity has stabilized.

The amount of methane produced from a landfill can be estimated using the following empirical relationship:[2]

$$CH_aO_bN_c + \tfrac{1}{4}(4 - a - 2b + 3c)H_2O \rightarrow \tfrac{1}{8}(4 - a + 2b + 3c)CO_2$$
$$+ \tfrac{1}{8}(4 + a - 2b - 3c)CH_4$$

This equation is useful only if the chemical composition of the waste is known. An empirical formula for gas production from municipal landfills in California has been stated as:[3]

$$Q = 18.77 \times 10^6 \left(\frac{Ah}{R^2} \right)$$

[1]Tenn, D. G., K. J. Haney and T. V. Degeare. *Use of the Water Balance Method for Predicting Leachate Generation from Solid Waste Disposal Sites,* U.S. EPA OSWMP, SW-168, (1975).

[2]Chian, E. S. K., F. B. DeWalle and E. Hammerberg. "Effect of Moisture Regime and Other Factors on Municipal Solid Waste Stabilization," in *Management of Gas and Leachate,* S. K. Banerji, Ed. U.S. EPA 600/9-77-026 (1977).

[3]Carlson, J. A. as reported in *Treatment and Utilization of Landfill Gas,* U.S. EPA OSWMP SW-583 (1977).

Figure 12-2   Pollution of groundwater from a sanitary landfill.

EXPLANATION

45 C   WELL AND NUMBER—Showing specific
conductance value at indicated screen

—200— LINE OF EQUAL SPECIFIC CONDUCTANCE—
Interval, 100, 500, and 1000 micromhos per
centimetre at 20°Celsius

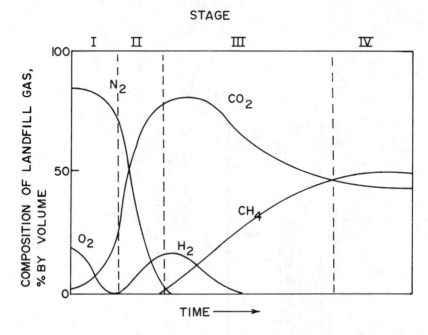

Figure 12-3   States in the decomposition of organic matter in landfills.

where Q = production rate of methane, standard ft^3/day
          A = area landfill, acre
          h = depth of landfill, ft
          R = radius of influence of extraction wells, ft

The rate of gas production from sanitary landfills can be controlled by varying the particle size of the refuse (shredding the refuse before placing it in the landfill) and by changing the moisture content. Figure 12-4 represents general relationships showing these variables. Gas production can be minimized with the combination of low moisture, large particle size and high density. Unwanted migration of the gas can be prevented by escape vents installed in the landfills. These escape vents are sometimes lit and the gas is burned off.

Since landfills produce considerable quantities of methane, interest has recently been shown in the capture and use of landfill gases. The gas must however be cleaned (CO_2 and other contaminants removed) before it can be used as pipeline gas. Such cleaning is both expensive and troublesome. A more reasonable use of landfill gas is to burn it as is, perhaps in industrial applications such as brick kilns.

The biological aspects of landfills as well as the structural properties of

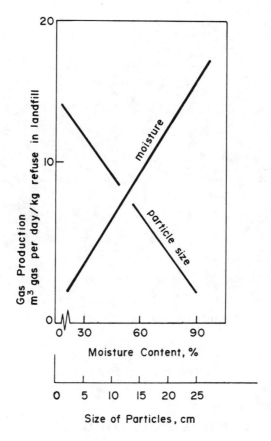

Figure 12-4   Effect of moisture and particle size of refuse on gas products in landfills (adapted from Reference 3).

compacted refuse limit the ultimate uses of landfills. Uneven settling is often a problem, and it is generally suggested that nothing be constructed on a landfill for at least two years after completion. With poor initial compaction, it is not unreasonable to expect 50% settling within the first five years. The owners of the motel shown in Figure 12-5 learned this fact the hard way.

Landfills should never be disturbed. Not only will this cause additional structural problems, but trapped gases can be a hazard. Buildings constructed on landfill sites should have spread footings (large concrete slabs) as foundations, although some have been constructed on pilings which extend through the fill and onto rock or other adequate strong material.

Figure 12.5.   A motel that was built on a landfill and experienced differential settling.

The cost of operating a landfill varies from about $3 to $15 per ton of refuse and usually represents the least-cost method of acceptable solid waste disposal.

VOLUME REDUCTION PRIOR TO DISPOSAL

Refuse is a bulky material which does not compact easily and thus the volume requirements in landfills are significant. Where land is expensive, the costs of landfilling can be high. Some of our larger cities, for example, pay as much as $20/ton of refuse for their landfill. Accordingly, various methods of reducing the volume of refuse to be disposed of have been found to be effective.

Incineration is often touted as the most effective solution to the solid waste disposal problem. Actually, incineration only reduces the volume of waste to 20 to 30% of the original volume and makes the product stable, but the residue must still be disposed of in some manner.

Incinerators have high capital costs, usually exceeding $5000/ton of 24-hr capacity.* Similarly, operating expenses can be high. Air pollution

---

*Incinerators are sized on the basis of 24 hours of operation. For example, if an incinerator is fed 20 tons of refuse per day during a normal 8-hr shift, the incinerator capacity is rated as 60 tons.

control has increased the cost of incineration to anywhere from \$15 to \$50 per ton of refuse. Understandably the United States is littered with incinerators that are too expensive to operate.

A schematic of a typical large incinerator is shown in Figure 12-6. The grapple bucket lifts the refuse from a storage pit and drops it into the charging chute. The stoker (in this case a traveling grate) moves the

Figure 12-6   Schematic of a typical solid waste incinerator.

refuse to the furnace area. Combustion occurs both on the stoker and in the furnace. Air is fed under and over the burning refuse. The walls of the furnace are cooled by pipes filled with water. The flue-gases exit through an electrostatic precipitator for controlling particulates and then up the stack.

Smaller incinerators, known popularly as modular incinerators, have been widely used. These units (Figure 12-7) have two chambers, the first chamber being a "smoldering" pit, where the refuse is combusted with limited air. The fly ash carried off is combusted in an afterburner, using oil or natural gas as supplemental fuel. This process produces a clear stack gas and can reduce the volume of refuse by 90%. It is, however, expensive and dependent on the availability of oil and natural gas for its

Figure 12-7   Modular incinerators used for solid waste processing (courtesy Consumat).

operation. Modular incineration in larger facilities is composed of a series of smaller units, each operating independently.

Shredding solid wastes (also known as pulverizing) and then spreading the material on fields has been successful in a number of places. The organics are too ground up to interest rats, and spreading dries the refuse, thus avoiding odor and fly problems.

The shredded material does not have to be covered with dirt—a significant advantage over the landfill. Landfill operations are at best difficult during wet or freezing weather, and the capacities of many landfill sites are limited not by available volume but by the dirt available for covering the refuse.

Pyrolysis is combustion in the absence of oxygen. One stated advantage of pyrolysis is that the residues have economic value. However, these products—combustible gas, tar and charcoal—have thus far not found acceptance as a raw material. Most of the gas is in fact used in making the process go. The tar is full of water and must be refined. The charcoal is full of glass and metal which must be separated before the charcoal can be useful. This separation, however, makes the charcoal too expensive when compared to charcoal derived from wood.

Pyrolysis reduces the volume considerably, produces a stable end product, and has fewer air pollution problems (small volume of gas to treat). On a large scale, such as for some of our larger cities, pyrolysis as a method of volume reduction has significant advantages over incineration. It also can be combined with sludge disposal, thus solving two major solid waste problems for a community. Such systems however, are still to be proven in full-scale operation.

CONCLUSION

We started this chapter by defining solid waste disposal so it no longer impacts society. This was at one time fairly easy to achieve. In fact, dumping solid waste over the city walls was quite adequate as a method of disposal. In our modern civilization, however, this is no longer possible, and it is becoming increasingly difficult to get rid of the stuff so it no longer impacts society.

One potential solution would be to simply redefine solid waste as a resource, and use it for the production of goods for people. This idea is explored in the next chapter.

PROBLEMS

12.1 Suppose the municipal garbage collectors in a town of 10,000 go on strike, and as a gesture to the community your college or university decides to accept all the city refuse temporarily and pile it on the football field. If all the people did indeed dump the refuse into the stadium, how many days must the garbage men be on strike before the stadium is filled to 1 yard deep? (Note: Assume density of refuse as 300 lb/yd, and assume the dimensions of the stadium as 120 yd long, and 100 yd wide).

12.2 If a town has a population of 100,000, what is the daily production of wastepaper?

12.3 Describe how you would run a composting operation for leaves in your community. Be sure to include the technical, economic and psychological considerations.

12.4 What would some environmental impacts and effects of depositing dewatered (but sloppy wet) sludge from a wastewater treatment plant into a sanitary landfill?

12.5 If all of the organic refuse in a landfill were cellulose, $C_6H_{10}O_5$, how much gas (CO_2 and CH_4) would be produced (m^3/kg refuse)?

12.6 Estimate how much gas you might obtain by capturing the CO_2 and CH_4 from the landfill in your community.

12.7 Choose a place for a 25-acre landfill on the map shown in Figure 12-8. What other information would you need? Justify your selection of the site.

LIST OF SYMBOLS

A = area, m^2
C = total percolation of rain into the soil, mm
E = evaporation, mm
h = depth of landfill, m
P = precipitation, mm
Q = flow rate of gas, m^3/sec
R = radius of influence of a gas withdrawal well, m
R = runoff coefficient
S = storage, mm

Figure 12-8   Map for siting a landfill.

SUGGESTED READING

Breidenbach, A. W. *Composting Municipal Solid Wastes in the U.S.,*
SW–47r, U.S. EPA (1971).

Brunner, D. R., and D. J. Keller. *Sanitary Landfill Design and Operation,* U.S. EPA OSWMP SW–65ts (1972).

Corey, R. C., Ed. *Principles and Practices of Incineration* (New York: Wiley-Interscience, 1969).

Golueke, C. G. *Composting* (Emmaus, PA: Rodale Press, 1972).

Golueke, C. G., and P. H. McGauhey. *Comprehensive Studies of Solid Waste Management,* U.S. Department of Health, Education and Welfare (1970).

Office of Local Government, New York State (Albany). *Municipal Refuse Collection and Disposal* (1964).

Reindl, J. "Managing Gas and Leachate Production on Landfills," *Solid Waste Management* (July 1977).

Remson, I., A. A. Fungaroli and A. W. Lawrence. "Water Movement in an Unsaturated Sanitary Landfill," *J. San. Eng. Div. ASCE* 94(SA2) (1968).

(See also Suggested Reading, Chapter 11)

Chapter 13

Resource Recovery

It is becoming increasingly difficult to find new sources of energy and materials to feed our industrial society. Concurrently, we are finding it more and more difficult to locate solid waste disposal sites, and mainly because of transportation requirements, the cost of disposal is escalating exponentially. These two factors—energy and material shortages, and fewer and more expensive disposal options—have fueled a new technology called resource recovery.

Except for the process of mass burning of raw refuse, discussed in the previous chapter, the recovery of resources from refuse is primarily a quest for purity. Pure materials can be obtained from mixed municipal solid waste in one of two ways:

1. Separation of the materials is performed by the user, the person who decides to discard the various "consumer products."*
2. Separation is performed after the mixed refuse is collected at a central processing facility.

SOURCE SEPARATION

Materials separation by the public, commonly known as source separation, has not been very successful in the United States. There are only two incentives which could be used to convince the public to undertake source separation. The first is regulatory, where the governmental agency *dictates* that only separated material will be picked up. Unfortunately, this is not a successful approach in a democracy, since public officials advocating unpopular regulations can be removed from office.

---

*The word consumer is clearly a misnomer. If we all were true consumers, and not users, there would be no solid waste.

Source separation in totalitarian regimes on the other hand is easy to implement, but this is hardly a compelling argument to abolish democracy.

The second means of achieving cooperation in source separation programs is to appeal to the sense of community spirit and the ethics of environmental concern. Indeed, studies conducted on the feasibility of community source separation programs, in which householders were asked if they would participate in such programs, attained a 90 to 95% positive response. Unfortunately, the *active* response, or the participation in a source separation project, has seldom exceeded 5% of the households. There is a wide gap between what people say they will do (especially if they perceive that the question contains a value component) and how they will actually perform. A further complication is that in some inner cities it is difficult to convince people to put refuse in trash cans, much less convince them to separate the refuse into components. Although source separation appears to be the least expensive and least energy-intensive method of resource recovery, it has not been successfully implemented on a large scale in the United States, and effort therefore has been directed toward separation of mixed and collected refuse.

SOLID WASTE SEPARATION PROCESSES

Most processes for separation of the various materials in refuse rely on a characteristic or property of the specific material as a code and this code is used to separate the material from the rest of the mixed refuse. Before such separation can be achieved, however, the material must be in separate and discrete pieces, a condition clearly not met by most components of mixed refuse. A common "tin can," for example, contains steel in its body, zinc on the seam, a paper wrapper on the outside, and perhaps an aluminum top. Other common items in refuse provide equally or even more challenging problems in separation.

One means of assisting in the separation process is to decrease the particle size of refuse, thus increasing the number of particles and achieving a greater number of "clean" particles. This size reduction step, although not strictly materials separation, is commonly a first step in a solid waste processing facility.

Size Reduction

Commonly called *shredding,* the process of size reduction usually consists of a brute force breakage of particles by swinging hammers in an enclosure. Two types of shredders are common in solid waste processing: the vertical and horizontal hammermills, as shown in Figure 13-1. In the former, the refuse enters the top and must work its way past the rapidly swinging hammers, clearing the space between the hammer tips and the enclosure. Particle size is controlled by adjusting this clearance. In the horizontal hammermill, the hammers swing over a grate which can be changed depending on the size of product required.

Figure 13-2 shows a solid waste processing facility with a conveyor belt leading up to a vertical shredder. The control room is above and to the left of the conveyor. The hammers inside the shredder are shown in Figure 13-3. As the hammers reduce the size of the refuse components, they are themselves worn down. Typically, a set of hammers such as those shown can process 20,000 to 30,000 tons of refuse before having to be replaced.[1]

Shredder Performance and Design

The performance of a shredder is measured with respect to the degree of particle size reduction achieved, and then compared to the cost in power and shredder wear.

The particle size distribution of a nonhomogeneous material like MSW is conveniently expressed by the Rosin–Rammler distribution function,[2] defined as

$$Y = 1 - \exp(-x/x_c)^n$$

where Y = the cumulative fraction of particles (by weight) less than size x,
n = a constant
x_c = the *characteristic particle size,* defined as the size at which 63.2% ($1 - 1/e = 0.632$) of the particles (by weight) are smaller.

This function plots as a straight line on log-log coordinates if $\ln[1/(1-y)]$ is plotted against x, as in Figure 13-4. Often, a shredder performance is specified on the basis of "90% of the particles (by weight) passing a given size." The conversion from characteristic size to 90% passing is possible by recognizing that

[1]Vesilind, P. A., A. E. Rimer and W. A. Worrell. "Performance Testing of a Vertical Hammermill Shredder," *Proc. National Conference on Solid Waste Processing* (New York: American Society of Mechanical Engineers, 1980).

[2]Rosin, P., and E. Rammler. "Laws Covering the Fineness of Powdered Coal." *J. Inst. Fuel,* 7:29–36 (1933).

Figure 13-1   Vertical and horizontal hammermills.

Figure 13-2 A shredding facility, showing the conveyor belt leading to a vertical shredder.

$$x_c = \frac{x}{[\ln(1/1 - Y)]^{1/n}}$$

If $Y = 90\%$, or 0.90, this expression reduces to

$$x_c = \frac{x_{90}}{(2.3)^{1/n}}$$

where x_{90} is the size at which 90% pass. Commonly $n = 1.0$, in which case $x_{90} = 2.3\, x_c$. Some typical values of x_c and n are shown in Table 13-1.

In addition to particle size reduction, shredder performance must be measured in terms of power usage, expressed as specific energy (kWh/ton refuse processed). Figure 13-5 illustrates how solids feed rate, shredder speed, moisture content and size of particles in the feed are interrelated.

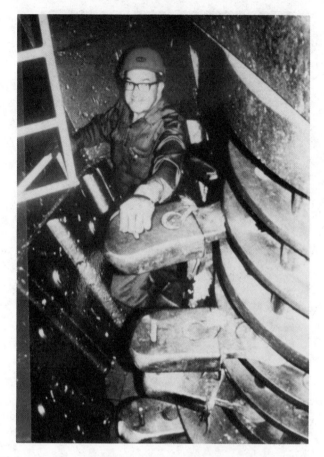

Figure 13-3   Inside a vertical hammermill. The hammers have been worn down by the shredding process. (Photo courtesy of W. A. Worrell.)

A partially empirical expression often used to estimate the power requirements for shredders was developed by Bond.[3] The specific energy, W, required to reduce a unit weight of material with 80% finer than some diameter L_F in micrometers, to a product with 80% finer than some diameter L_P in micrometers, is expressed as

$$W = 10\,W_i \left(\frac{1}{\sqrt{L_P}} - \frac{1}{\sqrt{L_F}} \right)$$

[3] Bond, F. C. "The Third Theory of Comminution," *Trans. AIME* 193:484 (1952).

Figure 13-4 Rosin-Rammler particle size distribution equation. In this sample, n = 1.0 and x_c = 1.6.

Table 13-1. Rosin-Rammler Exponents for Typical Shredder Installation[a]

Location	n (dimensionless)	x_c (cm)
Washington, DC	0.689	2.77
Wilmington, DE	0.629	4.16
Charleston, SC	0.823	4.03
St. Louis, MO	0.995	1.61
Pompano Beach, FL	0.587	0.67

[a] Source: Stratton, F. E., and H. Alter. "Application of Bond Theory to Solid Waste Shredding," *J. Environ. Eng. Div., ASCE,* 104 (EE1)(1978).

where W_i is the *Bond work index,* a factor that is a function of the material processed for a given shredder, and a function of shredder efficiency for a given material. W has the dimension of kWh/ton if L_P and L_F are in micrometers. The factor 10 also corrects for the dimensions. Typical values of the Bond work index for municipal solid waste (MSW) and other specific materials are listed in Table 13-2.

It is convenient to combine the Rosin–Rammler particle size distribution equation with the Bond work index concept, by first noting that

$$(1 - Y) = \exp(-x/x_c)^n$$

where $(1 - Y)$ is the cumulative fraction *larger* than some size x. If $(1 - Y) = 0.2$, $x = L_P$ the screen size through which 80% of the product passes. Solving for L_P

Figure 13-5   (a) The effect of feed rate and rotor speed on specific energy used in shredding. (b) The effect of feed rate and feed particles size on specific energy. (c) The correlation between moisture content in the refuse and rotor speed.

$$0.2 = \exp\left(\frac{L_P}{x_c}\right)^n$$

$$L_P = x_c(1.61)^{1/n}$$

Substituting this back into the Bond work index equation,

$$W = \frac{10\,W_i}{[x_c(1.61)^{1/n}]^{1/2}} - \frac{10\,W_i}{[L_F]^{1/2}}$$

Table 13-2.   Typical Bond Work Indexes[a]

Shredder Location	Material	Bond Work Index (kWh/ton)
Washington, DC	MSW	463
Wilmington, DE	MSW	451
Charleston, SC	MSW	400
St. Louis, MO	MSW	434
Pompano Beach, FL	MSW	405
Washington, DC	paper	194
Washington, DC	steel cans	262
Washington, DC	aluminum cans	654
Washington, DC	glass	8

[a]Source: Stratton, F. E., and H. Alter. "Application of Bond Theory to Solid Waste Shredding," *J. Environ. Eng. Div., ASCE,* 104 (EE1)(1978).

Example 13.1

Assuming $W_i = 400$ and x_c of the product is 1.62 cm, $n = 1$ and $L_F = 25$ cm (about 10 in., a realistic estimate of raw refuse), find the expected power requirements for a shredder processing 10 ton/hr.

$$W = \frac{10\,W_i}{[x_c(1.61)^{1/n}]^{1/2}} - \frac{10\,W_i}{[L_F]^{1/2}}$$

$$= \frac{10\,(400)}{[16,200\,(1.61)^{1/1}]^{1/2}} - \frac{10\,(400)}{[250,000]^{1/2}} = 16.7 \text{ kWh/ton}$$

(Note: x_c and L_F must be in micrometers, μm.)

The power requirement is 16.7 kWh/ton × 10 tons/hr = 167 kW.

General Expressions for Materials Recovery

In the separation of any one material from a mixture, the separation is termed *binary,* in that only two outputs are required. When a device is to separate more than one material from a mixture, the process is termed *polynary* separation.

Figure 13-6 shows a binary separator receiving a mixed feed of x_0 and y_0. The objective is to separate out the x fraction. The first exit stream is to have the x component, but since the separation is not perfect, it also

Figure 13-6   Definition sketch of a binary separator.

contains contamination in the amount of y_1. This stream is called the *product* or *extract,* while the second stream, containing mostly the y, but also some x, is known as the *reject*. The recovery of x can be expressed as

$$R_{(x_1)} = \left(\frac{x_1}{x_0} \right) 100$$

when $R_{(x_1)}$ is the recovery of x in the first output stream, as percent.

This expression, however, is not an adequate descriptor of the performance of the binary separator. Imagine a situation where the separator is turned *off,* i.e., all of the feed stream goes to the first output (extract). That is, $x_0 = x_1$, since there is no x_2. This would make $R_{(x_1)} = 100\%$. But obviously nothing has been done to the feed. Accordingly, a second requirement is the *purity* of the extract stream, defined as

$$P_{(x_1)} = \left(\frac{x_1}{x_1 + y_1} \right) 100$$

Similarly, purity is not an adequate descriptor of binary separator performance, since it might be possible to extract only a very small amount of x, in a pure state, but have the recovery ($R_{(x_1)}$) be very small, a clearly unacceptable condition from the process engineering standpoint. It is thus necessary to describe a materials separation device by *both* the recovery and purity.

This requirement, of course, makes it difficult to express the performance of a separation device as a single-valued parameter. This problem can be resolved by defining a term for binary separator *efficiency* as[4]

---

[4] Worrell, W. A., and P. A. Vesilind. "Testing and Evaluation of Air Classifier Performance," *Resource Recovery and Cons.,* 4:247–259 (1979).

$$E_{(x,y)} = \left(\frac{x_1}{x_0} \right)\left(\frac{y_2}{y_0} \right) 100$$

Example 13.2

A binary separator (magnet) is to separate a product (ferrous materials) from a feed stream (shredded refuse). The feed rate to the magnet is 1000 kg/hr and it contains 50 kg ferrous (product x) and 950 kg of other material (reject y). Of the 40 kg in the product stream, 35 kg is ferrous. What is the percent recovery of ferrous, the purity of the product, and the overall efficiency?

$$R_{(x_1)} = \left(\frac{x_1}{x_0} \right) 100 = \left(\frac{35}{50} \right) 100 = 70\%$$

$$P_{(x_1)} = \left(\frac{x_1}{x_1 + y_1} \right) 100 = \left(\frac{35}{40} \right) 100 = 88\%$$

$$E_{(x,y)} = \left(\frac{x_1}{x_0} \right)\left(\frac{y_2}{y_0} \right) 100 = \left(\frac{35}{50} \right)\left(\frac{[1000-40] - [50-35]}{950} \right) 100 = 70\%$$

Screens

Screens separate materials solely by size and do not identify the material by any other property. Consequently, screens are most often used in resource recovery as a classification step prior to a materials separation process. For example, it is possible technically (if not economically) to sort glass into clear and colored fractions by optical coding. This process, however, requires that the glass be of a given size, and screens can be used to produce such a feed to an optical sorter.

The most widely used screen in resource recovery is the *trommel,* pictured in Figure 13-7. The charge inside a trommel behaves in three distinctly different ways, depending on the speed of rotation. At slow speeds, the trommel material is *cascading,* i.e., not being lifted, but simply rolling back. At higher speed, *cataracting* occurs, where the centrifugal force carries the material up to the side and then it falls back. At even higher speeds, *centrifuging* occurs, where the material adheres to the inside of the trommel. Obviously, the efficiency of a trommel is enhanced when the particles have the greatest opportunity to drop through the holes, and this occurs during the cataracting type of operation. Trommel speed is thus often designed as some fraction of the criti-

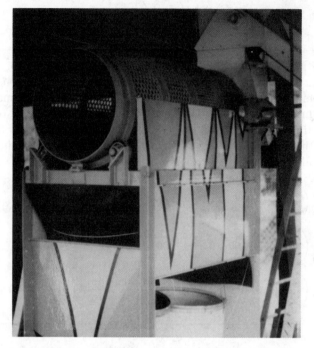

Figure 13-7   Trommel screen.

cal speed, defined as that rotational speed at which the materials will just begin centrifuging, and calculated as

$$n_c = \sqrt{\frac{g}{4\pi^2 r}}$$

where n_c = critical speed, rotations/sec
        g = gravitational acceleration, cm/sec^2
        r = radius of drum, cm

Example 13.3
   Find the critical speed for a 3-m-diameter trommel.

$$n_c = \left(\frac{g}{4\pi^2 r}\right)^{1/2}$$

$$= \left(\frac{980}{4(3.14)^2(150)}\right)^{1/2} = 0.407 \text{ rotation/sec}$$

Air Classifiers

Materials can be separated according to their aerodynamic properties. In shredded MSW, most of the aerodynamically light materials are organic, and most of the heavy materials are inorganic, thus air classification can produce a refuse-derived fuel (RDF) superior to unclassified shredded refuse.

Most air classifiers are similar to the unit pictured in Figure 13-8. The

Figure 13-8   Air classifier.

fraction escaping with the air stream is the product (or the overflow), the fraction falling out the bottom is the reject (or underflow). The recovery of organics by air classification is adversely influenced by two factors:

1.   Not all organics are aerodynamically light, and not all inorganics are aerodynamically heavy.
2.   Perfect separation of heavies and lights is difficult due to the stochastic nature of material movement in the classifier.

The first factor is illustrated in Figure 13-9, where the terminal settling velocity (that air velocity at which the particle will just begin to rise with the air stream) is plotted against the fraction of particles of various materials.[5] Regardless of the air velocity chosen, there can never be a com-

[5]McNabb, M. "A Simple Test for Air Classifier Design," MS Thesis, Duke University, Durham, NC (1980).

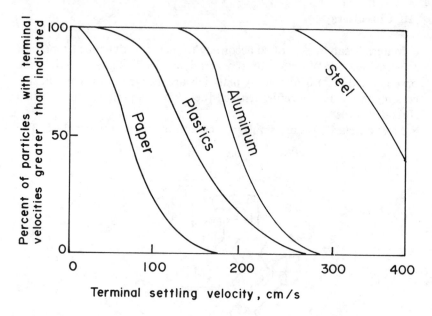

Figure 13-9   Terminal settling velocities of various components of MSW. Note that no single air speed will result in complete separation of organics (plastics and paper) from inorganics (steel and aluminum). Idealized curves. (Source: Ref. 5.)

plete separation of the organics (plastics and paper) from the inorganics (aluminum and steel).

The second factor is illustrated in Figure 13-10, which shows the efficiency (as defined above) versus the feed rate to the classifier. As solids loading increases, more particles that would have exited with the light fraction might get caught in the underflow stream, and vice versa.

Recovery of organics in the product stream can be plotted against the recovery of inorganics in the reject stream. The ideal performance curve can be obtained from terminal settling plots like Figure 13-9 and by calculating $R_{(x_1)}$ and $R_{(y_2)}$ at different air velocities. The actual performance curve is obtained by measuring $R_{(x_1)}$ and $R_{(y_2)}$ for a continuous classifier at different air speeds. The closer the two curves are, the more effective is the air classifier. In addition, as noted in Figure 13-10, the performance deteriorates as the loading on the air classifier is increased. With high feed rates the throat of the classifier becomes clogged and a clean separation is no longer possible.

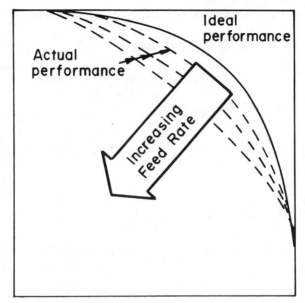

Recovery of inorganics in the
reject stream

Figure 13-10   Actual and ideal performance of air classifiers (idealized curves).
(Source: Reference 5.)

Example 13.4
If an air classifier operates with an air velocity of 200 cm/sec, and the feed contains equal amounts of paper, plastics, aluminum and steel having terminal settling curves as shown in Figure 13-9, what would be the recovery of organics and the purity of the product?

At 200 cm/sec, the fractions of the components that report as the overflow (product) are (from Figure 13-9):

Paper	100%
Plastics	90%
Aluminum	50%
Steel	0%

The total organics in the product is

$$(100)(1/4) + (80)(1/4) = 45\% \text{ of feed}$$

The total inorganics in the product is

$$(50)(1/4) + (0)(1/4) = 12\% \text{ of the feed}$$

The recovery of organics is thus

$$R_{org} = \frac{45}{25 + 25} \times 100 = 90\%$$

The purity is

$$P_{product} = \frac{45}{45 + 12} \times 100 = 79\%$$

Magnets

Ferrous material is removed using magnets which continually extract the ferrous and reject the remainder. Two types of magnets are shown in Figure 13-11. With the belt magnet, recovery of ferrous is enhanced by placing the belt close to the refuse, but this also decreases the purity of the product. A major problem in using belt magnets is the depth of the refuse on the conveyor belt. The heavy ferrous particles tend to settle to the bottom of the refuse carried on a conveyor, and these are then the farthest away from the magnet.

Other Separation Equipment

Countless other unit operations for materials handling and storage have been tried. Jigs have been used for removing glass, froth flotation has been successfully employed for separating glass from ceramics, eddy current devices have recovered aluminum in commercial quantities, etc. As resource recovery operations evolve, more and better materials separation and handling equipment will be introduced. A schematic of a typical materials separation facility is presented in Figure 13-12.

Energy Recovery from the Organic Fraction of MSW

The organic fraction of refuse is, as mentioned earlier, a useful secondary fuel. This shredded and classified product can be used in exist-

Figure 13-11 Two types of magnets used in resource recovery: (a) belt magnet, (b) drum magnet.

ing power plants either as a supplemental fuel with coal, or fired as the sole fuel in separate boilers. A cross section of a typical heat recovery boiler is shown in the previous chapter, Figure 12-6. In smaller installations, modular heat recovery boilers, also described in Chapter 12, have found wide use.

In the combustion of an organic material, the presumed reaction is

$$(HC)_x + O_2 \rightarrow CO_2 + H_2O + heat$$

The process is much more complex, however, since not all the hydrocarbons are converted to carbon dioxide and water, and other components of the fuel such as nitrogen and sulfur also are oxidized. In the case of nitrogen, the product of combustion is nitrogen oxides

$$N_2 + O_2 \rightarrow 2NO$$

$$2NO + O_2 \rightarrow 2NO_2$$

Figure 13-12   Schematic of typical materials separation facility for refuse processing.

Nitrogen dioxide is an important component in the formation of photochemical smog.

Sulfur is also oxidized to sulfur oxides

$$S + O_2 \rightarrow SO_2$$

$$2SO_2 + O_2 \rightarrow 2SO_3$$

and SO_3 combines with water to form sulfuric acid,

$$SO_3 + H_2O \rightarrow H_2SO_4$$

which is, of course, highly corrosive and a hazard to health and vegetation. (See Chapter 17.)

The theoretical amount of oxygen required for combustion is known as *stoichiometric oxygen* (in terms of air it is *stoichiometric air*) and is calculated from the chemical reaction, as in the following example.

Example 13.5

If carbon is combusted, as

$$C + O_2 \rightarrow CO_2 + heat$$

how much air is required? One mole of oxygen is required for each mole of carbon used. The molecular weight of carbon is 14 and of oxygen is $2 \times 16 = 32$. Hence one gram of C requires

$$\frac{32}{14} = 2.28 \text{ g oxygen}$$

Since air is about 21% oxygen by volume and 23.15% by weight, the total amount of air required to combust one gram of carbon is

$$\frac{2.28}{0.2315} = 9.87 \text{ g air}$$

The yield of energy from combustion is measured as the calories of heat liberated per unit weight of material burned. This is known as the *heat of combustion,* or in engineering shorthand as the *heat value,* and is measured in *calorimeters,* in which a small sample of the fuel is placed in a stainless steel bomb under a high pressure of pure oxygen and then fired. The heat is transferred to the water in which a bomb is placed and the rise in temperature measured. With knowledge of the mass of the water, the energy liberated during combustion can be calculated. In SI units the heat value is expressed as kJ/kg; common engineering units express it as Btu/lb. Table 13-3 lists some heats of combustion for common hydrocarbons and gives some typical values for refuse and RDF.

Table 13-3.   Typical Values of Heats of Combustion

Fuel	Heat of Combustion	
	(kJ/kg)	(Btu/lb)
Carbon (to CO_2)	32,800	14,100
Hydrogen	142,000	61,100
Sulfur (to SO_2)	9,300	3,980
Methane (CH_4)	55,500	23,875
Raw Refuse	9,300	4,000
RDF (Air Classified)	18,600	8,000

In the design of any combustion operation, the system needs to be analyzed for thermal as well as materials balance. The latter requires simply that the mass of various inputs (air and fuel) must equal the mass of the outputs (stack emissions and bottom ashes). In the case of a thermal balance, the heat input must equal the output plus losses. Typical thermal balance calculations are shown in Example 13.6.

Example 13.6

Assume a processed refuse containing 20% moisture and 60% organics. It is to be fed to a boiler at a rate of 1000 kg/hr. From a calorimetric analysis of the refuse, it was determined that a dry sample has a heat value of 19,000 kJ/kg.

The heat from the combustion of the RDF is

$$H_{comb} = 19,000 \text{ kJ/kg} \times 1000 \text{ kg/hr}$$

$$= 19 \times 10^6 \text{ kJ/hr}$$

The RDF organic fraction contains hydrogen, which becomes water. The heat of combustion value above includes the latent heat of vaporization (the hydrogen in the sample all condenses to water, thus yielding heat). Since this does not occur in full-scale combustion (i.e., the water exists as steam), the latent heat of vaporization must be considered a heat loss. Assuming the organics had 50% hydrogen, the latent heat loss due to vaporization is

$$H_{vap} = 1000 \text{ kg/hr} \times 0.6 \text{ (organics)} \times 0.5 \text{ (hydrogen)}$$

$$\times 2420 \text{ kJ/kg (latent heat of vaporization for water)}$$

$$= 0.726 \times 10^6 \text{ kJ/hr}$$

The RDF also contains moisture which is vaporized.

$$H_{mois} = 1000 \text{ kg/hr} \times 0.2 \text{ (moisture)} \times 2420 \text{ kJ/kg} = 0.484 \times 10^6 \text{ kJ/hr}$$

There is heat loss due to radiation, usually assumed as 5% of the heat input, or

$$H_{rad} = 19 \times 10^6 \text{ kJ/hr} \times 0.05 = 0.95 \times 10^6 \text{ kJ/hr}$$

Not all of the organics will combust. Assume that the ashes contain 10% of the organics, so that the heat loss is

$$H_{noncom} = (10/60) \times 19 \times 10^6 \text{ kJ/hr} = 3.17 \times 10^6 \text{ kJ/hr}$$

The stack gases also contain heat, and since the amount is unknown, it is usually calculated as the difference, as

heat input = heat output

$$H_{comb} = (H_{vap} + H_{mois} + H_{rad} + H_{noncom}) + H_{stack}$$

$$19 \times 10^6 = (0.726 + 0.484 + 0.95 + 3.17) \times 10^6 + H_{stack}$$

$$H_{stack} = 13.67 \text{ kJ/hr}$$

So the heat lost in the stack gases is 13.67 kJ/hr. This heat can be recovered by running cold water into the boiler (the water wall tubes) and producing steam. If 2000 kg/hr of steam, at a temperature of 300°C and a pressure of 4×10^3 kPa is required, and the temperature of the boiler water is 80°C, calculate the heat loss in the stack gases. Heat in the boiler water is

$$H_{wat} = 2000 \text{ kg/hr (water)} \times (80 + 273)°\text{K} \times 0.00418$$

$$\text{kJ/kg-}°\text{K} = 2951 \text{ kJ/hr}$$

(Note: the specific heat of water is 0.00418 kJ/kg-°K.)
The heat in the steam, at 300°C and 4×10^3 kPa (see steam tables in any thermodynamics textbook) is 2975 kJ/kg

$$H_{steam} = 2000 \text{ kg/hr} \times 2975 \text{ kJ/kg} = 5.95 \times 10^6 \text{ kJ/hr}$$

The heat balance then yields

In = Out

Heat in combustion
Heat in boiler water
$\Big\} = \Big\{$
Latent heat due to vaporization
Heat loss due to radiation
Heat in ashes
Heat in steam
Heat in stack gases

By subtraction, the heat in the stack gases is

$$H_{comb} + H_{wat} = (H_{vap} + H_{mois} + H_{rad} + H_{noncom} + H_{steam}) + H_{stack}$$

$$(19 + 0.0002) \times 10^6 = (0.726 + 0.484 + 0.95 + 3.17 + 5.95) \times 10^6 + \text{heat in stack gases}$$

$$\text{Heat in stack gases} = 7.72 \times 10^6 \text{ kJ/hr}$$

Combustion of the organic fraction of refuse is not the only means of extracting useful energy from it. This extraction can also be by chemical or biochemical means.

Using a process known as *acid hydrolysis,* the cellulose fraction of RDF can be treated so as to produce methane gas. Other chemical processes are presently being developed for producing alcohol from RDF. Alcohol can be added to gasoline and thus serve as substitute liquid fuels. A mixture of 80% gasoline and 20% alcohol will not require readjustments in an internal combustion engine.

The two types of biochemical processes used for extracting useful products from refuse are anaerobic and aerobic decomposition. In the anaerobic system, refuse is mixed with sewage sludge and the mixture digested. Operational problems have made this process impractical on a large scale, although single household units which combine human excreta with refuse have been used.

Aerobic decomposition of refuse is better known as *composting* and results in the production of a useful soil conditioner which has moderate fertilizer value. The process is exothermic, and in a household level has been used as a means of producing hot water for heating homes. On a community scale, composting can be a mechanized operation using an aerobic digester (Figures 13-13 and 13-14) or it may be a low-technology operation, using long rows of shredded refuse known as *windrows.*

Windrows are normally 3 m (10 ft) wide at the base and 1.5 m (4 to 6 ft) high. Under these conditions, sufficient moisture and oxygen are available to support aerobic life. The piles must be turned periodically to allow sufficient oxygen to penetrate to all parts of the pile. A modification of this system is to blow air into the piles. This is commonly known as *static pile composting* (Figure 13-15).

Temperatures within a windrow approach 140°F, due entirely to biological activity. The pH will approach neutrality after an initial drop. With most wastes, additional nutrients are not needed. The composting of bark and other materials, however, is successful only with the addition of nitrogen and phosphorus nutrients.

Moisture must usually be controlled. Excessive moisture makes it difficult to maintain aerobic conditions while a dearth inhibits biological life. A 40–60% moisture content is considered desirable.

Figure 13-13   Schematic of a mechanical composting operation.

There has been some controversy over the use of inoculants, freeze-dried cultures, used to speed up the process. Once the composting pile is established, which requires about two weeks, the inoculants have not proven to be of any significant value. Most municipal refuse contains all the organisms required for successful composting, and "mystery cultures" are thus not needed.

The end point of a composting operation can be measured by noting a drop in temperature. The compost should have an earthy smell, similar to peat moss, and should have a dark brown color.

Although compost is an excellent soil conditioner, it is not widely used by U.S. farmers. Inorganic fertilizers are cheap and easy to apply and most farms are located where soil conditions are good. The plentiful food supply in most developed countries does not dictate the use of marginal lands where compost would be of real value.

CONCLUSION

It is obvious that we must attack the solid waste problem from both ends—from the source as well as from the disposal methods. Solid wastes, often called the "third pollution," are only now being considered a problem equal in magnitude to that of air and water pollution.

It has been suggested that one solution to the solid waste problem is the development of truly biodegradable forms of materials such as plastics and glass. A scientist in England has developed a plastic which will disintegrate when exposed to ultraviolet light. Direct sunlight will thus cause

Figure 13-14   A community composting operation in Auckland, New Zealand, showing the rotary digester and windrows.

Compost Pile

Figure 13-15   Static pole composting.

disintegration, while light filtered through a window pane will not. The materials thus broken down will degrade biologically.

We are still many years away from the development and use of fully recyclable or biodegradable materials. The only truly disposable package available today is the ice cream cone.

PROBLEMS

13.1 Estimate the critical speed for a trommel screen, 3 m in diameter, processing refuse so as to run 25% full.

13.2 What air speed is required in an air classifier to suspend a piece of glass 1 mm in diameter?

13.3 Refer to the curves shown in Figure 13-16. By replotting the curves, estimate Rosin–Rammler constants n and x_c. Compare these to the data in Table 13-1.

13.4 If the Bond work index is 400 kWh/ton, what is the power requirement to process 100 ton/hr of the refuse described in Figure 13-15?

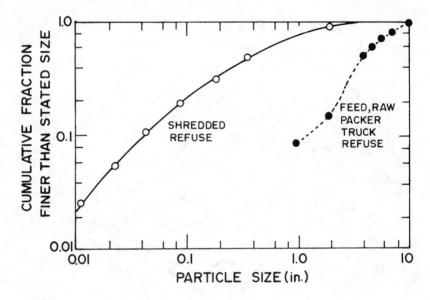

Figure 13-16 Feed and product curves in shredding (Problem 13.3).

13.5 An air classifier performance is shown below

	Organics (kg/hr)	Inorganic (kg/hr)
Feed	80	20
Product	60	10
Reject	20	10

Calculate the recovery, purity and efficiency.

13.6 Estimate the heating value of MSW based on the ultimate analysis shown in Table 11-1.

13.7 A power plant burns 100 ton/hr of coal. How much air is needed if 50% excess air is used?

13.8 Using the data in Example 13.6, what percent of the heating value can be saved by reducing the moisture content of refuse to zero?

13.9 In Example 13.6, what fraction of the coal is "wasted" due to the loss of heat in the stack gases (if steam is produced)?

13.10 What is the "wasted" coal in Problem 13.9 heat equivalent in barrels of oil?

LIST OF SYMBOLS

E = efficiency of materials separation
L_F = feed particle size, 80% finer than, microns
L_P = product size, 80% finer than, microns
MSW = municipal solid waste
n_c = critical speed of a trommel, rotations/sec
P_x = purity of a product x, %
r = radius, cm
RDF = refuse-derived fuel
R_x = recovery of a product x, %
W = specific energy, kWh/ton
W_i = Bond work index, kWh/ton
x = particle size, m
x_c = characteristic particle size, m
x_0, y_0 = mass per time of feed to a materials separation device
x_1, y_1 = mass per time of components x and y exiting from a material separation device through exit stream 1
x_2, y_2 = mass per time of components x and y exiting from a material separation device through exit stream 2
Y = cumulative fraction of particles (by weight) less than some specified size

SUGGESTED READING

Alter, H., and J. J. Dunn, Jr. *Solid Waste Conversion to Energy* (New York: Marcel Dekker, Inc., 1980).

Barton, A. F. M. *Resource Recovery and Recycling* (New York: John Wiley & Sons, 1979).

Cheremisinoff, P. N., and A. C. Morresi. *Energy from Solid Waste* (New York: Marcel Dekker, 1976).

Combustion Fundamentals for Waste Incineration (New York: American Society of Mechanical Engineers, 1974).

"Fourth Report to Congress: Resource Recovery and Waste Reduction," U.S. EPA, OSWMP (1976).

Frankiewicz, T. C. Ed. *Energy from Waste* (Ann Arbor, MI: Ann Arbor Science Publishers, Inc., 1980).

Haug, R. T. *Compost Engineering* (Ann Arbor, MI: Ann Arbor Science Publishers, Inc., 1980).

Kesler, S. E. *Our Finite Mineral Resources* (New York: McGraw-Hill Book Company, 1976).

Kirov, N. Y. *Waste Management, Control Recovery and Reuse* (Ann Arbor, MI: Ann Arbor Science Publishers, Inc., 1975).

Stratton, F. E., and H. Alter. "Application of Bond Theory to Solid Waste Shredding," *J. Environ. Eng. Div. ASCE* 104 (EEI) (1978).

Taggart, A. F. *Handbook of Mineral Dressing* (New York: John Wiley & Sons, Inc., 1974).

Trezek, G. J., and G. Savage. "Report on a Comprehensive Refuse Comminution Study," *Waste Age* (July 1975).

Vesilind, P. A., and A. E. Rimer. *Unit Operations in Resource Recovery Engineering* (Englewood Cliffs, NJ: Prentice-Hall, 1981).

Chapter 14

Hazardous Waste

For centuries, chemical wastes have been the necessary by-products of developing societies. Here a disposal site, there a disposal site, everywhere a disposal site—all with little or no attention to potential impacts on groundwater quality, runoff to streams and lakes, and skin contact as children played hide-and-seek in a forest of abandoned 55-gallon drums. Engineering decisions here historically were made by default; little or no handling/processing/disposal planning at the corporate or plant level necessitated quick and dirty decision by mid- and entry-level engineers at the end of production processes. These production engineers solved disposal problems by simply piling or dumping these waste products "out back."

Attitudes began to change in the 1960s and 1970s. As other chapters of this text indicate, air, water and land were no longer viewed as commodities to be polluted with the problems of cleanup freely passed to neighboring towns or future generations. Individuals responded with court actions against pollution, and governments responded with revised local zoning ordinances, updated public health laws and new major Federal Clean Air and Clean Water Acts. In 1976, the Federal Resource Conservation and Recovery Act (RCRA) was enacted to give the U.S. Environmental Protection Agency specific authority to regulate the generation, transport and disposal of hazardous waste. As the 1980s begin, we find that engineering knowledge and expertise has not kept pace with this awakening to the necessity to adequately manage hazardous wastes. This chapter discusses the state of knowledge in the field of hazardous waste engineering, tracing the quantities of wastes generated in the nation from handling and processing options through transportation controls to resource recovery and ultimate disposal alternatives.

MAGNITUDE OF THE PROBLEM

Over the years, the term "hazardous" has evolved in a confusing setting as different groups advocate many criteria for classifying a waste as "hazardous." Within the federal government, different agencies used such descriptions as toxic, explosive and radioactive to label a waste as hazardous. Different states had other classification systems, as did the National Academy of Sciences and the National Cancer Institute. These past systems are displayed in Table 14-1. Selected classification criteria are described in more detail in Table 14-2.

The federal government attempted to impose a nationwide classification system under the implementation of RCRA, wherein a hazardous waste is defined by the degree of ignitability, corrosivity, reactivity and/or toxicity. This definition includes acids, toxic chemicals, explosives and other harmful or potentially harmful waste. In this chapter, this will be the applicable definition of hazardous waste. Radioactive wastes are excluded. Such wastes obviously are hazardous, but because their generation, handling, processing and disposal differ so drastically from nonnuclear hazards, the radioactive waste problem is addressed separately in Chapter 15.

Given this somewhat limited definition, more than 60 million metric tons, by wet weight, of hazardous waste are generated annually throughout the United States. More than 60% is generated by chemical and allied products industry, and the machinery, primary metals, paper and glass products industries each generate between 3 and 10% of the nation's total. Approximately 60% of the hazardous waste is liquid or sludge. Major generating states, including New Jersey, Illinois, Ohio, California, Pennsylvania, Texas, New York, Michigan, Tennessee and Indiana, contribute more than 80% of the nation's total production of hazardous waste, and most waste is disposed of on the generator's property. More than 80% of all disposal is in inadequately designed and operated pits, ponds, landfills and incinerators.

A hasty reading of these hazardous waste facts points to several interesting, though shocking, conclusions. First, most hazardous waste is generated and inadequately disposed of in the eastern portion of the country. In this region, the climate is wet with patterns of rainfall that permit infiltration and/or runoff to occur. Infiltration permits the transport of hazardous waste into groundwater supplies, and surface runoff leads to the contamination of streams and lakes. Secondly, most hazardous waste is generated and disposed of in areas where people rely on aquifers for drinking water. Major aquifers and well withdrawals underlie areas where the wastes are generated. Thus, the hazardous waste problem is compounded by two considerations: the wastes are generated and disposed of

Table 14-1.   Historical Definitions of Hazardous Waste

System	Toxicological	Flammability	Explosives	Corrosive	Reactivity	Oxidizing material	Radioactive	Irritant	Strong sanitizer	Bioconcentration	Carcinogenic, mutagenic, teratogenic	Sufficient quantity
Title 15, U.S. Code, Sec. 1261	x	x		x			x	x	x			x
CPSC-Title 16, CFR, Part 1500	x	x		x			x	x	x			x
Food, Drug, and Cosmetic Act	x									x	x	x
DOT-Title 49, CFR, Parts 100–199	x	x	x	x		x	x	x				x
Pesticides-Title 40, CFR, Part 162	x	x								x	x	
Ocean Dumping-Title 40, CFR, Part 227	x									x	x	x
NIOSH-Toxic Substances List	x										x	
Drinking Water Standards	x									x	x	
FWPCA Sec. 304 (a)(1)	x									x	x	
Sec. 307 (a)	x									x	x	x
Sec. 311 (b)(2)(A)	x											x
Clean Air Act-Sec. 112	x								x		x	
California State List	x	x	x	x					x	x		
National Academy of Sciences	x	x			x							
TRW Systems Group	x	x	x	x			x			x		x
Batelle Memorial Institute	x	x	x		x	x	x	x		x	x	
Booz-Allen Applied Research, Inc.	x	x	x		x							x
Dept. of the Army	x											
Dept. of the Navy	x	x	x	x	x	x	x					
National Cancer Institute	x									x	x	x

Table 14-2.   Classification Criteria for Hazardous Waste

CRITERIA	DESCRIPTION
1. Bioconcentration (Bioaccumulation)	The process by which living organisms concentrate an element or compound in levels in excess of the surrounding environment.
2. LD_{50} (Lethal Dose 50)	A calculated dose of a chemical substance which is expected kill to 50% of a population exposed through a route other than respiration (mg/kg of body weight).
3. LDC_{50} (Lethal Dose Concentration 50)	A calculated concentration of a chemical substance which, when following the respiratory route, will kill 50% of a population during a 4-hour exposure period (ambient concentration in mg/l).
4. Phytotoxicity	The ability of a chemical substance to cause poisonous reactions in plants.

in areas where it rains and in areas where people rely on aquifers for supplies of drinking water.

WASTE PROCESSING AND HANDLING

Waste processing and handling are key concerns as a hazardous waste begins its journey from the generator site to a secure long-term storage facility. Ideally, the waste can be stabilized or detoxified or somehow rendered harmless in a treatment process similar to those outlined briefly below.

Chemical Stabilization/Fixation. In these processes, chemicals are mixed with waste sludges, the mixture is pumped onto land, and solidification occurs in several days or weeks. The result is a chemical nest that entraps the waste, and pollutants such as heavy metals may be chemically bound in insoluble complexes. Proponents of these processes have argued for building roadways, dams and bridges using a selected cement as the fixing agent. The environmental adequacy of the processes has not been documented, however, as long-term leaching and defixation potentials are not well understood.

Volume Reduction. Volume reduction is usually achieved in an incineration process. This process takes advantage of the large organic fraction of waste being generated by many industries, but may lead to secondary problems for hazardous waste engineers: air emissions in the stack of the incinerator and ash production in the base of the incinerator.

Both by-products of incineration must be addressed in terms of legal, cost and ethical constraints. Because incineration is often considered a very good method for the ultimate disposal of hazardous waste, we discuss it in some detail later in this chapter.

Waste Segregation. Prior to shipment to a processing or long-term storage facility, wastes are segregated by type and chemical characteristics. Similar wastes are grouped in a 55-gallon drum or group of drums, segregating liquids like acids from solids such as contaminated lab clothing and animal carcasses. Waste segregation is generally practiced to prevent undesirable reactions at disposal sites, and may lead to economies of scale in the design of detoxification or resource recovery facilities.

Detoxification. Numerous thermal, chemical and biological processes are available to detoxify chemical wastes. Options include:

- neutralization
- ion exchange
- incineration
- pyrolysis
- aerated lagoons
- waste stabilization ponds

These technologies are extremely waste-specific; ion exchange obviously doesn't work for every chemical and some forms of heat treatment can be prohibitively expensive for sludges that have a high water content. It's time to call in the chemical engineers whenever detoxification technologies are being considered.

Degradation. Methods exist which chemically degrade some hazardous wastes and render them safer, if not completely safe. Chemical degradation processes, which are very waste-specific, include hydrolysis, to destroy organophosphorus and carbonate pesticides, and chemical dechlorination, to destroy some polychlorinated pesticides. Biological degradation generally involves incorporating the waste into the soil. Landfarming, as it has been termed, relies on healthy soil microorganisms to metabolize the waste components. Such landfarming sites must be strictly controlled for possible water and air pollution that results from overactive or underactive organism populations. For the most part, degradation of hazardous waste is in the research and development stages.

Encapsulation. A wide range of material is available to encapsulate hazardous waste. Ranging from the basic 55-gallon steel drums utilized throughout the nation, options include:

- concrete
- asphalt
- plastics

Several layers of different materials are often recommended, such as a steel drum coated with an inch or more of polyurethane foam to prevent rust.

TRANSPORTATION OF HAZARDOUS WASTES

Hazardous wastes are transported across the nation on trucks, rail flatcars and barges. Because many hazardous wastes are often generated in relatively small quantities, truck transportation, and often small-truck transportation, is a highly visible and constant threat to public safety and the environment. There are four basic elements in the control strategy for the movement of hazardous waste from a generator.

1. *Haulers.* Major concerns over hazardous waste haulers include operator training, insurance coverage and special registration of transport vehicles. Handling precautions include gloves, face masks and coveralls for workers, as well as registration of handling equipment to control future use of the equipment and avoid situations where hazardous waste trucks today are used to carry produce to market tomorrow. Schedules for relicensing haulers and checking equipment are part of an overall program for ensuring proper transport of hazardous wastes.

2. *Hazardous Waste Manifest.* The concept of a cradle-to-grave tracking system has long been considered key to proper management of hazardous waste. This "bill of lading" or "trip ticket" ideally accompanies each barrel of waste and describes the content of each barrel to its recipient. Copies of the manifest are submitted to generators and state officials so all parties know each waste has reached its desired destination in a timely manner. This system serves four major purposes: (1) provides the government with a means of tracking waste within a given state and determining quantities, types and locations where the waste originates and is ultimately disposed; (2) certifies that wastes being hauled are accurately described to the manager of the processing/disposing facility; (3) provides information for recommended emergency response if a copy of the manifest is not returned to the generator; and (4) provides a data base for future planning within a state. Figure 14-1 illustrates one possible routing of copies of a selected manifest. In this example, the original manifest and five copies are passed from the state regulatory agency to the generator of the waste. Copies accompany each barrel of waste that leaves the generating site and are signed and mailed to the respective locations to indicate the transfer of the waste from one location to another.

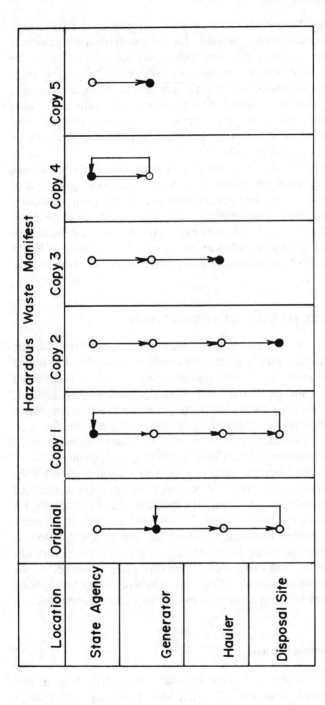

KEY:

O = Transshipment Point: for signature and relay with waste   shipment to next location

● = Final Destination: to file

Figure 14-1   Possible routing of copies of a hazardous waste manifest.

3. *Labeling and Placarding.* Before a waste is transported from a generating site, each container is labeled and the transportation vehicle is placarded. Announcements that are appropriate include warnings for explosives, flammable liquids, corrosive material, strong oxidizers, compressed gases and poisonous and/or toxic substances. Multiple labeling is desirable if, for example, a waste is both explosive and flammable. These labels and placards warn the general public of possible dangers, and assist emergency response teams as they react in the event of a spill or accident along a transportation route.

4. *Accident and Incident Reporting.* Accidents involving hazardous wastes must be reported immediately to state regulatory agencies and local health officials. Accident reports that are submitted immediately and indicate the amount of materials released, the hazards of these materials, and the nature of the failure that caused the accident can be instrumental in containing a waste and cleaning the site. For example, if liquid waste can be contained, groundwater and surface water pollution may be avoided.

RESOURCE RECOVERY ALTERNATIVES

Resource recovery alternatives are based on the premise that one man's waste is another man's prize. What may be a worthless drum of electroplating sludge to the plating engineer may be a silver mine to an engineer skilled in metals recovery. In hazardous waste management, two types of systems exist for transferring this waste to a location where it is viewed as a resource: *hazardous waste materials transfers* and *hazardous waste information clearinghouses.* In practice one organization may display characteristics of both of these pure systems.

The rationale behind both transfer mechanisms is illustrated in Figure 14-2. An industrial process typically has three outputs: (1) a principal product, which is sold to a consumer; (2) a useful by-product available for sale to another industry, and (3) waste, historically destined for ultimate disposal. Waste transfers and clearinghouses act to minimize this flow of waste to a landfill or to ocean burial by directing it to a heretofore unidentified industry or firm that perceives the waste as a resource. As the regulatory and economic climate of the nation evolves, these perceptions can continue to change and more and more waste can be economically recycled.

Information Clearinghouses

The pure clearinghouse has limited functions. Basically, these institutions offer a central point for collecting and displaying information

Figure 14-2   Rationale for hazardous waste clearinghouses and exchanges.

about industrial wastes. Their goal is to introduce interested potential trading partners to each other through the use of anonymous advertisements and contacts. Clearinghouses generally do not seek customers, negotiate transfers, set prices, process materials or provide legal advice to interested parties. One major function of a clearinghouse is to keep all data and transactions confidential so trade secrets are not compromised.

Clearinghouses are also generally subsidized by sponsors, either trade or governmental. Small clerical staffs are organized in a single office or offices spread throughout a region. Little capital is required to get these operations off the ground, and annual operation expenses can range from $20,000 to $90,000.

The value of clearinghouse operations should not be overemphasized. Often they are only able to operate in the short term; they evolve from an organization with many listings and active trading to rapid drop-offs in activity as plant managers make their contacts with waste suppliers and short-circuit the system by eliminating the clearinghouse.

Materials Exchanges

In comparison to the clearinghouse concept, a pure materials exchange has many complex functions. A transfer agent within the exchange typically identifies generators of waste and potential users of the waste. The exchange will buy or accept waste, analyze its chemical and physical properties, identify buyers, reprocess the waste as needed, and sell it at a profit.

The success of an exchange depends on several factors. Initially, a highly competent technical staff is required to analyze waste flows and design and prescribe methods for processing the waste into a saleable resource. Capitalization can easily exceed $500,000 in laboratory equipment and supplies, and a yearly cash flow, including debt service, can exceed $200,000 for even small operations. The ability to diversify is critical to the success of an exchange. Its management must be able to identify local suppliers and buyers of their products. Additionally, an exchange may even enter the disposal business and incinerate or landfill waste.

Clearinghouses and exchanges have been attempted with some success in the United States. However, a longer track record exists in Europe. Belgium, Switzerland, West Germany, most of the Scandinavian countries, and the United Kingdom all have experienced some success with exchanges. The general characteristics of European waste exchanges include:

- operation by the national industrial associations
- services offered without charge
- waste availability made known through published advertisements
- advertisements discuss chemical and physical properties, as well as quantities, of waste
- advertisements are coded to keep the identity of offerer confidential

Much can be learned in the United States from these experiences.

Five wastes are generally recognized as having transfer value: (1) wastes having high concentrations of metals, (2) solvents, (3) concentrated acids, (4) oils and (5) combustibles for fuel. That is not to say these wastes are the only transferable items. Transformed from waste to resource in one European exchange were:

- foundry slag, 50–60% metallic Al, 400 ton/yr
- methanol, 90% with trace mineral acids, 150 m^3/yr
- cherries, deep frozen, 4 tons

One man's waste can truly be another man's valued resource.

HAZARDOUS WASTE MANAGEMENT FACILITIES

Siting Considerations

A wide range of factors must be considered as hazardous waste management facilities are sited across the nation. Important inputs to site selection include studies of an area's "ologies": hydrology, climatology, geology, and ecology, as well as current land use patterns, environmental health, and transportation factors. Socioeconomic factors may be the key to any effort to site a facility.

Hydrology. Hazardous waste landfills should be located well above historically high groundwater tables. Care should be taken to ensure that a location has no surface or subsurface connection, such as a crack in confining strata, between the site and a water course. Hydrological considerations limit direct discharge of wastes into groundwater or surface water supplies.

Climatology. Hazardous waste management facilities should be located outside the paths of recurring severe storms. Hurricanes and tornadoes disrupt the integrity of landfills and incinerators and cause immediate catastrophic effects on the surrounding environment and public health in the region of the facility. In addition, areas of high air pollution potential should be avoided in site selection processes. These areas include valleys where winds and/or inversions act to hold pollutants close to the surface of the earth, as well as areas on the windward side of mountain ranges, i.e., areas similar to the Los Angeles area where longterm inversions are prevalent.

Geology. A disposal or processing facility should only be located on stable geologic formations. Impervious rock, which is not littered with cracks and fissures, is an ideal final liner for hazardous waste landfills.

Ecology. The ecological balance must be considered as hazardous waste management facilities are located in a region. Ideal sites in this respect include areas of low fauna and flora density, and efforts should be made to avoid wilderness areas, wildlife refuges and migration routes. Areas with unique plants and animals, especially endangered species, should also be avoided.

Alternative Land Use. Areas with low ultimate land use should receive prime consideration as facilities are sited in a region. Areas with high recreational use potential should be avoided because of the increased possibility of direct human contact with the wastes.

Environmental Health. Landfills and processing facilities should be located away from private wells, away from municipal water supplies, and away from high population densities. Flood plains should be avoided, at least up to the 100-year storm level.

Transportation. Transportation routes to facilities are a major consideration in siting hazardous waste management facilities. Such facilities should be accessible by all-weather highways to avoid spills and accidents during periods of rain and snowfall. Ideally, the closer a facility is to the generators of the waste, the less likely are spills and accidents as the wastes move along the countryside.

Socioeconomic Factors. Factors which could make or break an effort to site a hazardous waste management facility fall under this major heading. Such factors, which range from citizen acceptance to long-term care and monitoring of the facility, are:

1. Citizen acceptance and public education programs: Will local towns-people permit it?
2. Land use changes and industrial development trends: Does the region wish to experience the industrial growth that is induced by such facilities?
3. User fee structures and recovery of project costs: Who will pay for the facility, can user changes be used to induce industry to reuse, reduce or recover the resources in the waste?
4. Long-term care and monitoring: How will post-closure maintenance be guaranteed and who will pay?

All are critical concerns in a hazardous waste management scheme.

Incinerators

Incineration is a controlled process that uses combustion to convert a waste to a less bulky, less toxic or less noxious material. The principal products of incineration from a volume standpoint are carbon dioxide, water and ash, but the products of primary concern due to their environmental effects are compounds containing sulfur, nitrogen and halogens. When the gaseous combustion products from an incineration process contain undesirable compounds, a secondary treatment such as afterburning, scrubbing or filtration is required to lower concentrations to acceptable levels prior to atmospheric release. The solid ash products from the incineration process are a major concern and must reach adequate ultimate disposal.

Figure 14-3 shows the schematic of a generalized incineration system which includes the components of a waste incineration system. The actual system, which may contain one or more of the components, is usually dependent on the individual waste's application requirement. The advantages of incineration as a means of disposal for hazardous waste are:

1. Burning wastes and fuels in a controlled manner has been carried on for many years and the basic process technology is available and reasonably well developed. This is not the case for some of the more exotic chemical degradation processes.
2. Incineration is broadly applicable to most organic wastes and can be scaled to handle large volumes of liquid waste.
3. Large expensive land areas are not required.

Figure 14-3   Waste incineration system.

The disadvantages of incineration include:

1. The equipment tends to be more costly to operate than many other alternatives.
2. It is not always a means of ultimate disposal in that normally an ash remains which may or may not be toxic, but which in any case must be disposed of properly.
3. Unless controlled by applications of air pollution control technology, the gaseous and particulate products of combustion can be hazardous to health or damaging to property.

The decision to incinerate a specific waste will therefore depend, first, on the environmental adequacy of incineration as compared to other alternatives, and second, on the relative costs of incineration and the environmentally sound alternatives.

The variables which have the greatest effect on the completion of the oxidation of wastes are waste combustibility, dwell time in the combustor, flame temperature, and the turbulence present in the reaction zone of the incinerator. The combustibility is a measure of the ease with which a material can be oxidized in a combustion environment. Materials with a low flammability limit, low flash point, and low ignition and auto ignition temperatures may be combusted in a less severe oxidation environment, i.e., at a lower temperature and with less excess oxygen.

Of the three "T's" of good combustion—time, temperature, and turbulence—only the *temperature* may be readily controlled after the incinerator unit is constructed. This can be done by varying the air-to-fuel ratio. If solid carbonaceous waste is to be burned without smoke, a minimum temperature of 760°C (1400°F) must be maintained in the combustion chamber. Upper temperature limits in the incinerator are dictated by the refractory materials available. Above 1300°C (2400°F) special refractories are needed.

The degree of *turbulence* of the air for oxidation with the waste fuel will affect the incinerator performance significantly. In general, both mechanical and aerodynamic means are utilized to achieve mixing of the air and fuel. The completeness of combustion and the time required for complete combustion are significantly affected by the amount and the effectiveness of the turbulence.

The third major requirement for good combustion is *time*. Sufficient time must be provided to the combustion process to allow slow-burning particles or droplets to completely burn before they are chilled by contact with cold surfaces or the atmosphere. The amount of time required depends on the temperature, fuel size and degree of turbulence achieved.

The type and form of waste will dictate the type of combustion unit required. A number of control methods have been successfully developed for applications where the pollutants are in the form of fumes or gas. If the waste gas contains organic materials which are combustible, then incineration should be considered as a final method of disposal. When the amount of combustible material in the mixture is below the lower flammable limit, it may be necessary to add small quantities of natural gas or other auxiliary fuel to sustain combustion in the burner. Thus economic considerations are critical in the selection of incinerator systems because of the high costs of these additional fuels.

Incineration is also a possibility for the destruction of liquid wastes. Liquid wastes may be classified into two types from a combustion standpoint: combustible liquids and partially combustible liquids. Combustible liquids include all materials having sufficient calorific value to support combustion in a conventional combustor or burner. Noncombustible liquids cannot be treated or disposed of by incineration and include materials that would not support combustion without the addition of auxiliary fuel and would have a high percentage of noncombustible constituents such as water.

When starting with a waste in liquid form, it is necessary to supply sufficient heat for vaporization in addition to raising it to its ignition temperature. In order that a waste may be considered combustible, several rules of thumb should be used. The waste should be pumpable at ambient temperature or capable of being pumped after heating to some reasonable temperature level. Since liquids vaporize and react more rapidly when finely divided in the form of a spray, atomizing nozzles are usually employed to inject waste liquids into incineration equipment whenever the viscosity of the waste permits atomization. If the waste cannot be pumped or atomized, it cannot be burned as a liquid but must be handled as a sludge or solid.

In order to support combustion in air without the assistance of an auxiliary fuel, the waste must generally have a calorific value of 18,500–23,000 kJ/kg (8000–10,000 Btu/lb) or higher. Liquid waste having a heating value below 18,500 kJ/kg (8000 Btu/lb) is considered a partially combustible material and requires special treatment.

Several basic considerations are important in the design of an incinerator for a partially combustible waste. First, the waste material must be atomized as finely as possible to present the greatest surface area for mixing with combustion air. Second, adequate combustion air to supply all the oxygen required for oxidation or incineration of the organics present should be provided in accordance with carefully calculated requirements. Third, the heat from the auxiliary fuel must be sufficient to raise the temperature of the waste and the combustion air to a point above the ignition temperature of the organic material in the waste.

Incineration of wastes which are not pure liquids but which might be considered sludges or slurries is also an important waste disposal problem. Incinerator types applicable for this kind of waste would be fluidized bed incinerators, rotary kiln incinerators and multiple hearth incinerators.

Incineration is not a total disposal method for many solids and sludges because most of these materials contain noncombustibles and have residual ash. Complications develop with the wide variety of materials which must be burned. Controlling the proper amount of air to give combustion of both solids and sludges is difficult, and with most currently available incinerator designs this is impossible.

The types of incinerators which are applicable to solid wastes are open pit incinerators and closed incinerators such as rotary kilns and multiple hearth incinerators. Generally, the incinerator design does not have to be limited to a single combustible or partially combustible waste. Often it is both economical and feasible to utilize a combustible waste, either liquid or gas, as the heat source for the incineration of a partially combustible waste which may be either liquid or gas.

Experience indicates that wastes which contain only carbon, hydrogen, and oxygen and which can be handled in power generation systems can be destroyed in a way that reclaims some of their energy content. These types of wastes may also be judiciously blended with wastes having low energy content, such as the highly chlorinated organics, in order to minimize the use of purchased fossil fuel. On the other hand, rising energy costs will not be a significant deterrent to the use of thermal destruction methods where they are clearly indicated to be the most desirable method on an environmental basis.

If the gaseous emissions of the hazardous waste incinerator contains undesirable compounds, air pollution control equipment is required to remove these compounds. The solid and liquid effluents from the pollution control system may require further treatment prior to their ultimate disposal. The selection of air pollution control equipment is a function of the chemical constituents in the waste.

Landfills

Landfills must be adequately designed and operated if public health and the environment are to be protected. The general components that go into the design of these facilities, as well as the correct procedure to follow during the operation and post-closure phase of the facility's life, are discussed below.

Design

Three levels of safeguard must be incorporated into the design of a hazardous landfill. These levels are displayed in Figure 14-4. The primary system is an impermeable liner, either clay or synthetic material, coupled with a leachate collection and treatment system. Infiltration can be minimized with a cap of impervious material overlaying the landfill, and sloped to permit adequate runoff and to discourage pooling of the

LEVEL 1: Liner plus leachate collection/treatment

LEVEL 2: Back up liner plus leachate collection/treatment

LEVEL 3: Wells to monitor and, if needed, control leachate plume

Figure 14-4   Three levels of safeguard in hazardous waste landfills.

water. The objectives are to prevent rainwater and snow melt from entering the soil and percolating to the waste containers and, in case water does enter the disposal cells, to collect and treat it as quickly as possible. Side slopes of the landfill should be a maximum of 3:1 to reduce stress on the liner material. Research and testing of the range of synthetic liners must be viewed with respect to a liner's strength, compatibility with wastes, costs, and life expectancy. Rubber, asphalt, concrete and a variety of plastics are available, and such combinations as polyvinyl chloride overlaying clay may prove useful on a site-specific basis.

A leachate collection system must be designed by contours to promote movement of the waste to pumps for extraction to the surface and subsequent treatment. Plastic pipes, or sand and gravel, similar to systems in municipal landfills and used on golf courses around the country, are adequate to channel the leachate to a pumping station below the landfill. One or more pumps direct the collected leachate to the surface, where a wide range of waste-specific treatment technologies are available, including:

- sorbent material: carbon and fly ash arranged in a column through which the leachate is passed.
- packaged physical-chemical units, including chemical addition and flash mixing, controlled flocculation, sedimentation, pressure filtration, pH adjustment, and reverse osmosis.

The effectiveness of each method is highly waste-specific and tests must be conducted on a site-by-site basis before a reliable leachate treatment system can be designed. All methods produce waste sludges which must reach ultimate disposal.

A secondary safeguard system consists of another barrier contoured to provide a backup leachate collection system. In the event of failure of the primary system, the secondary collection system conveys the leachate to a pumping station, which in turn relays the wastewater to the surface for treatment.

A final safeguard system is also advisable. This system consists of a series of discharge wells up-gradient and down-gradient to monitor groundwater quality in the area and to control leachate plumes if the primary and secondary systems fail. Up-gradient wells act to define the background levels of selected chemicals in the groundwater and to serve as a basis for comparing the concentrations of these chemicals in the discharge from the down-gradient wells. This system thus provides an alarm mechanism if the primary and secondary systems fail.

Operation

As waste containers are brought to a landfill site for burial, specific precautions should be taken to ensure the protection of public health, worker safety and the environment. Wastes should be segregated by physical and chemical characteristics and buried in the same cells of the landfill. Three-dimensional mapping of the site is useful for future mining of these cells for resource recovery purposes. Observation wells with continuous monitoring should be maintained, and regular core soil samples should be taken around the perimeter of the site to verify the integrity of the liner materials.

Site Closure

Once a site is closed and does not accept more waste, the operation and maintenance of the site must continue. The impervious cap on top of the landfill must be inspected and maintained to minimize infiltration. Surface water runoff must be managed, collected and possibly treated. Continuous monitoring of surface water, groundwater, soil and air quality is necessary, as ballooning and rupture of the cover material may occur if gases produced and/or released from the waste rise to the surface. Waste inventories and burial maps must be maintained for future land use and waste reclamation. A major component of post-closure management is maintaining limited access to the area.

CONCLUSION

Hazardous waste is a relatively new concern of environmental engineers. For years, the necessary by-products of an industrialized society were piled "out back" on land which had little value. As time passed and the rains came and went, the migration of harmful chemicals moved hazardous waste to the front page of the newspaper and into the classroom. Engineers employed in all public and private sectors must now face head-on the processing, transport and disposal of these wastes.

PROBLEMS

14.1 Assume you are an engineer working for a hazardous waste processing firm. Your vice president thinks it would be profitable to locate a new regional facility near the state capitol. Given what you know about that region, rank the factors which distinguish a good site from a bad site. Discuss the reasons for this ranking; i.e., why, for example, are hydrologic considerations more critical in that region than, for example, the geology.

14.2 If you were a town engineer just informed of a chemical spill on Main Street, sequence your responses. List and describe the actions your town should take for the next 48 hours if the spill is relatively small (100–500 gallons) and is confined to a small plot of land.

14.3 The manifest system, through which hazardous waste must be tracked from generator to disposal site, is expensive for industry. Make the following assumptions about a simple electroplating operation: 50 barrels of waste per day, 1 "trip ticket" per barrel, $25/hr labor charge. Assume the generator's lab technician can identify the contents of each barrel at zero additional time for no additional cost to the company because he routinely has done that for years anyway. What is the cost in man-hours and in dollars for this generator to comply with the manifest system illustrated in Figure 14-1? Document your necessary assumptions about the time required for the generator to complete each step of each trip ticket.

14.4 Compare and contrast the design considerations of a hazardous waste landfill with the design considerations of a conventional municipal refuse landfill.

14.5 As town engineer, design a system to detect and stop the movement of hazardous wastes into your municipal refuse landfill.

LIST OF SYMBOLS

LD_{50} = Lethal Dose, at which 50% of the subjects are killed
LDC_{50} = Lethal Dose Concentration, at which 50% of the subjects are killed
RCRA = Resource Conservation and Recovery Act

SUGGESTED READING

Land and Natural Resources Division. *The Superfund Concept,* U.S. Department of Justice (1979).

Municipal Environmental Research Laboratory. *Survey of Solidification/Stabilization Technology for Hazardous Industrial Waste,* U.S. Environmental Protection Agency (1979).

National Wildlife Federation. *The Toxic Substances Dilemma, A Plan for Citizen Action* (Washington, DC: National Wildlife Federation, 1980).

Office of Solid Waste. *Incineration in Hazardous Waste Management,* U.S. Environmental Protection Agency (1975).

Office of Water and Waste Management. *Electrical Resistivity Evaluations at Solid Waste Disposal Facilities,* U.S. Environmental Protection Agency (1978).

Office of Water and Waste Management. *Siting of Hazardous Waste Management Facilities and Public Opposition,* U.S. Environmental Protection Agency (1979).

Office of Water and Waste Management. *Siting of Waste Impoundment Disposal Facilities,* U.S. Environmental Protection Agency (1980).

Office of Water and Waste Management. *Status of State Programs for Hazardous and Solid Waste Management,* U.S. Environmental Protection Agency (1978).

Office of Water and Waste Management. *Toxicity of Leachates,* U.S. Environmental Protection Agency (1980).

Chapter 15
Radioactive Waste

In the late 1800s, French scientists determined that uranium minerals routinely emitted invisible radiation capable of passing through apparently solid objects. Building on the research, Pierre and Marie Curie were able to isolate two new chemical elements from the uranium minerals: polonium and radium; and these newly discovered elements were observed to produce radiation that was even more intense than uranium emissions. Subsequently, it was determined that the radiation was a result of an atomic disintegration, as atoms of the minerals somehow exploded spontaneously through time. Three atomic decay chains are found in nature: the uranium chain, the actinium chain and the thorium chain.

Subsequently, it was discovered that other artificial transmutations are achievable in the lab. Each natural chain is in a series of decay, with a parent nuclide succeeded by offspring nuclides as the parent transforms into a stable, nonexplosive element through a series of alpha, beta and gamma emissions. The movement of these emissions through our environment can have significant impact on our quality of life, and is a major concern to the environmental engineer.

This chapter presents a general background discussion of this radiation process, highlights radioactive wastes, maps the radiation as it moves from the waste deposits through the environment and impacts public health, and summarizes options available to environmental engineers to protect public health. Many of the technologies and management issues addressed in Chapter 14 are applicable to these radioactive wastes as well, e.g., facility siting and packaging and transportation considerations.

RADIATION

In the early 1900s, the Curies and their contemporaries were able to identify three major types of radiation that were emitted from exploding atoms: *alpha* (α), *beta* (β) and *gamma* (γ) radiations. Modern physics has subsequently identified many more types, including neutrons and pions, but not all are of general concern to the environmental engineer. Those that present public health problems only during and after the explosion of an atomic bomb, or possibly as they exist in beams of selected cancer therapy machines, are not discussed here. Environmental engineers must develop a basic understanding of alpha, beta and gamma radiation if they are to deal with the significant problems associated with the management of radioactive wastes.

Radioactive Decay

An atom that is radioactive has an unstable nucleus that moves to a more stable condition either by emitting a nuclear particle (alpha and beta emissions) or an amount of energy (gamma emissions). In cases where a charged particle is emitted, the atom ends up with a different atomic number and is thus defined as a different element. These disintegrations follow the laws of chance, and the decrease in the number of radioactive atoms within a material is defined as:

$$\Delta N = -K_b N \Delta t$$

where ΔN = the change in the number of radioactive atoms during the selected time period Δt
   K_b = the proportionality factor termed the disintegration constant with unit 1/time
   K_b = the fraction of atoms which disintegrate per unit time

Integrating this equation, letting $N = N_0$ at time $= 0$, we get the classical exponential radioactive decay equation:

$$\int_{N_0}^{N} \frac{dN}{N} = -\int_0^t K_b \, dt$$

Using calculus, we get:

$$\ln \frac{N}{N_0} = -K_b t$$

Rearranging gives us:

$$\frac{N}{N_0} = \exp(-K_b t)$$

The data points in Figure 15-1 correspond to this equation.

After a specific time period, $t = {}_{1/2}L$, the value equals one-half of the previous N. That is, at the end of some time period defined as ${}_{1/2}L$, half of the radioactive atoms have disintegrated. This half-life is determined from the above equation:

$$_{1/2}L = \frac{\ln 2}{K_b} = \frac{0.693}{K_b}$$

The half-lives of selected radioactive products are presented in Table 15-1.

Looking at Figure 15-1, we see that N does not become zero at $t = 2 \cdot {}_{1/2}L$. In fact N becomes $\frac{1}{4} N_0$. We therefore see that the equation

Figure 15-1   General description of radioactive decay.

$$\frac{N}{N_0} = \exp(-K_b t)$$

is so constructed that N never becomes zero for any finite time period; for every half-life that passes, the number of atoms is halved.

Table 15-1.   Dangerous Radioactive Products

Product	Type of Radiation	Half-life
Krypton-85	beta and gamma	10 years
Strontium-90	beta	20 years
Iodine-131	beta and gamma	8 days
Cesium-137	beta and gamma	30 years
Tritium	beta	12 years
Cobalt-60	beta and gamma	5 years
Carbon-14	beta	5770 years

Example 15.1

An engineer, with the assistance of a chemist, prepares 10.0 g of pure $_6C^{11}$. This notation represents carbon with 6 protons and a total mass of 11 atomic mass units (amu, the number of protons plus the number of neutrons). (Conversion: 1 amu $= 1.66 \times 10^{-24}$ g.) The radioactive decay of this atom is

$$_6C^{11} \rightarrow {}_1e^0 + {}_5B^{11}$$

If the half-life is 21 minutes, how many grams of carbon are left 24 hours after the preparation?

The equations above describing the half-life refer to number of atoms, so we must initially calculate the number of atoms in 1 g of $_6C^{11}$

$$\frac{1 \text{ atom } _6C^{11}}{11.0 \text{ amu}} \times \frac{1 \text{ amu}}{1.66 \times 10^{-24} \text{ g } _6C^{11}} = \frac{1 \text{ atom } _6C^{11}}{18.3 \times 10^{-24} \text{ g } _6C^{11}}$$

Now, compute the number of atoms of $_6C^{11}$ in 10 g:

$$\frac{1 \text{ atom } _6C^{11}}{18.3 \times 10^{-24} \text{ g } _6C^{11}} \times 10 \text{ g } _6C^{11} = 55 \times 10^{22} \text{ atoms } _6C^{11}$$

Now, apply the equation for $_{1/2}L$:

$$K_b = \frac{0.693}{_{1/2}L} = \frac{0.693}{21 \text{ min}} = 33 \times 10^{-3} \text{ min}^{-1}$$

Following convention to avoid dealing with negative logarithms, we can rearrange our half-life equation as:

$$-K_b t = \ln \frac{N}{N_0} = -\ln \frac{N_0}{N}$$

or

$$K_b t = \ln \frac{N_0}{N}$$

Now, realizing that we must express t and K_b so that $(K_b t)$ is unitless, we must include the number of minutes in a day in our calculations:

$$K_b t = (33 \times 10^{-3} \text{ min}^{-1})(1440 \text{ min}) = 47.5$$

and

$$\ln \frac{N_0}{N} = 2.303 \log \frac{N_0}{N}$$

So we have

$$2.303 \log \frac{N_0}{N} = 47.5$$

or

$$\log \frac{N_0}{N} = 20.6$$

Applying the properties of logarithms:

$$\frac{N_0}{N} = (\text{antilog } 0.6) \times (\text{antilog } 20)$$

where we get:

$$\frac{N_0}{N} = 4.0 \times 10^{20} \quad \text{or} \quad N = \frac{N_0}{4.0 \times 10^{20}}$$

This gives us

$$N = \frac{55 \times 10^{22}}{4.0 \times 10^{20}} = 14 \times 10^2 \text{ atoms } {_6}C^{11}$$

After 1 day there are approximately 1400 atoms of ${_6}C^{11}$ remaining. To convert this to grams:

$$(14 \times 10^2 \text{ atoms } {_6}C^{11}) \times \left(\frac{18.3 \times 10^{-24} \text{ g}}{\text{atoms } {_6}C^{11}} \right) = 250 \times 10^{-22} \text{ g } {_6}C^{11}$$

Often, the original radionuclide, called the parent, decays to a nucleus which is also unstable, called the daughter; the daughter often decays even further. Chains where radioactive daughters produce radioactive second daughters which produce radioactive third daughters and on and on and on are not uncommon. For example, for the decay chain beginning with U^{238} we observe ten steps before the stable Pb^{207} is reached.

Alpha, Beta and Gamma Radiations

The three basic types of radiation can be categorized using research apparatus diagrammed in Figure 15-2. Here, a beam of exploding atoms is aimed with a lead barrel at a fluorescent screen designed to glow when hit by the radiation. Two alternately charged probes direct the positively charged α-radiation and negatively charged β-radiation accordingly. The γ-radiation is seen to be "invisible light," a stream of neutral particles that passes through the electromagnetic force field. Alpha and beta emissions are typically classified as particulates; gamma emissions consist of electromagnetic radiation (waves).

Alpha radiations, identified as being physically identical to the nuclei of helium atoms, stripped of their planetary electrons with only 2 protons and 2 neutrons remaining, are exploded from the nucleus of selected radioactive atoms with a kinetic energy of somewhere between 4 and 10 MeV.* The particles have a mass of about 4 amu (6.642×10^{-4} g) and a positive charge of 2.** As these charged particles travel along at approxi-

---

*Mass is designated as energy in the equation $E = mc^2$. One MeV $= 10^6$ electron volts (ev). One electron volt $= 1.603 \times 10^{-12}$ erg.

**This charge, in atomic mass units (amu), is expressed in units relative to an electron with a negative charge of 1.

Figure 15-2   Controlled measurement of alpha (α), beta (β) and gamma (γ) radiation.

mately 10,000 miles per second, they interact with other atoms. Each interaction transfers a portion of the energy to the electrons of these other atoms, and results in the production of an ion pair; i.e., a negative electron with an associated positive ion. The huge number of interactions produces between 30,000 and 100,000 of these ion pairs per centimeter of air traveled, resulting in rapid depletion of the alpha energy. Subsequently, the range of the alpha particle is only 1–8 cm in air. If such a particle runs into a solid object, notably human skin cells, the energy is rapidly dissipated. Thus, alpha particles present no direct problem of external radiation damage to humans, but could cause health problems when emitted inside the body where protective layers are not present to diffuse the energy. Even the strongest alpha particles are stopped by the epidermal layer of the skin and rarely reach the sensitive layers. Humans are typically contaminated with material that emits alpha particles only if the material is inhaled, ingested or absorbed in a skin wound.

Beta radiation is electrons that are also emitted from the nucleus of a radioactive atom at a velocity approaching the speed of light, with kinetic energy between 0.2 and 3.2 MeV. Given their lower mass of approximately 5.5×10^{-4} amu (9.130×10^{-28} g), the interactions between beta particles and the atoms of pass-through materials are much less frequent than alpha particle interactions. Fewer than 200 ion pairs are formed in each centimeter of flight through air, and the resulting slower rate of energy loss enables beta particles to travel several meters in air and several centimeters through human tissue. Thus, although beta-radiation can damage tissue under the human skin, internal organs are generally protected. However, exposed organs such as eyes are sensitive to beta damage.

Gamma radiation is identified as invisible, electromagnetic rays emitted from the nucleus of radioactive atoms. These rays are like medical X-rays, in that they are composed of photons. Because of their neutral charge, gamma photons collide randomly with the atoms of the material as they pass through. Thus, the *relaxation length,* the distance to decrease the intensity of the ray by a factor of $1/e$, is much greater than for α and β particles. A typical gamma ray, with an energy of about 0.7 MeV, has a unique relaxation length for different pass-through materials: for example, lead, water and air have relaxation lengths of 5, 50 and 10,000 cm respectively. The dose of gamma radiation received by unprotected human tissue can be significant because the dose is not greatly impacted by air molecules.

Energy is lost as the alpha, beta and gamma radiations pass through different materials. If the material is human tissue, the energy gained (by the tissue) may cause a disordering in the chemical or biological structure of the tissue and subsequent dysfunction of the cells. These effects are discussed later in this chapter.

Units for Measuring Radiation

Standard units have evolved which are used to measure radiation and its impacts on material. As we see above, the damage to human tissue is directly related to the amount of energy deposited in the tissue by the alpha, beta and gamma particles. This energy, in the form of ionization and excitation of molecules, results in heat damage to the tissue or even radiation burn. For this reason, many of the units are related to energy deposits.

A *rad* is a unit of absorbed dose defined as the disposition of 100 erg/g of any absorbing material. The amount of energy delivered to a material by radiation is a function of how frequently the radiation interacts with the material. When the same masses of lead and human tissue are exposed to an identical gamma flux, for example, rates of interaction are vastly different and energy transfer is different. More interactions occur per unit time in the lead than in the human tissue and different absorbed doses or rads are measured for these two materials.

The *Roentgen* is a standard unit of exposure corresponding to a quantity of gamma rays that deposit 87.7 ergs, per gram of air, at standard temperature and pressure. This unit is equivalent to the ejection of 1 electrostatic unit (esu) of charge in 1 cm^3 of air, and describes the electromagnetic field associated with radioactive decay. However, the unit does not indicate where the gamma rays deposit their energy. For the environmental engineer interested in health effects of radiation, the locations of their energy deposit is also of concern.

The *dose equivalent* addresses this issue: all types of radiation do not produce identical biological effects for a given amount of energy delivered to the human tissue. In general, radiations with high specific ionization along their tracks will produce a greater effect, but the quantitative difference will depend on the tissue or organ and biological change selected for study. The dose equivalent, measured as a *rem* (Roentgen equivalent man) or millirem (1/1000 of a rem), is equal to the product of the dose in rads and a conventional *quality factor* that describes the biological effects of concern. These units thus take into account the effect on living tissue of the alpha, beta and gamma radiation. Because all radiations do not produce identical biological effects for a given amount of energy deposited in, for example, human tissue, the standard for comparison is defined as gamma radiation having a linear energy transfer in water of 3.5 keV/μm and a rate of 10 rad/min. For example, for radiation from emitters located inside human tissue, the quality factors have been developed to convert the absorbed dose to doses equivalent as shown in Table 15-2. By standard, the quality factors are selected on the basis of long-term effects that may occur during the lifetime of an individual. Thus, this definition of dose equivalent is applicable to long-term impacts and not to the more obvious immediate impacts from a large absorbed dose. Table 15-3 illustrates the average annual doses to reproductive organs from radiation sources in the United States.

The *curie* is a unit which measures the activity of radioactive material. This activity is defined as

$$\frac{dN}{dt}$$

where dN is the average number of nuclear transformations which occur in a quantity of material in time period dt. One curie is defined to equal the quantity of a radioactive nuclide where the number of disintegrations

Table 15-2.   **Sample Quality Factors for Internal Radiation**[a]

Internal Radiation	Quality Factor
α	10.0
β and γ	
$E_{max} > 0.03$ Mev	1.0
$E_{max} < 0.03$ Mev	1.7

[a]E_{max} refers to the maximum level of emissions by the β and γ source.

Table 15-3.   Typical Average Annual Doses of Radiation
in the United States

Radiation Type	Sample Source	Average Annual Dose to Reproductive Organs (mrem/year)
Alpha (and Gamma)	Natural radioactivity (uranium) in rocks and minerals	30.0
Beta	Natural radioactivity (potassium-40) in rocks and minerals	20.0
	Natural radioactivity in the atmosphere (tritium)	2.0
	Television, 1 hr/day	0.5
Gamma	Medical X-rays	20.0
	Cosmic radiation at sea level	40.0

is 3.7×10^{10} per seconds. The curie is also related to the mass of the radioactive material through the half-life and the number of atoms per unit weight of the material, i.e., the atomic weight of the material.

This mass of material which disintegrates 3.7×10^{10} times per second varies widely from material to material. Note the data in Table 15-4. The relationship between the curies and the mass of a given radionuclide is given by:

$$Q = \frac{K_T \, M \, N^0}{W \, (3.7 \times 10^{10})}$$

where
Q = number of curies
K_T = disintegration constant = $0.693/_{1/2}L$; i.e., the fraction of atoms which decay per second
$_{1/2}L$ = half-life of the radionuclide, in seconds
M = mass of the radionuclide, in grams
N^0 = Avogadro's number, 6.03×10^{23} grams/gram atomic weight
W = atomic weight of the radionuclide

A complex relationship exists between the quantity of radiation, the curie, and the radiation dose rate, rad/sec. This relation depends on the energy of the radiation, the type of radiation, the path of the radiation, and the amount of absorbing material between the emittor and the receptor.

Table 15-4.   Description of Selected Radioactive Material

Material	Half-Life	Weight of 1 Curie (g)
Na^{24}	14.8 hr	10^{-15}
Ra^{226}	1622 yr	1
U^{238}	4.5×10^9 yr	3×10^6

Measuring Radiation

Devices have been developed to measure the radiation dose, dose rate, or the quantity of active material that is present. The particle counter, the ionization chamber, and photographic film are three methods widely used in the field.

Counters are designed to note the movement of single particles through a defined volume. Gas-filled counters collect the ionization produced by the radiation as it passes through the gas, and amplify it to produce an audible pulse. Counters are typically used to determine the radioactivity present by measuring the number of particles of protons that are emitted.

Ionization chambers basically consist of a pair of charged electrodes which collect ions formed within their respective electrical fields. Chambers are generally designed to determine dose or dose-rate measurements because they provide an indirect representation of the energy deposited in the chamber.

Photographic film darkens if exposed to radiation and is a useful indicator of the presence of radioactivity. Such film is often used for determining personnel exposure and making other dose measurements where a long record of dose is necessary or if a permanent record of dose is required.

HEALTH EFFECTS

When alpha and beta particles and gamma radiation penetrate living cells, or any other matter, they transfer their energy through a series of collisions with the atoms or nuclei of the receiving material. Many molecules are damaged in the process, as chemical bonds are broken and electrons are lost (ionization). In fact, energy is lost all along the path of the radiation, and a measure of this rate of *linear energy transfer* (LET) is the density of ionization activity along this path. The more ionization that is observed, the higher the intensity of biological damage to the cells.

As discussed earlier, alpha particles have a high LET and potentially cause significant damage. However, because their LET is high, i.e., because they lose their energy quickly, they cannot penetrate and are characteristically stopped by 1 in. of air or the outer layer of human skin. Beta particles and gamma radiation, on the other hand, lose energy slowly and penetrate deeper into living organisms.

Biological effects from these penetrations can be grouped as *somatic* and *genetic*. Somatic effects are the impacts on individuals directly exposed to the radiation, and include a change in the incidence of cancer and/or a change in life expectancy. Genetic effects, on the other hand, are impacts of the radiation which are transferred from the receiving individual to offspring through mutations in genes.

Radiation sickness and possible death are examples of early somatic effects of large doses of radiation exposure lasting a few seconds to a few minutes. Such doses typically could come from a nuclear bomb or an accident at a nuclear power plant. Waste products, if properly handled, do not pose these problems. For man, it is estimated that a gamma-radiation dose of 400–500 rads delivered to a major portion of the body in a short time will result in death within a few months.[1] Most mammals appear to have the same degree of sensitivity to the LD_{50} (lethal dose resulting in a 50% kill within 30 days). Lower doses, but above 100 rads, will lead to vomiting, diarrhea and nausea in humans; hair loss is generally observed within two weeks after an individual has been exposed to 300 rads or more.[2,3]

Delayed somatic effects often become evident after massive exposure or as a result of long-term exposure to low-level radiation. These effects are generally only evident in a statistical analysis of the incidence of the effect among those exposed compared to the incidence of the same effect in a control group. No one can predict who will contract what cancer as a result of low-level exposures which continue over several years. Such effects are discussed in terms of the probability of an individual contracting the specific ailment. This probability is generally seen as a function of radiation dose, exposure time, the organs exposed, and many other factors not fully understood.

---

[1] Braestrup, C. G., and H. O. Wyckoff. *Radiation Protection* (Springfield, IL: Charles C. Thomas, 1958).

[2] Glasstone, S. "Effects of Nuclear Weapons," U.S. Government Printing Offices, (1962).

[3] Nickson, J. J., and H. N. Baine. "Physiological Effects of Radiation," in *Radiation Hygiene Handbook* (New York: McGraw-Hill Book Company, 1959).

Genetic effects are as little understood as are delayed somatic effects. Evidence points to the fact that mutations in sperm cells and egg cells are passed to next generations, but no one understands the mutation process to the point where predictions can be made and corrective actions taken.

A complete listing of the radiation mechanisms that damage cells, tissues or organs is not available at this time. Many theories focus on alterations of the DNA, possibly by the deposit of energy from a direct hit. Suffice it to conclude here that the health effects of a given dose of radiation depend on a large number of factors, including

- magnitude of the absorbed dose
- type of radiation
- penetrating power of the radiation
- sensitivity of the receiving cells and organs
- rate at which the dose is delivered
- proportion of the cell/organ/human body exposed

To protect public health, the environmental engineer must focus on the factors which are most in his control: the magnitude of potential doses, the type of radiation available to impact public health in an area, and the rate at which doses are delivered. The environmental engineer must, in short, act to minimize unnecessary radioactive exposure.

SOURCES OF RADIOACTIVE WASTE

Radioactive wastes are produced in a range of forms, concentrations and quantities. The wastes may be gaseous, liquid or solid, may be soluble or insoluble in water, and may emit several types of radiation. The concentration of the wastes varies from the highly radioactive spent fuel wastes of a nuclear power plant to the low-level residues found in many mineral processing technologies. Between these extremes is the broad range of wastes from medical, industrial and scientific applications.

High-level waste traditionally refers only to the liquid or solid derivatives of the waste stream from the reprocessing of spent reactor fuel. These wastes, which contain the products of nuclear fission discussed below, emit the gamma radiation which significantly impacts human tissues. These wastes also tend to produce significant heat from radioactive decay which can lead to uncontrolled release of the emissions into the environment. Dilution would generally reduce the direct radiation hazard from such a release, but inhalation and ingestion of the material could result through time. Radioactive wastes, separated from reactant coolant as solids, generally have a very small volume and are sealed in special containers.

Low-level waste is generally used to describe radioactive wastes that do not present an acute threat of radiation sickness even to individuals routinely exposed to its emissions, i.e., all radioactive wastes not classified as "high-level" waste. Another definition of low-level radioactive waste is commonly proposed: unwanted radioactive material not produced as a by-product in the fission reaction. A potential public health hazard from low-level waste is radiation exposure to water that comes in contact with the waste and is subsequently ingested by humans through the food chain or drinking water. Exposure in such cases depends on the rate of transfer of the radionuclides through the surrounding soil to the groundwater/surface water as well as the dilution of the material.

Long-lived wastes are particularly bothersome to the environmental engineer because they represent continued risks to public health and the environment far into the future. Transuranic (TRU) contaminated wastes, isotopes with atomic numbers greater than 92 like Np, Pu and Am, are especially critical because of their long half-lives. TRU wastes are primarily produced by the U^{238} in fuel elements during the operation of a nuclear reactor. Reprocessing of spent fuel elements would remove plutonium, but because the separation is not complete, the effluent high-level liquids still contain plutonium ($<0.5\%$) as well as other TRU material. Similarly, TRU contamination of low-level waste occurs when the TRU materials are handled or processed, mostly at government facilities that produce nuclear weapons.

Nuclear Fuel Cycle

The nuclear fuel cycle is a major generator of radioactive waste. The cycle starts with the mining of uranium and proceeds through the refining of the ore which concentrates the raw uranium to the desired uranium compound. At this point the fuel elements for reactors are fabricated with the natural uranium. After the elements are used in the reactor, the remaining uranium and plutonium can be separated, sent to fuel fabrication, and thus recycled. Figure 15-3 displays this movement of radioactive material in the cycle, and pinpoints the locations where waste is produced.

We will focus here on the nuclear reactor: the solid, liquid and gaseous radioactive wastes produced by nuclear power plants. Some solid wastes, including parts of contaminated machinery, work gloves and shoe covers, generally have a very low level of activity. Other solid wastes, particu-

Figure 15-3   Sources of radioactive waste in the nuclear fuel chain.

larly spent fuel, are more hazardous. Liquid wastes include corrosion products, iron, cobalt and zinc resulting from the chemical attack on radioactive metal parts in the reactor, some fission products, and tritium. Tritium is a radioactive form of hydrogen produced within the reactor and removed as either a liquid or a gas. Because it can become part of a water molecule, and because of its long 12.3-year half-life, tritium is a potential biological hazard. Gaseous wastes contain some gaseous fission products as well as tritium. Generally speaking, gaseous and liquid wastes from nuclear power plants contain very low levels of radioactivity. The solid derivatives from the processing of the coolants pose the real problems, as do the spent fuel rods.

The fission products alluded to above are the result of a special type of nuclear reaction in which the nucleus of the bombarded atom splits into two pieces and releases energy in the form of heat and radiations as:

$$U^{235} + 1 \text{ neutron} \rightarrow \text{fission products} + 2 \text{ (or 3) neutrons} + \text{energy}$$

Fission can be induced in most elements having an atomic number above 30 by bombardment with particles which have sufficient energy. U^{238}, for example, requires neutrons with an energy of 1 MeV or greater. Of special interest is the fact that this fission process produces other neutrons which themselves produce further fissions. It is therefore possible to arrange a mass of fissionable material so that these additional neutrons lead to a chain reaction. Figure 15-4 illustrates a nuclear power plant where such chain reactions can take place to produce heated water and then electricity.

In the fission process, the uranium splits into two new, lighter atoms. These atoms are called fission products, most of which are stable (nonradioactive), but others of which are very radioactive. These products build up in the fuel rods during their use and are held in place until the rod is removed for reprocessing or long-term storage. Most of these products pose no threat to public health because either they are produced in extremely small quantities or they have very short half-lives. On the other hand, several radioactive products are troublesome because of

Figure 15-4   Diagram of three major circuits in a nuclear plant (courtesy American Lung Association).

their long half-lives, their high yield, or sometimes because of their chemical properties. Strict regulations control the release of the radioactive products listed in Table 15-1.

Other Sources of Radioactive Waste

The increasingly widespread use of radioisotopes in research, medicine and industry has created a lengthy list of potential sources of other radioactive waste. Sources range from a large number of laboratories using small quantities of a few isotopes to large medical and research laboratories where many different isotopes are produced, used and wasted in large volumes.

In addition, naturally occurring radionuclides exist and become threats to public health. Phosphate mining, for example, produces a waste that, by virtue of its natural origin and methods of extraction, acts as a diffuse source of radiation. Similar wastes, much lower in radioactivity, result from copper mining and coal combustion. These wastes tend to contain the naturally occurring radioisotopes of uranium, thorium and radium.

MOVEMENT THROUGH THE ENVIRONMENT

Radioactive wastes contaminate the air, the water and the land. This contamination can have a significant impact on public health. Figure 15-5 pictures the pathways of contamination from a hypothetical mound of radioactive waste.

From an *air transport* standpoint, all nuclear waste handling and disposal facilities fall into two general categories: (1) those that have a planned and predictable discharge to the atmosphere and (2) those where any discharge would be purely accidental. Boiling-water reactors, fuel processing plants, and a large number of atomic experimental systems—particularly military—fit into the first category, and, on occasion, also into the second category. Good examples of the second category are well contained pressurized water reactors and long-term storage facilities.

The atmospheric pathways from both types of discharges have received a great deal of attention because of the analogies between radioactive emission and sulfur dioxides and inert gases generated by automobiles and smoke stacks across the nation. Airborne radioactivity is available for inhalation and/or disposition on water supplies or food chain land and thus impacts public health. Chapters 17 and 18 discuss air

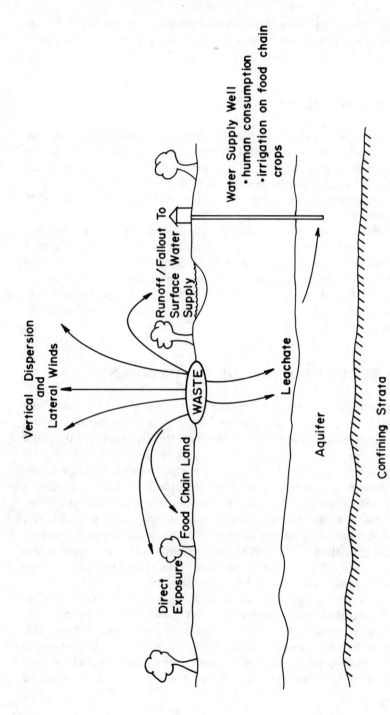

Figure 15-5   Potential movement of radioactivity from waste storage and "disposal" areas.

pollution and meteorology, and much of that discussion is applicable to the study of radioactive particles.

Water transport occurs whenever radionuclides in surface or subsurface soil erode or leach into a watercourse or whenever fallout occurs from the atmosphere. The radioactive waste is transferred from its solid or liquid state into groundwater or surface water supplies where the radionuclides can enter the human food chain or drinking water. However, little is known about the leachability of radionuclides, and only gross estimates can be made about surface water transport and disposition. As with any such radioactive material transport equations, the half-life of the isotopes must be considered. Chapters 4 and 9 address water transport considerations that are somewhat applicable to the problems associated with radioactive waste.

Radioactive isotopes also move *from the land* into the human food chain. Plant uptake mechanisms have virtually no way of screening an element which is radioactive; thus, plants generally collect a deposit of radionuclides in their structure if the radionuclides are present with the water, phosphorus, nitrogen and trace elements necessary for plant growth. Rain and wind also act to deposit the hazardous material on the roots and leaves of the plants, all to be passed along to whatever or whomever eats the plants. Cows eating contaminated corn, for example, can pass radioactivity along to milk drinkers.

Transuranic wastes last for long periods of time within the environment, but for the most part are strongly held by soil particles. They are not easily translocated through most food chains, although some concentration does pass through certain aquatic food systems. Additionally, they pose only a slight biological hazard to humans because adsorption in the gastrointestinal track is limited. The greater potential hazard from TRU wastes is from inhalation of dust particles containing the materials since a large fraction is generally retained in the human lung.

In summary, very little is known or understood about the transport of radioactive materials through the environment, making the management of such wastes especially troublesome.

RADIOACTIVE WASTE MANAGEMENT

The objective of the environmental engineer is to prevent radiation from being introduced into the biosphere over the lifeline of the radioactive waste. Control of the potential direct impact on humans is necessary, but, as we have seen, it is not sufficient because water, air and land pathways transmit radioactivity to current and future generations. A "lifetime" is defined by the half-life of the radioisotopes, but also by current and future reprocessing technologies which can and will retrieve all or part of the waste as a recycled resource. Recycling is an option now and will be in the future, but massive amounts of waste are produced today that are destined not for reprocessing but for long-term storage.

Engineers charged with radioactive waste control must, because of current technical, economic and political factors outside their control, focus on long-term storage technologies, i.e., *disposal*. With the given half-lives of many radioisotopes, it is difficult to imagine any technology which truly offers ultimate disposal for these wastes; thus, we will think in terms of long-term storage: 10, 100 or 10,000 years or longer. Many of the issues discussed in Chapter 14 are applicable to the radioactive waste problem as well.

Many options have been suggested for the long-term storage of radioactive waste. These options can be grouped into four major categories:

1. land disposal
    - liquid injection into geological formations
    - burial in deep caverns
    - burial in deep holes
2. ocean disposal
    - direct ocean dumping
    - disposal in trenches
    - sea bed disposal
3. polar ice disposal
4. spaceship technology

Alternatives have been proposed with varying degrees of forethought. Each option must be ranked by a list of criteria which includes: retrievability of the waste, length of time available for undisturbed storage, probable storage capacity, and cost of disposal.

Several types of geological formations exist under land masses which could be used as storage facilities: bedded salt, granite deposits, shales, metamorphic rock. The ideal formation is one in which rock units are homogeneous, thick, and deep, with adequate hydraulic and mechanical stability to maintain isolation of radioactive material. For many years, bedded salt has been favored; the Federal Republic of Germany has an existing policy of burying toxic materials (nuclear wastes as well as mercurides and arsenic) in salt deposits. The major salt deposit regions in the United States are illustrated in Figure 15-6.

For several reasons, salt deposits are particularly attractive for long-term storage of radioactive waste:

SCALE

| 500 miles |
| 300 km |

KEY

Area over rocksalt deposits

Area over salt dunes

Figure 15-6   Salt storage potential in the United States.

- Salt mining technology is well developed and storage sites could be constructed.
- Salt deposits tend to have a high plasticity, and thus have a tendency to seal themselves if fractures are created by major movements in the earth's crust.
- Salt deposits have low permeability, and are essentially sealed from groundwater and surface water supplies.
- Salt has a high thermal conductivity, which helps dissipate heat that builds in waste containers.
- Salt formations have a high structural strength, with the ability to withstand effects of heat and radiation.

A major drawback in many regions is the lack of historical data on oil drilling operations. Generally, no records were kept and puncture wounds from these operations could still scar many salt deposits.

Granite crystalline rocks have been considered as alternative formations for radioactive waste disposal. Such rocks tend to have greater mechanical strength than salt, but also have little if any plasticity and are often scarred with cracks and fissures.

Ocean disposal is not attractive for several reasons. Direct dumping would immediately endanger public health, and the integrity of any container system located along the ocean floor could not be guaranteed because of the corrosiveness of seawater. In addition, the residence time of deep seawater is only in the range of 100 to 1000 years, so isotopes with long half-lives would escape and be transported upward to ocean layers where the human food chain could be impacted. Little is understood about the transport of sediment along the ocean floor; thus the risks associated with burial in the sediment or in trenches are open to speculation.

Polar ice disposal is not a good alternative for several reasons. Most of the ice in the Antarctic, for example, was deposited less than 100,000 years ago and is not particularly stable. If a radioactive mass were placed on the surface of the ice and allowed to melt down to the bedrock, the hole could seal itself (temperatures in many regions of Antarctica have remained below freezing for more than 1 million years). However, the continued heating after the mass reached bedrock might result in a pool of water under the ice pack that eventually could find its way to the open sea and the human food chain.

Spaceship disposal of solidified waste is very expensive. The expense could be justified, particularly in the era of the space shuttle, but probably only then if smaller volumes of long-lived elements like iodine-129 could be separated from other waste. In addition, the risks associated with suborbital and orbital flight prior to the space-dump are not negligible.

CONCLUSIONS

Radiation is energy traveling in the form of particles or waves. Examples include microwave ovens, X-rays in medicine, and sunlight. This energy flows from a natural and spontaneous process in which unstable atoms of an element emit excess energy from their nuclei; it also results from man-induced chain reactions in nuclear power facilities.

This energy flux can have adverse impacts on biological matter. Radiation sickness, cancer, shortened life or immediate death can result from varying exposures. The most useful units for measuring radiation doses to humans are the *rem* and *millirem* (1/1000 of a rem, abbreviated *mrem*). These units take into account the effects on living tissue of the three types of radiation: alpha, beta and gamma radiations.

The general concern of the environmental engineer is radiation from man-made sources, particularly the wastes from nuclear power plants and laboratory research studies. These wastes must be handled with care exceeding the care given the hazardous wastes discussed in Chapter 14, but in much the same manner. Long-term storage is a problem that is not fully resolved at this writing. In 1981, high-level radioactive wastes from nuclear power plants are being stored onsite in dry swimming-pool-like structures. Policy formulation is taking place at the national level to determine what should be done next:

- no nukes?
- no medical research?
- no more nuclear weapons?

The future for all three major sources of man-made radioactive waste is out of the hands of the scientists and engineers and in the political arena. Only time and Chapter 25 will tell.

PROBLEMS

15.1 Qualitatively rank the four types of long-term storage (land, ocean, ice and spaceship) according to each of the following criteria:

- environmental protection (water, air and surface land)
- direct protection of public health
- retrievability of waste
- time available for undisturbed storage
- storge capacity
- cost of disposal

Discuss your rankings in detail and make general recommendations for the Governor's Task Force on radioactive waste disposal in your home state.

15.2 Construct a mass balance of radioactive material through your neighborhood hospital. What are the relative magnitudes of the respective outputs? How much do patients carry home with them?

15.3 Design a fallout shelter to support a family (2 adults, 2 children, no pets) for a period of 10 weeks. Draw on your knowledge gained in other chapters of this text. Discuss your selection of building materials and design considerations for the various environmental support systems.

15.4 Discuss the *environmental* impacts of a nuclear power plant, including water, air and land quality considerations. Include discussions of thermal pollution and impacts on the aquatic environment.

15.5 Assume your sister-in-law, 7-months pregnant, has a toothache. Outline the benefits and costs associated with these three courses of action:

- no dentist, no treatment
- dentist visit, treatment aided with X-rays
- dentist visit, treated without the aid of X-rays

What would you advise?

LIST OF SYMBOLS

α = alpha radiation
β = beta radiation
γ = gamma radiation
amu = atomic mass unit
esu = electrostatic unit
K_b = fraction of atoms which disintegrate per unit time
K_T = fraction of atoms which disintegrate per second = $0.693/_{1/2}L$
$_{1/2}L$ = half-life of a radionuclide, in seconds
LD_{50} = lethal dose 50
LDC_{50} = lethal dose concentration 50
LET = linear energy transfer
M = mass of a radionuclide, grams
MeV = 10^6 electron volts
N_0 = number of radioactive atoms within a material
N^0 = Avogadro's number, 6.02×10^{23} grams/gram atomic weight
Q = number of curies
rem = Roentgen equivalent man
t = time
TRU = transuranic (long-lived isotopes)
W = atomic weight of a radionuclide

SUGGESTED READING

Andrews, D. H., and Kokes, R. J. *Fundamental Chemistry* (New York: John Wiley & Sons, 1965).

Ames, L. L., and D. Rai. "Radionuclide Interactions with Soil and Rock Media," U.S. EPA Report 520/6-78-007, (1978).

Bredehoeft, J. D., et al. "Geological Disposal of High-Level Radioactive Wastes—Earth Science Perspectives," Geological Survey Circular, 779, U.S. Geological Survey (1978).

Committee on Radioactive Waste Management. "Radioactive Wastes at Hanford Reservation, A Technical Review," Commission on Natural Resources, National Research Council (Washington, DC: National Academy of Sciences, 1978).

Meteorology and Atomic Energy, National Technical Information Service, U.S. Department of Commerce (1966).

Pojasek, R. B. *Toxic and Hazardous Waste Disposal* (Ann Arbor, MI: Ann Arbor Science Publishers, Inc., 1980).

U.S. Environmental Protection Agency. "Criteria for Radioactive Wastes," *43, Federal Register,* 53262 (1978).

Chapter 16

Solid and Hazardous Waste Law

Laws controlling environmental pollution are discussed in this text in terms of their evolution from the courtroom through Congressional committees to administrative agencies. The gaps in common law are filled by statutory laws adopted by Congress and state legislatures, and implemented by administrative agencies such as the U.S. Environmental Protection Agency (EPA) and state departments of natural resources. For several reasons, this evolutionary process was particularly rapid in the area of solid waste law.

For decades, and in fact centuries, solid waste was disposed of on land no one really cared about. Municipal refuse was historically trucked to a landfill in the middle of a woodland, and industrial waste—often hazardous—was generally "piled out back" on land owned by the industry itself. In both cases, environmental protection and public health were not perceived as issues. Solid wastes were definitely out of sight and neatly out of mind.

Only relatively recently has public interest in solid waste disposal sites reached the level of concern that equals that for water and air pollution. Some local courtrooms and zoning commissions have dealt with siting selected disposal facilities over the years, but most decisions simply resulted in the city or industry hauling the waste a little farther away from the complaining public. At these remote locations, solid waste was not visible like the smokestacks that emitted pollutants into the atmosphere and the pipes discharging wastewater into the rivers. Pollution of the land from solid waste disposal facilities was and is a much more subtle type of pollution.

The passage of federal and state environmental statutes reflects this naivety. Initially, the highly visible problem of air pollution was addressed in a series of clean air acts. Then, the next most visible pollution was confronted in the water pollution control laws. Finally, as researchers dug deeper into environmental and public health concerns, it was realized that even obscure landfills and holding lagoons had the potential to significantly pollute the land and impact public health. Subsurface and surface water supplies, and even local air quality, are threatened by solid waste.

This chapter addresses solid waste law in two major sections: nonhazardous waste and hazardous waste. The division reflects the regulatory philosophy for dealing with these two very distinct problems. Realizing the general lack of common law in this area, we move directly into statutory controls of solid waste.

NONHAZARDOUS SOLID WASTE

The most significant solid waste disposal regulations were developed under the Resource Conservation and Recovery Act (RCRA) of 1976. This federal statute, which amended the elementary Solid Waste Disposal Act of 1965, reflected the concerns of the public in general and Congress in particular about: (1) protecting public health and the environment from solid waste disposal, (2) filling the loopholes in existing surface water and air quality laws, (3) ensuring adequate land disposal of residues from air pollution technologies and sludge from wastewater treatment processes, and (4) promoting resource conservation and recovery.

The EPA implemented RCRA in a manner that reflected these concerns. Disposal sites were cataloged as either landfills, lagoons or landspreading operations and the adverse effects of improper disposal were grouped into eight categories of impacts:

- *Floodplains* were historically prime locations for industrial disposal facilities because many firms elected to locate along rivers for power generation and/or transportation of process inputs or production outputs. When the rivers flood, the disposed wastes are washed downstream, impacting water quality.
- *Endangered species* may be threatened directly by habitat destruction or kills as the disposal site is being developed and operated.
- *Surface water quality* can also be impacted by certain disposal practices. Without proper controls of runoff and leachate, rainwater has the potential to transport pollutants from the disposal site to nearby lakes and streams.
- *Groundwater* impacts are of great concern because about half of the nation's population depends on groundwater for water supply.

- *Food chain crops* can be adversely affected by landspreading or solid waste which can impact on public health and agricultural productivity. Food chain crops often accumulate trace chemicals and increase the dangers to public health.
- *Air quality* can be degraded by pollutants emitted from waste decomposition, such as methane, and can cause serious pollution problems downwind from a disposal site.
- *Health* and *safety* of onsite workers and the nearby public can also be directly impacted by explosions, fire and bird hazards to aircraft.

The EPA guidelines spell out the operational and performance requirements to eliminate these eight types of impacts from solid waste disposal.

Under operational standards, technologies, designs and/or operating methods are specified to a degree that theoretically ensures the protection of public health and the environment. Any or all of a long list of operational considerations could go into a plan to build and operate a disposal facility: type of waste to be handled, facility location, facility design, operating parameters, and monitoring and testing procedures. The advantage of operational standards is that the best practical technologies can be utilized for solid waste disposal, and the state agency mandated with environmental protection can easily determine compliance with a specified operating standard. The major drawback is that compliance is generally not measured by monitoring actual effects on the environment, but rather as an either/or situation where the facility either does or doesn't meet the required operational requirement.

On the other hand, performance standards are developed to provide given levels of protection to land/air/water quality around the disposal site. Determining compliance with performances standard is not easy because the actual monitoring and testing of groundwater, surface water, and land and air quality are costly, complex undertakings.

Because solid wastes are heterogeneous and because site-specific considerations are all-important in their adequate disposal, the federal regulatory effort realized the necessity of allowing state and local discretion in protecting the environmental quality and public health. Therefore, the EPA developed both operational and performance standards to minimize the effect of each of the eight potential impacts of solid waste disposals.

Floodplains. The protection for floodplains focuses on limiting disposal facilities in such areas unless the local area has been protected against being washed out by the base flood. A base flood is defined as a flood that has a 1% or greater chance of occurring in any year, or a flood equaled or exceeded only once in every 100 years on the average.

Endangered and Threatened Species. Solid waste disposal facilities must be constructed and operated so they will not contribute to the "taking" of endangered and threatened species. "Taking" means the harming, pursuing, hunting, wounding, killing, trapping or capturing of species so listed by the U.S. Department of the Interior. Operators of a solid waste disposal facility may be required to modify its operation to comply with these rules if nearby areas are designated as critical habitats.

Surface Water. Solid waste disposal leads to the pollution of surface waters whenever rainwater percolates through the refuse and runoff occurs, or whenever spills take place during solid waste shipments. Point source discharges from disposal sites are regulated by the NPDES permit system discussed in detail in Chapter 10. Nonpoint source, or diffused water, movement is regulated by the requirements implementing an area-wide or statewide water quality management plan approved by the EPA. Generally, these plans regulate the degradation of surface water quality by facility design, operation and maintenance. Artificial and natural runoff barriers like liners, levees and dikes are often required by states. If runoff waters are collected the site becomes by definition a point source of pollution, and the facility must have a NPDES permit if this new point source discharges into surface watercourses.

Groundwater. To prevent contamination of groundwater supplies, the owners and operators of a solid waste disposal facility must comply with one or more of five design and operational requirements: (1) utilize natural hydrogeologic conditions like underlying confining strata to block flow to aquifers, (2) collect and properly dispose of leachate by installing natural or synthetic liners, (3) reduce the infiltration of rainwater into the solid waste with correct cover material, (4) divert contaminated groundwater or leachate from groundwater supplies, and (5) conduct groundwater monitoring and testing procedures.

Land Used for Food Chain Crops. The federal guidelines attempt to regulate the movement of heavy metals and synthetic organics, principally polychlorinated biphenyls (PCB), from solid waste into soils used to produce human food chain crops. Typically, the solid waste of concern here is sludge from municipal wastewater treatment facilities. To control heavy metals, the regulations give the operator of such disposal facilities two options to follow: the controlled application approach and the dedicated facility approach. If the operator elects the controlled application approach, he must conform with an annual application rate of cadmium from solid waste. Cadmium is controlled basically as a function of time and the type of crop to be grown. A cadmium addition of 2.0 kg/ha (1.8 lb/acre) is allowed for all food-chain crops other than tobacco, leafy vegetables and root crops grown for human consumption. An application rate of 0.5 kg/ha (0.45 lb/acre) is in force for these accumulator crops. Under this option, the operator must conform to a cumulative limit on the amount of cadmium which can be disposed of on food chain land. This cumulative limit is a function of the cation exchange capacity (CEC) of the soil and background soil pH. At low pH and low CEC, a maximum cumulative application rate of 5.0 kg/ha (4.50 lb/acre) is allowed; whereas at high CEC and soils with high or near neutral pH levels, a 20 kg/ha (18 lb/acre) amount is allowed. The theory is that in soils of high CEC and high pH, the cadmium remains bound in the soil and is not taken up by the plants. All operators electing this option must ensure that the pH of the solid waste and soil mixture is 6.5 or greater at the time of application.

The second option open to food chain landspreaders is the dedicated facility approach. This option is distinguished from the first option by its reliance on output control, or crop management, as opposed to input control, or limiting the amounts of cadmium that can be applied to the soil. This second cadmium control option is designed specifically for facilities with the resources and the capabilities to closely manage and monitor the performance of their respective operations. The requirements under this option include: (1) only growing animal feed crops, (2) the pH of the solid waste and soil mixture is 6.5 or greater at the time of application, (3) there is a facility operating plan which demonstrates how the animal feed will be distributed to preclude direct ingestion by humans and (4) there are deed notifications for future property owners that the property has received solid wastes at high cadmium application rates and that food-chain crops should not be grown due to a possible health hazard.

Solid waste with PCB concentrations greater than or equal to 10 mg/kg dry weight must be incorporated into the soil when applied to land used for producing animal feed, unless the PCB concentration is found to be less than 1.5 mg/kg fat basis in the produced milk. By requiring soil incorporation, the solid waste regulations attempt to provide that the Food and Drug Administration levels for PCB in milk and animal feed will not be exceeded.

Health. The federal rules require that the operation of a solid waste disposal facility protect the public health from disease vectors. Disease vectors are any routes by which disease is transmitted to humans (birds, rodents, flies and mosquitoes are good examples). This protection must be achieved by minimizing the availability of food for the vectors. At landfills, an effective means to control vectors, especially rodents, is the application of cover material at the end of each day of operation. Other techniques include poisons, repellents and natural controls like the supply predators. Treating sewage sludge with pathogen-reduction processes serves to control the spread of disease from landscaping activities.

Air Quality. A solid waste disposal facility must comply with the Clean Air Act and the relevant State Implementation Plans developed by state and local air emission control boards. Open burning of residential, commercial, institutional and industrial solid waste is generally prohibited. However, several special wastes are excluded from this prohibition: diseased trees, land clearing debris, debris from emergency cleanup operations, and wastes from silviculture and agriculture operations. Open burning is defined as the combustion of solid waste without: (1) control of combustion in the air to maintain adequate temperatures for efficient combustion, (2) containment of the combustion reaction in an enclosed device to provide adequate residence time, or (3) control of the emissions from the combustion process.

Safety. Fires at solid waste disposal facilities are a constant threat to public safety. Death, injury and vast property damage have resulted in the past from fires breaking out in open dumps. The ban on open burning also reduces the chance of accidental fires. In terms of aircraft safety, facilities should not be located between airports and bird feeding, roosting or watering sites. This hazard is important when birds fly across an airport or landing/takeoff pattern.

HAZARDOUS WASTE

Common law is particularly not well developed in the hazardous waste area. The issues discussed in Chapter 14 are newly recognized by society and little case law has had a chance to develop. Since the public health impacts of solid waste disposal in general and hazardous waste disposal in particular were so poorly understood, few plaintiffs ever bothered to take a defendant into the common law courtroom to seek payment for damages or injunctions against such activities. Hazardous waste law is therefore discussed here as statutory law, specifically in terms of the compensation for victims of improper hazardous waste disposal and efforts to regulate the generation, transport and disposal of hazardous waste.

Historically, federal statutory law was generally lacking in describing how victims should be compensated for improper hazardous waste disposal. A complex, repetitious, confusing list of federal statutes was the only recourse for victims.

The federal Clean Water Act only covers wastes that are discharged to navigable waterways. Only surface water and ocean waters within 200 miles from the coast are considered, and a $5 million revolving fund is set up by the act and administered by the Coast Guard. Fines and charges are deposited into the fund to compensate victims of discharges, but the funds are only available if the discharger of the waste is clearly identifiable. The fund is used most often for compensating states for cleanup of spills from large tanker ships.

The Outer Continental Shelf (OCS) Lands Act sets up two funds to help pay for hazardous waste cleanup and to compensate victims. Under the act, OCS leaseholders are required to report spills and leaks from petroleum producing sites, and the Offshore Oil Spill Pollution Fund exists to finance cleanup costs and compensate injured parties for loss of the use of their property, loss of the use of natural resources, loss of profits, and a state or local government's loss of tax revenue. The U.S. Department of Transportation (DOT) is responsible for the administration of this fund, with a balance maintained between $100 and $200 million provided by a 3% tax on oil produced at the OSC sites. If the operator of the site cooperates with DOT once a spill occurs, his liability is limited to a maxiumum of $35 million for a facility spill and $250,000 for a ship spill at the site.

The Fisherman's Contingency Fund also exists under the OCS Lands Act to repay fishermen for loss of profit and equipment due to oil and gas exploration, development and production. The Secretary of the U.S. Department of Commerce is responsible for the administration of this fund. The balance of under $1 million is funded by assessments collected by the U.S. Department of the Interior from holders of permits and pipeline easements. If a fisherman cannot pinpoint the site responsible for the discharge of hazardous waste, his claim against the fund can still be acceptable if his boat was operated within the vicinity of OCS activity and if he filed his claim within 5 days of his injury. The fact that two agencies are involved in the administration makes this fund especially confusing to the victims of hazardous waste incidents.

The Price-Anderson Act was passed to provide compensation to victims of discharge from nuclear facilities. Nuclear incidents including radioactive material/toxic spills and emissions are covered, as are explosions. The licensees of a nuclear facility are required by the Nuclear Regulatory Commission (NRC) to obtain insurance protection. If damages from an accident are greater than this insurance coverage, the federal government indemnifies the licensee up to $500 million. In no case does financial liability exceed $560 million. If a major meltdown had occurred at Three Mile Island in Pennsylvania, the inadequacy of this fund would have been painfully obvious.

Under the Deepwater Ports Act, the Coast Guard acts to remove oil from deepwater ports. The DOT administers a liability fund which helps pay for the cleanup and which compensates injured parties. Financed by a 2¢ per barrel tax on oil loaded or unloaded at such a facility, the fund takes effect once the required insurance coverage is exceeded. The deepwater port itself must hold insurance for $50 million for claims against its waste discharges, and vessels using the port must hold insurance for $20 million for claims against their waste spills. Once these limits are exceeded, the fund established by the Deepwater Ports Act takes over. Again, the usefulness of the fund is dependent on how well two agencies work together and how well insurance claims are administered. This situation is again confusing at best to injured parties.

Statutes at the state level have generally paralleled these federal efforts. New Jersey has a Spill Compensation and Control Act for hazardous wastes listed by the EPA. The fund covers cleanup costs, losses of income, losses of tax revenues, and the restoration of damaged property and natural resources. A tax of 1¢ per barrel of hazardous substance finances a fund that pays injured parties if they file a claim within 6 years of the hazardous waste discharge and within 1 year of the discovery of the damage.

The New York Environmental Protection and Spill Compensation Fund is similar to this New Jersey fund, but with two major differences. In New York, only petroleum discharges are covered, and the generator may not blame a third party for a spill or discharge incident. Thus, if a trucker or handler accidentally spills hazardous waste, the generator of the waste may still be held liable for damages.

Other states have limited efforts in the compensation of victims of hazardous waste incidents. Florida has a Coastal Protection Trust Fund of $35 million which compensates victims for spills, leaks and dumping of waste. However, the coverage is limited to injury due to oil, pesticides, ammonia and chlorine, which provides a loophole for many types of hazardous waste discussed in Chapter 14. Other states that do have compensation funds generally spell out limited compensatable injuries and provide limited funds, often less than $100,000.

In the past, the value of these federal and state statutes has been questioned. Even when taken as a group, they do not provide for complete compensation strategy for dealing with hazardous waste. Few funds address personal injuries, and abandoned disposal sites are not considered. Huge administrative problems also exist as several agencies attempt to respond to a hazardous waste spill. Questions of which fund applies, which agency has jurisdiction, and what injuries can receive compensation linger.

To clean up their acts, the federal government passed a law labeled the

Superfund Act. This act evolved out of several years of battles between congressional committees, the EPA staff, and special interest groups. Originally, the Superfund Act was to serve two purposes: (1) provide money for the cleanup of abandoned hazardous waste disposal sites, and (2) establish liability so dischargers could be made to pay for injuries and damages. Three considerations went into the development of this act:

- *What type of incidents should be covered?* The focus here was on decisions to include spills and abandoned sites under the Superfund Act, as well as onsite toxic pollutants in harbors and rivers across the country. Fires and explosions caused by hazardous materials were also a focus of this consideration as the Superfund concept evolved through the years.
- *What type of damages should be compensated?* As alluded to above, three types of damages could be covered by a superfund: environmental cleanup costs; economic losses associated with property use, income and tax revenues; and personal injury in the form of medical costs for acute injuries, chronic illness, death and general pain and suffering.
- *Who pays for the compensation?* Choices here include federal appropriations into the fund, industry contributions from sales tax, or income tax surcharges. Federal and state cost sharing is also possible, as are fees for disposal of hazardous waste at permitted disposal facilities.

The version adopted by Congress addressed each consideration and reflected the mood of compromise that surrounded the Superfund's evolution. A fund of between $1 billion and $4 billion was established that is financed 87% by a tax on the chemical industry and 13% by general revenues of the federal government. Small payments are permitted out of the fund, but only for out-of-pocket medical costs and partial payment for diagnostic services. The active Superfund Act does not set strict liability for spills and abandoned hazardous wastes.

Compensation for damages is one concern; regulations that control generators, transporters and disposers of hazardous waste are the other side of the hazardous waste law coin. The Federal Resource Conservation and Recovery Act (RCRA), which is discussed at length in this chapter as it relates to solid waste, also is the principal statute that deals with hazardous waste. The RCRA requires the EPA to establish a comprehensive regulatory program to control hazardous waste. This solid and hazardous waste act offers a good example of the types of reporting and recordkeeping requirements that are mandated in federal environmental laws. The Clean Water Act and Clean Air Act place requirements on water and air polluters similar to the types of requirements discussed below for generators of hazardous waste.

A generator of hazardous waste must meet these EPA requirements:

1. determine if the waste is hazardous, as defined by an EPA listing of hazardous waste, by EPA testing procedures, or as indicated by the materials and processes used in production
2. obtain an EPA general identification number
3. obtain a facility permit if hazardous waste is stored at the generating site for 90 days or longer
4. use appropriate containers and labels prior to shipment offsite
5. prepare a manifest for tracking the waste shipment offsite, as described in Chapter 14 (see Figure 14-3).
6. assure arrival of the waste at the disposal site
7. submit annual summaries of activities to federal and/or state regulatory agencies

The key to the regulatory system is this manifest system of "cradle to grave" tracking (see Figure 14-3).

A transporter of hazardous waste must meet several requirements under RCRA:

1. obtain an EPA Transporter Identification Number
2. comply with the provisions of the manifest system
3. deliver the entire quantity of waste to the disposal/processing site
4. keep a copy of the manifest for 3 years
5. follow DOT rules for responding to spills of hazardous waste.

Again, the key element of transportation controls is the implementation and administration of the manifest system.

Facilities which treat, store and/or dispose of hazardous waste also have requirements placed on them by RCRA:

1. The owner of the facility must apply for an operator's permit and supply data on the proposed site and waste to be handled. This permit will spell out the terms of compliance: construction and operating schedules, as well as monitoring and recordkeeping procedures.
2. As permits are granted by the federal or state agency, minimum operating standards will be placed on the facility. Design and engineering standards for containing, neutralizing and destroying the wastes are addressed in these permits, as are safety and emergency measures in the event of accident. Personnel training is included in each permit's requirements.

3.   The financial responsibility of the facility is defined, and a trust fund generally must be established at site closure. This fund enables the monitoring of groundwater and surface discharges to continue after the site is closed, and enables the executor of the trust to maintain the site for years to come.

The key to the successful regulation of hazardous waste generation, transportation and disposal is the manifest system and the manner in which the disposal processing facilities are regulated. The responsibilities for administering RCRA are passed from the EPA to state agencies as these agencies show they have the authority and expertise to effectively regulate hazardous waste.

In addition to the control of hazardous waste, Congress has passed the Toxic Substances Control Act (TOSCA) in an attempt to control hazardous substances before they become hazardous waste. One facet of this regulatory program is known as the Premanufacturing Notification (PMN) System. Under the system, all manufacturers must notify the EPA prior to marketing any substance not included in the EPA's 1979 inventory of toxic substances. In this notification, the manufacturer must analyze the predicted effect of the substance on workers, on the environment and on consumers. This analysis must be based on test data and all relevant literature. A Compliance Checklist is presented in Table 16-1 where specific steps are spelled out for manufacturers who wish to produce a new product.

Table 16-1.   Checklist to Comply with TOSCA's Premanufacture Notification (PMN) System

Step	Requirement
1	Inspect EPA inventory of toxic substances to see if new product is listed.
2	Assess EPA rules for testing procedures to determine if substances not listed by the EPA are, in fact, "toxic."
3	Prepare and submit the PMN form at least 90 days prior to manufacture.
4	Obtain an EPA ruling: • *No ruling:* begin manufacturing at end of 90-day period • *EPA Order:* respond to EPA order for more data • *EPA Ruling:* EPA may prohibit or limit the proposed manufacturing, processing, use, or disposal of the substance.

CONCLUSION

Solid waste statutory law developed rather rapidly once scientists and engineers realized the real and potential impacts of improper disposal. Air, land and water quality from such disposal became a key concern to local health officials and federal and state regulators. Hazardous and nonhazardous solid wastes are now regulated under a complex system of federal and state statutes which place operating requirements on facility operators.

The newness of these statutes leaves their effectiveness in doubt. How will the EPA implement RCRA? If states are given the responsibility to control hazardous and nonhazardous waste within their boundaries, how will they respond? In the end, what will be the public health, environmental quality, and economic impacts of RCRA and the Superfund?

PROBLEMS

16.1 Local land use ordinances often play key roles in limiting the number of sites available for a solid or hazardous waste disposal facility. Discuss the types of these zoning restrictions which apply in your home town.

16.2 Assume you work for a firm which contracts to clean the inside and outside of factories and office buildings in the state capitol. Your boss thinks he could make more money by expanding his business to include the handling and transportation of hazardous wastes generated by his current clients. Outline the types of data you must collect in order to advise him to expand/not expand.

16.3 Laws which govern refuse collection often begin with controls on the generator; i.e., rules which each household most follow if city trucks are to pick up their solid waste. If you were asked by a town council to develop a set of such "household rules," what controls would you include? Emphasize public health concerns, minimizing the cost of collection, and even resource recovery considerations.

16.4 The oceans have long been viewed as a bottomless pit into which the solid wastes of the world may be dumped. Engineers, scientists and politicians are divided on the issue. Should ocean disposal be banned? Develop arguments pro and con.

LIST OF SYMBOLS

DOT = U.S. Department of Transportation
EPA = U.S. Environmental Protection Agency
NRC = Nuclear Regulatory Commission
OCS = outer continental shelf
PMN = Premanufacture Notification
RCRA = Resource Conservation and Recovery Act
TOSCA = Toxic Substances Control Act

SUGGESTED READING

Office of Solid Waste. *Environmental Impact Statement: Criteria for Classification of Solid Waste Disposal Facilities and Practices,* U.S. Environmental Protection Agency (1979).

U.S. Congress. *Resource Conservation and Recovery Act: Public Law 94-580,* U.S. Government Printing Office (1976).

U.S. Environmental Protection Agency. "Criteria for Classification of Solid Waste Disposal Facilities and Practices," in *Federal Register 53438,* U.S. Government Printing Office (September 13, 1979).

U.S. Environmental Protection Agency. "Hazardous Waste and Consolidated Permit Regulations," 44, *Federal Register 33066,* U.S. Government Printing Office (May 19, 1981).

U.S. Congress. *Toxic Substances Control Act, Public Law 94-469,* U.S. Government Printing Office (1976).

Chapter 17

Air Pollution

> Be it known to all within the sound
> of my voice, whosoever shall be
> found guilty of burning coal shall
> suffer the loss of his head.
> King Edward II, ca. 1300

Obviously, air pollution is not a new problem. King Edward II (1284–1327) tried to solve the problem of what Eleanor of Aquitaine called "the unendurable smoke" by prohibiting the burning of coal while Parliament was in session. His successors, Richard III (1377–1422) and Henry V (1413–1422) both took action against smoke, the former by taxation and the latter by forming a commission to study the problem. This was the first of a plethora of commissions, none of which helped reduce the level of air pollution in England. Under Charles II (1630–1685) a pamphlet was authored in 1661 by John Evelyn, entitled "Fumifugium or the Inconvenience of Aer and Smoak of London Dissipated, together with some Remedies Humbly Proposed." His suggestions included moving industry to the outskirts of town and establishing green belts around the city. None of his proposals were implemented. A subsequent commission in 1845 suggested, among other solutions to the smoke problems, that locomotives "consume their own smoke." In 1847 this requirement was extended to chimneys and in 1853 it was decreed that offending chimneys be torn down.

In fact, for all the rhetoric and commissions, there was little action in England, or anywhere else in the industrially developing world, until after World War II. Action was finally prompted in part by two major air pollution episodes where human deaths were directly attributed to high levels of pollutants.

The first of these occurred in Donora, a small steel town (pop. 14,000)

in western Pennsylvania. Donora is located in a bend of the Monongahela River, and in 1948 had three main industrial plants—a steel mill, a
wire mill and a zinc plating plant.

During the last week of October 1948, a heavy smog settled in the area,
and a weather inversion prevented the movement of pollutants out of the
valley. (See Chapter 18 for a definition of inversion.) On Wednesday, the
smog became especially intense. It was reported that streamers of carbon
appeared to hang motionless in the air and visibility was so poor that
even natives of the area became lost.[1] By Friday, the doctors' offices and
hospitals were flooded with calls for medical help. Yet no alarm had been
sounded. The Friday Halloween parade was well attended, and a large
crowd watched the Saturday afternoon football game between Donora
and Monongahela High Schools.*

The first death had however occurred at 2 a.m. More followed in quick
succession, and by midnight 17 persons were dead. Four more died
before the effects of the smog abated. By this time the emergency had
been recognized and special medical help was rushed in.

Although the Donora episode helped focus attention on air pollution
problems in the United States, it took four more years before England
suffered a similar disaster and action was finally taken. The "Killer
Smog" of 1952 occurred in London, with meteorological conditions similar to those during the Donora episode. A dense fog at ground level
coupled with bitter cold and the smoke from coal burners caused the
formation of another infamous "pea souper." This one was more severe
than the usual smog however, and lasted for over a week. The smog was
so heavy that visibility during daylight hours was cut to only a few
meters. Bus conductors had to walk in front of their vehicles to guide
them through the streets (Figure 17-1). Two days after the fog set in, the
death rate in London began to soar (Figure 17-2). Sulfur dioxide concentrations increased to nearly seven times their normal levels and carbon
monoxide was twice the normal. It is important, however, to not conclude from the apparent relationship in Figure 17-2 that sulfur dioxides

---

*It has been reported that the Monongahela coach protested the game and asked that it
not be recorded in the books, because he alleged that the Donora coach contrived to have a
pall of smog hanging over the field so that on kicking and passing plays, the ball would disappear and his players did not know where it would reappear.[2]

[1]Schrenk, H. H. et al. "Air Pollution in Donora, Pa." U.S. Public Health Service Bulletin, No. 306 (1949).

[2]Chanlett, E. "History of Sanitation" in *History of Environmental Sciences and
Engineering.* P. A. Vesilind, Ed., University of North Carolina, Department of E.S.E.
(1975).

Figure 17-1   London smog during the 1952 episode (courtesy BBC Hulton Picture Library).

caused the deaths. Many other factors may have been responsible, and the various air pollutants probably acted synergistically to affect the death rate.

Primarily due to these acute episodes in London and Donora, public opinion and concern forced the initial attempts to clean up urban air.

Interestingly, we have difficulty in defining what *is* clean air. From the scientific standpoint, clean air is composed of the constituents listed in Table 17-1. If we accept this as a definition of clean air, however, we are in trouble, since any naturally occurring suspended material can be called a pollutant, and one never finds such "clean air" in nature. It may thus be more appropriate to define air pollutants as those substances which exist in such concentrations as to cause an unwanted effect. These pol-

Figure 17-2   Deaths and pollutant concentration during the London 1952 air pollution episode. [Source: Wilkins, E. T., *Journal of the Royal San. Inst.* 74 (1) (1954).]

lutants can be natural (such as smoke from forest fires) or man-made (such as automobile exhaust) and can be in the form of gases or particulates (liquid or solid particles larger than 1 micrometer).*

TYPES OF AIR POLLUTANTS

Gaseous Pollutants

In the context of air pollution control, gaseous pollutants include substances that are gases at normal temperature and pressure as well as vapors of substances that are liquid or solid at normal temperature and pressure. Among the gaseous pollutants of greatest importance in terms of present knowledge are carbon monoxide, hydrocarbons, hydrogen sulfide, nitrogen oxides, ozone and other oxidants, and sulfur oxides.

---

*In air pollution control parlance, a micrometer is often referred to as a *micron* (μ). We will adopt this usage here.

Table 17-1.   **Components of Normal Dry Air**

	Concentration (ppm)
Nitrogen	780,900
Oxygen	209,400
Argon	9,300
Carbon Dioxide	315
Neon	18
Helium	5.2
Methane	1.0–1.2
Krypton	1.0
Nitrous Oxide	0.5
Hydrogen	0.5
Xenon	0.008
Nitrogen Dioxide	0.02
Ozone	0.01–0.04

Carbon dioxide should be added to this list because of its potential effect on climate. These and other gaseous air pollutants are listed in Table 17-2.

Pollutant concentrations are commonly expressed as micrograms per cubic meter ($\mu g/m^3$). An older and still used method is to express gaseous pollutant concentrations as parts per million, where

$$1 \text{ ppm} = \frac{1 \text{ volume of gaseous pollutant}}{10^6 \text{ volumes, pollutant + air}}$$

and 1 ppm = 0.0001 percent by volume.

At 25°C and 760 mm Hg (1 atmosphere) the relationship between ppm and $\mu g/m^3$ is

$$\mu g/m^3 = \frac{\text{ppm} \times \text{molecular weight} \times 10^3}{24.5}$$

The constant 24.5 is the volume in liters occupied by one gram-mole of an ideal gas (at 25°C and 760 mm Hg). For different temperatures and pressures, this constant of course changes.

The relationship between temperature, pressure and volume of an ideal gas is expressed as Boyle's and Charles' Laws:

At constant temperature, $P_1 V_1 = P_2 V_2$

Table 17-2.   Gaseous Air Pollutants

Name	Formula	Properties of Importance	Significance as Air Pollutant
Sulfur Dioxide	SO_2	Colorless gas, intense choking odor, highly soluble in water to form sulfurous acid H_2SO_3	Damage to vegetation, property and health
Sulfur Trioxide	SO_3	Soluble in water to form sulfuric acid H_2SO_4	Highly corrosive
Hydrogen Sulfide	H_2S	Rotten egg odor at low concentrations, odorless at high concentrations	Highly poisonous
Nitrous Oxide	N_2O	Colorless gas, used as carrier gas in aerosol bottles	Relatively inert. Not produced in combustion
Nitric Oxide	NO	Colorless gas	Produced during high-temperature, high-pressure combustion. Oxidizes to NO_2
Nitrogen Dioxide	NO_2	Brown to orange gas	Major component in the formation of photochemical smog
Carbon Monoxide	CO	Colorless and odorless	Product of incomplete combustion. Poisonous
Carbon Dioxide	CO_2	Colorless and odorless	Formed during complete combusion. Possible effects in producing changes in global climate
Ozone	O_3	Highly reactive	Damage to vegetation and property. Produced mainly during the formation of photochemical smog
Hydrocarbons	C_xH_y or HC	Many	Some hydrocarbons are emitted from automobiles and industries, others are formed in the atmosphere

$$\text{At constant pressure,} \quad \frac{V_1}{T_1} = \frac{V_2}{T_2}$$

Combined, we have the ideal gas law:

$$PV = nRT$$

where   P = pressure, N/m^2
          V = volume, m^3
          n = number of moles of a gas
          R = universal gas constant, 8.31 J/degree-mole
          T = absolute temperature, °K

For air at lower altitudes, this equation is more conveniently expressed in terms of mass density:

$$P = \rho R_a T$$

where ρ = air density in kg/m^3
          R_a = 287 J/kg-degree.

Example 17.1

An automobile emission contains 2% by volume of carbon monoxide and has a temperature of 80°C. What is the concentration in ppm and $\mu g/m^3$?

Since 1 ppm = 0.0001%,

$$2\% = 20,000 \text{ ppm}$$

From Charles' Law;

$$\frac{V_1}{T_1} = \frac{V_2}{T_2}$$

$$\frac{24.5}{273° + 25°} = \frac{V_2}{273° + 80°}$$

$$V_2 = 29.0 \text{ liters}$$

This one g-mole of a gas at 80°C occupies 29.0 liters.
The molecular weight of CO is 12 + 16 = 28, and

$$\mu g/m^3 = \frac{20,000 \text{ ppm} \times 28 \times 10^3}{29.0}$$

$$= 19.3 \times 10^6 \ \mu g/m^3$$

Particulate Pollutants

Particulate pollutants are classified as follows:

1. *Dust*—solid particles which are: (a) entrained by process gases directly from the material being handled or processed, e.g., coal, ash and cement; (b) direct offspring of a parent material undergoing a mechanical operation, e.g., sawdust from woodworking; (c) entrained materials used in a mechanical operation, e.g., sand from sandblasting. Dusts from grain elevators and coal-cleaning plants typify this class of particulate. Dust consists of relatively large particles. Cement dust, for example, is about 100 μ in diameter.

2. *Fume*—a solid particle, frequently a metallic oxide, formed by the condensation of vapors by sublimation, distillation, calcination or chemical reaction processes. Examples of fumes are zinc and lead oxide resulting from the condensation and oxidation of metal volatilized in a high-temperature process. The particles in fumes are quite small, with diameters from 0.03 to 0.3 μ.

3. *Mist*—a liquid particle formed by the condensation of a vapor and perhaps by chemical reaction. An illustration of this process is the formation of sulfuric acid mist:

$$SO_3 \text{ (gas) } 22°C \rightarrow SO_3 \text{ (liquid)}$$

$$SO_3 \text{ (liquid) } + H_2O \rightarrow H_2SO_4$$

Sulfur trioxide gas becomes a liquid since its dew point is 22°C and SO_3 particles are hydroscopic. Mists typically range from 0.5 to 3.0 μ in diameter.

4. *Smoke*—solid particles formed as a result of incomplete combustion of carbonaceous materials. Although hydrocarbons, organic acids, sulfur oxides and nitrogen oxides are also produced in combustion processes, only the solid particles resulting from the incomplete combustion of carbonaceous materials are smoke. Smoke particles have diameters from 0.05 to approximately 1 μ.

5. *Spray*—a liquid particle formed by the atomization of a parent liquid.

Approximate size ranges of the various types of air pollutants are shown in Figure 17-3. The concentration of particles is always expressed as $\mu g/m^3$.

SOURCES OF AIR POLLUTION

Many of the pollutants of concern are formed and emitted through natural processes. For example, naturally occurring particulates include pollen grains, fungus spores, salt spray, smoke particles from forest fires, and dust from volcanic eruptions. Gaseous pollutants from natural

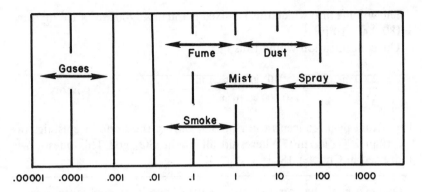

Figure 17-3   Approximate size ranges of air pollutants (μ).

sources include carbon monoxide as a breakdown product in the degradation of hemoglobin, hydrocarbons in the form of terpenes from pine trees, hydrogen sulfide resulting from the breakdown of cysteine and other sulfur-containing amino acids by bacterial action, nitrogen oxides and methane.

Man-made sources of pollutants can be conveniently classified as stationary combustion, transportation, industrial process and solid waste disposal sources. The principal pollutant emissions from stationary combustion processes are particulate pollutants, as fly ash and smoke, and sulfur and nitrogen oxides. Sulfur oxide emissions are, of course, a function of the amount of sulfur present in the fuel. Thus, combustion of coal and oil, both of which contain appreciable amounts of sulfur, yields significant quantities of sulfur oxide.

The amount of sulfur oxide produced can be calculated if the sulfur content of the fuel is known, as the example below illustrates.

Example 17.2

During the 1952 London episode, it is estimated that 25,000 metric tons of coal was burned which had an average sulfur content of about 4%. The mixing depth (height of the inversion layer or cap over the city which prevented the pollutants from escaping) was about 150 m, over an area of 1200 km². What is the expected SO_2 concentration at the end of this time?

Amount of sulfur emitted = 25,000 metric tons × 0.04 = 1000 metric tons. Sulfur has an atomic weight of 32. Sulfur dioxide has a molecular weight of 64. The amount of sulfur dioxide emitted was thus = (64/32) × 1000 metric tons = 2000 metric tons.

The volume into which this is mixed is 150 m × 1200 km^2 × 10^6 m^2/km^2 = 180,000 × 10^{6}m^3

The concentration would thus be =

$$\frac{2000 \text{ metric tons} \times 10^6 \text{ g/metric ton} \times 10^6 \text{ }\mu\text{g/g}}{180,000 \times 10^6 \text{m}^3} = 11,000 \text{ }\mu\text{g/m}^3$$

The actual peak concentration of SO$_2$ during the London episode was less than 2000 μg/m^3. Where did all of the SO$_2$ go? This question is addressed in Chapter 18.

One effect of the combustion of sulfur-containing coal is the formation of acid rain. Normal, uncontaminated rain has a pH of about 5.6, but acid rain can be as low as pH 2 or even below. Hundreds of lakes in North America and Scandinavia have become so acidic that they no longer can support fish life. In a recent study of Norwegian lakes, more than 70% of the lakes having a pH of less than 4.5 contained no fish, and nearly all lakes with a pH of 5.5 and above contained fish. The low pH not only affects fish directly, but contributes to the release of potentially toxic metals such as aluminum, thus magnifying the problem. In Norway, storms which travel over the industrial areas of Great Britain and continental Europe can be tracked and have been found to dump especially destructive precipitation. Figure 17-4 shows three such storms and the corresponding pH values of the rainwater. The recognition of this problem makes the use of the "tall stack" method of air pollution control highly questionable (see Chapter 20).

In North America, acid rain has already wiped out all fish and many plants in 50% of the high mountain lakes in the Adirondacks. The pH in many of these lakes has reached such levels of acidity as to replace the trout and native plants with acid-tolerant mats of algae.[3]

Nitrogen oxides are formed by the thermal fixation of atmospheric nitrogen in high-temperature processes. Accordingly, almost any combustion operation will produce nitric oxide (NO). Other pollutants of interest from combustion processes are organic acids, aldehydes, ammonia and carbon monoxide. The amount of carbon monoxide emitted relates to the efficiency of the combustion operation, i.e., a more efficient combustion operation will oxidize more of the carbon present to carbon dioxide, reducing the amount of carbon monoxide emitted. Of the fuels used in stationary combustion, natural gas contains practically no sulfur, and particulate emissions are almost nil.

---

[3]Schoefield, C. Cornell University, as reported in the *Washington Post* (7 January 1981).

	pH of rain water at sampling site
"Normal"	5.8
Storm of 25 Aug.	4.2
Storm of 26 Aug.	4.0
Storm of 27 Aug	4.1

Figure 17-4   Paths of three storms sampled in Kristiansand, Norway.

Transportation sources, particularly automobiles using the internal combustion engine, constitute a major source of air pollution. the various points of emission are shown in Figure 17-5. Particulate emissions from the automobile include smoke and lead particles, the latter usually as halogenated compounds. Smoke emissions, as in any other combustion operation, are due to the incomplete combustion of carbonaceous material. On the other hand, lead emissions relate directly to the addition of tetraethyl lead to the fuel as an antiknock compound. Gaseous pollutants from transportation sources include carbon monoxide, nitrogen oxides and hydrocarbons. Hydrocarbon emissions result from incomplete combustion and evaporation from the crankcase, carburetor and gasoline tank.

Significant progress has been made in controlling pollution from automobiles. Exhaust has been cleaned by better than 90% over 1963 models, and further improvement will be evident over the next few years.

Pollution emissions from industrial processes reflect the ingenuity of modern industrial technology. Thus, nearly every imaginable form of pollutant is emitted in some quantity by some industrial operation.

Although solid waste disposal operations need not be a major source of air pollutants, many communities still permit backyard burning and/or disposal of solid waste by open burning dumps. These techniques, which attempt to reduce the volume of waste, produce instead a variety of difficult-to-control pollutants. Among these are undesirable

Figure 17-5   Sources of automotive air pollution (courtesy General Motors).

odors, carbon monoxide, small amounts of nitrogen oxides, organic acids, hydrocarbons, aldehydes, and great quantities of smoke. (Both backyard burning and open burning dumps are prime examples of inefficient combustion.)

PRIMARY AND SECONDARY POLLUTANTS

An important approach to classification is that of primary and secondary pollutants. A primary pollutant is defined as one that is emitted as such to the atmosphere, whereas a secondary pollutant is one formed in the atmosphere. The components of automobile exhaust are particularly important in the formation of secondary pollutants. The well known and much discussed Los Angeles smog is a case of secondary pollution formation. Table 17-3 lists in simplified form some of the key reactions in the formation of photochemical smog.

The reaction sequence illustrates how nitrogen oxides formed in the combustion of gasoline and other fuels and emitted to the atmosphere are acted upon by sunlight to yield ozone (O_3), a compound not emitted as such from a source and hence considered a secondary pollutant. Ozone

Table 17-3.   Simplified Reaction Scheme
for Photochemical Smog

NO_2	+ Light	→ NO	+ O
Nitrogen Dioxide		Nitric oxide	Atomic oxygen
O	+ O_2 Molecular oxygen	→ O_3 Ozone	
O_3	+ NO	→ NO_2	+ O_2
O	+ HC Hydrocarbon	→ $HCO^•$ Radical	
$HCO^•$	+ O_2	→ $HCO_3^•$ Radical	
$HCO_3^•$	+ HC	→ Aldehydes, ketones, etc.	
$HCO_3^•$	+ NO	→ $HCO_2^•$ Radical	+ NO_2
$HCO_3^•$	+ O_2	→ O_3	+ $HCO_2^•$
$HCO_x^•$ Radical	+ NO_2	→ Peroxyacetyl nitrates	

in turn reacts with hydrocarbons to form a series of compounds which includes aldehydes, organic acids and epoxy compounds. Thus, the atmosphere can be viewed as a huge reaction vessel wherein new compounds are being formed while others are being destroyed.

The formation of photochemical smog is a dynamic process. Figure 17-6 is an illustration of how the concentrations of some of the components vary during the day. Note that as the morning rush hour begins the NO levels increase, followed quickly by NO_2. As the latter reacts with sunlight, O_3 and other oxidants are produced. The hydrocarbon level similarly increases at the beginning of the day and then drops off in the evening.

Figure 17-6   Formation of photochemical smog.

The reactions involved in photochemical smog remained a mystery for many years. Particularly baffling was the formation of high ozone levels. As seen from the first three reactions in Table 17-3, for every mole of NO_2 reacting to make atomic oxygen and hence ozone, one mole of NO_2 was created from reaction with the ozone. All of these reactions are fast. How, then, could the ozone concentrations build to such high levels?

One answer is that NO enters into other reactions, especially with various hydrocarbon radicals, and thus allows excess ozone to accumulate in the atmosphere (the seventh reaction in Table 17-3). In addition, some hydrocarbon radicals react with molecular oxygen and also produce ozone.

The chemistry of photochemical smog is still not clearly understood. This was witnessed by the recent attempt to reduce hydrocarbon emissions in order to control ozone levels. The thinking was that if HC is not available, the O_3 will be used to oxidize NO to NO_2, thus using the available ozone. Unfortunately, this control strategy was a failure, and the answer seems to be that all primary pollutants involved in photochemical smog formation must be controlled.

HEALTH EFFECTS

Much of our knowledge of the effects of air pollution on people comes from the study of acute air pollution episodes. As previously noted, the two most famous episodes occurred in Donora, Pennsylvania, and in London, England. In both episodes the illness appeared to be chemical irritation of the respiratory tract. The weather circumstances were also similar in that a high pressure system with an inversion layer was present. An inversion, which is discussed further in Chapter 18, is a layer of warm air aloft which prevents the pollutants from escaping and diluting vertically. In addition, inversions are usually accompanied by low surface winds leading to reduced horizontal dilution of pollutants.

In both episodes, the pollutants affected a specific segment of the public—those individuals already suffering from diseases of the cardio-respiratory system. Another observation of great importance is that it was not possible to blame the adverse effects on any one pollutant. This observation puzzled the investigators (industrial hygiene experts) who were accustomed to studying industrial problems where one could usually relate health effects to a specific pollutant. Today, after many years of study, it is thought that the health problems during the episodes were attributable to the combined action of a particulate matter (solid or liquid particles) and sulfur dioxide, a gas. No one pollutant by itself, however, could have been responsible.

Another episode of historical significance is one involving the accidental release of hydrogen sulfide gas from a refinery at Poza Rica, Mexico, in November 1950. Although the Poza Rica episode is not community air pollution in its usual sense, it is an important example of air pollution which resulted from an industrial accident. The accident occurred when the processes in a refinery went astray and a large quantity of hydrogen sulfide gas escaped. H_2S is a heavy, poisonous gas, and when it escaped, it hugged the ground around the refinery and crept into nearby homes. In all, 22 persons died and 320 became ill.*

Another historically important air pollution health effect is "Yokohama Asthma." Following the Second World War an unusual number of American military personnel and their dependents residing in the Yokohama area of Japan required treatment for attacks of an "asthma-like" condition. These attacks were first noted in the winter of 1948. Investigation showed that the attacks were coincident with conditions of low wind velocity and resulting high pollution levels. Interestingly, only people who were heavy smokers were affected, and the only solution to the problem was reassignment of the victims to another location.

In the United States similar problems affecting asthmatics have been reported in New Orleans. Studies of patients admitted for emergency treatment for asthma attacks were conducted over a period of years at New Orleans' Charity Hospital. These studies indicated that a sharply increased number of asthmatics required emergency treatment when winds of low speed from the south and southwest prevailed. Investigation further showed that the occurrence of these attacks was associated with a burning dump southwest of New Orleans. A measure of the severity of this problem is illustrated by the October 1962 episode which caused the deaths of nine asthmatics and required emergency treatment for 300.

Recently, several air pollution episodes have taken place in New York City and other metropolitan areas. The first well documented episode occurred in November 1953, and air pollution alerts occur every year. The morbidity and mortality** data from these episodes have demonstrated a statistical association between public health and the occurrence of temperature inversions and the resulting increased levels of particulate matter and sulfur dioxide.

Except for these episodes, scientists have very little information from which to evaluate the health effects of air pollution. Laboratory studies with animals are of some help, but the step from a rat to a man (anatomically speaking) is quite large.

Four of the most difficult problems in relating air pollution to health are unanswered questions concerning: (1) the existence of thresholds, (2) the total body burden of pollutants, (3) the time versus dosage problem, and (4) synergistic effects of various combinations of pollutants.

---

*Hydrogen sulfide, disguised as "sewer gas," has killed many unsuspecting municipal workers who ventured into large sewers.

**_Mortality_ is of course death (a statistic); _morbidity_ is illness not resulting in death.

1. *Threshold.* The existence of a threshold in health effects of pollutants has been debated for many years. With reference to Figure 17-7, there are three basic dose-response curves possible for a dose of a specific pollutant (e.g., carbon monoxide) and the response (e.g., reduction in the blood's oxygen-carrying capacity). Curve A shows that if this dose-response relationship holds, there is no effect on human metabolism until a critical concentration (the threshold) is reached. Curve B on the other hand suggests that there is a detectable response for *any* finite concentration of the pollutant. Curve C seems to be the most likely dose-response relationship for many pollutants and suggests no strict threshold, but also shows a minimal response up to a higher concentration, at which point the response becomes severe. The problem is that for most pollutants the shapes of these curves are unknown.

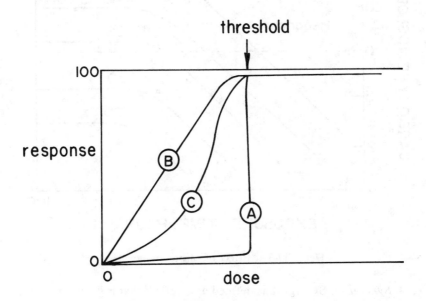

Figure 17-7   Possible dose-response curves.

2. *Total Body Burden.* Not all of our dose of pollutants comes from air. For example, although we breathe in about 50 μg/day of lead, we take in about 300 μg/day of lead in our water and food. In the setting of air quality standards for lead it must therefore be recognized that most of the lead intake is from food and water.

3. *Time Versus Dosage.* Most pollutants require time to react, and thus the time of contact is as important as the level. The best example of this is the effect of carbon monoxide, as illustrated in Figure 17-8. CO

reduces the oxygen-carrying capacity of the blood by combining with the hemoglobin and forming carboxyhemoglobin. At about 60% carboxy-hemoglobin concentration, death results from lack of oxygen. The effects of CO at sublethal concentrations are usually reversible. Because of the time-response problem, ambient air quality standards are set at maximum allowable concentrations for a given time (Chapter 21).

Figure 17-8   Effect of carbon monoxide.

4. *Synergism.* Synergism is defined as an effect that is greater than the sum of the parts. For example, black lung disease in coal miners occurs only when the miner is also a cigarette smoker. Coal mining by itself, or cigarette smoking by itself will not cause black lung, but the synergistic action of the two puts miners who smoke at high risk.

The Respiratory System

The major target of air pollutants is the respiratory system, pictured in Figure 17-9. Air (and entrained pollutants) enter the body through the throat and nasal cavities and pass to the lungs through the trachea. In the lungs, the air moves through bronchial tubes to the alveoli, small air sacks in which the gas transfer takes place. Pollutants are either absorbed into the bloodstream, or moved out of the lungs by tiny hair cells called cilia which are continually sweeping mucus up into the throat. The respiratory system can be damaged by both particulate and gaseous pollutants.

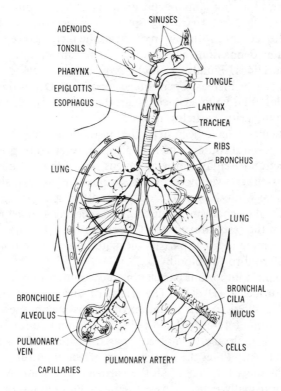

Figure 17-9   The respiratory system (courtesy American Lung Association).

Particulate Matter

The site and the extent of the deposition of particulates in the respiratory tract is a function of certain physical factors, the most important being particle size.

Alveolar deposition is especially important since that region of the lungs is not provided with cilia to remove particulates. Thus, particles deposited there would remain for a relatively greater length of time.

Very small particles, less than 0.1 μ in diameter, will be deposited in the alveoli due to their Brownian motion. Larger particles will be deposited only if they are able to negotiate the trip through the respiratory system without getting stuck on the cilia and expelled. Particles greater than 1 μ will generally be caught before they reach the alveoli.

Other factors that determine the amount of particulate deposition are respiratory frequency (breaths per unit time) and tidal volume (the volume moved in and out of the lungs with each breath). Low respiratory frequencies result in fairly high deposition percentages; somewhat higher frequencies cause a return to high-percentage deposition. This is explained by the fact that with low frequencies the increased residence time permits greater deposition, whereas with increasing frequency the residence time declines. Finally, at high frequencies increased turbulence may account for the increased percentage deposition. Deposition increases with increasing tidal volume since the greater volume of air inspired the greater the number of particles inspired.

Once particulates have been deposited in the lung, their removal can take place by several routes. The first of these, referred to previously, is due to the action of cilia in moving mucus out of the respiratory tract. The mucus is either swallowed or expectorated. Phagocytosis, wherein cells rather like white blood cells engulf and transport the particles, is another removal mechanism. Coughing and sneezing also act to dislodge particles from the respiratory tract. Of course, the solubility of the particle will determine in great part its ultimate fate, i.e., soluble particles will be transferred to the blood.

One of the most important particles man encounters in polluted atmospheres is sulfuric acid (H_2SO_4). The effects of sulfuric acid are basically those of sulfur dioxide, i.e., irritation of mucus membranes and bronchial constriction. Interestingly, if sulfuric acid and sulfur dioxide are compared on a molecule to molecule basis, sulfuric acid will produce from four to twenty times the physiological effect of sulfur dioxide.

Perhaps the most familiar of all the particulates is lead. Atmospheric lead results principally from the use of tetraethyl lead in motor fuel. Lead has been shown to interfere with the development and maturation of red blood cells; chronic exposure leads to a stippled red cell. Hemoglobin precursors, known as porphyrins, are excreted in the urine of those exposed to lead. Lead can therefore be detected in both blood and urine.

Investigations have shown that urinary and blood lead values are higher in urban residents than in their rural counterparts. Occupational exposure to auto exhaust also results in elevated blood lead levels. As might be expected, blood lead is higher in cigarette smokers than in nonsmokers. It must again be emphasized, however, that the total intake of lead (body burden) is not only from the atmosphere. If there is an effect resulting from lead exposures to urban atmospheres, it most certainly will be one of chronic toxicity since the levels experienced are not sufficient to produce acute lead intoxication.*

The toxicity of beryllium was first recognized when disease resulting from inhaling dust from broken fluorescent tubes was noted (beryllium was once used as the phosphor in the tubes). Beryllium can be a community air pollutant also. Berylliosis was discovered in a small community which had a beryllium plant and in which the majority of the stricken resided within 3/4 mile of the plant, where the estimated concentration of beryllium was about 0.1 $\mu g/m^3$.

Berylliosis is characterized by diffuse pulmonary fibrosis which interferes with the diffusion of gases from the lungs to the blood. The symptoms of berylliosis include weight loss, shortness of breath, cough, and sometimes changes in the bone. Beryllium can be detected in blood, urine and body tissues.

Gaseous Pollutants

Sulfur dioxide (SO_2) is a gas highly soluble in water and hence in body fluids. The primary effect of this gas is irritation of the tissues lining the upper respiratory tract resulting in increased resistance to air flow. Experiments have shown that the effect of sulfur dioxide on the flow of air in the respiratory system can be aggravated by the presence of an inert aerosol** such as sodium chloride. The effect of sulfur dioxide in the presence of the aerosol, which in itself has no effect, is much greater than the effect of the gas alone, a classical example of synergism. The aerosol probably adsorbs the sulfur dioxide, thus allowing deeper penetration of sulfur dioxide into the respiratory tract than would otherwise occur. It is also possible that the adsorbed sulfur dioxide undergoes oxidation to sulfuric acid which is a more potent irritant than sulfur dioxide. In general, it can be said that sulfur dioxide's effect is more acute than chronic.

Acute effect is short and usually severe (e.g., the common cold is an acute respiratory infection) whereas *chronic effect* is drawn out over a long time (e.g., chronic bronchitis). An example of an acute lead poisoning incident occurred in El Paso, Texas, where a number of school children were found to suffer from the effects of lead poisoning. A lead smelting plant has been blamed as the source.

** An aerosol is a suspension of particulates in a gas. Thus a suspension of tiny salt crystals in air is an aerosol.

The cilia (Figure 17-9) which protect the respiratory system by sweeping out particles are also affected by sulfur dioxide. Experiments using rabbits and other animals have shown that the frequency with which cilia beat is decreased in the presence of sulfur dioxide. Thus, sulfur dioxide, in addition to constricting the bronchi, also affects the protection mechanism of the respiratory tract.

Nitrogen dioxide (NO_2) is a pulmonary irritant. Although little is known of the specific toxic mechanism, it is known that nitrogen dioxide at concentrations admittedly greater than those found in community air is an edema producer and also results in pulmonary hemorrhage. Edema is an abnormal accumulation of fluid in body tissues. Pulmonary edema is excessive fluid in the lung tissues.

Nitric oxide (NO), the other common oxide of nitrogen, is not an irritant gas. *In vitro* nitric oxide combines readily with human hemoglobin to form a highly stable nitric oxide hemoglobin, but interestingly this hemoglobin effect has not been observed in living animals.

Ozone (O_3) is a highly irritating, oxidizing gas. Concentrations of a few parts per million produce pulmonary congestion, edema and hemorrhage. A one-hour exposure of human subjects to 2500 $\mu g/m^3$ can increase the residual lung volume and decrease maximum breathing capacity. The symptoms of ozone exposure are initially a dry throat, followed by headache, disorientation and altered breathing patterns.*

A fascinating observation from animal studies is the development of ozone tolerance. An animal exposed initially to a low concentration of ozone can survive a subsequent exposure to an ordinarily fatal concentration of ozone. To what extent ozone tolerance develops in humans is unknown.

Carbon monoxide (CO), discussed earlier, is an asphyxiant gas which combines readily with hemoglobin (Hb) to form carboxyhemoglobin (COHb). Human hemoglobin has a carbon monoxide affinity 210 times greater than its affinity for oxygen, thus preventing oxygen transfer. The

---

*Ozone in the air we breathe should not be confused with the so-called "ozone layer" in the stratosphere. The stratosphere begins at an altitude of seven to ten miles, depending on the latitude and season of the year. Ozone in this thin air absorbs a large part of the sun's ultraviolet radiation. It is believed that vapor from supersonic aircraft flights and fluorocarbon gases from spray cans will cause permanent reduction in stratospheric ozone. This could increase the ultraviolet radiation reaching the earth, raising the incidence of human skin cancer and probably affecting the earth's climate and ecological systems in unpredictable ways. The high ozone layer seems to be maintained by natural processes, mainly the sun's radiation. Lightning discharges are the principal natural source of ozone in the lower atmosphere.

formation of carboxyhemoglobin effectively reduces the amount of hemoglobin available to carry oxygen to the tissues, and death can occur by this asphyxiation.

Among the organic gases, formaldehyde (HCHO) is particularly important. Its effects are similar to those of sulfur dioxide, i.e., irritation of mucous membranes and bronchial constriction.

It has also been noted that a linear statistical association exists between aldehyde concentrations and eye irritation. An unsaturated aldehyde, acrolein (CH_2CHCHO), at concentratons of a tenth of a part per million will produce eye irritation comparable to that which occurs in Los Angeles photochemical smog. Acrolein is also irritating to mucous membranes and produces bronchoconstriction.

Among the chronic diseases thought to be causally related to air pollution are lung cancer, emphysema and asthma. The following facts have been cited in support of the hypothesis that lung cancer is related to air pollution: (1) lung cancer mortality is higher in urban than in rural areas, even when cigarette smoking is taken into account; (2) carcinogenic substances for experimental animals (both organic and metallic) are found in polluted atmospheres; (3) carcinogens are relatively stable in the atmosphere; (4) compounds extracted from air samples produce cancers in bioassay animals; (5) atmospheric irritants affect the protective action of cilia and mucus flow; and (6) both the carcinogens and irritants have physical characteristics compatible with the postulated effects. However, even this staggering statistical and experimental evidence does not conclusively prove that air pollution *does* cause cancer. The actual cause-and-effect relationship is still a medical mystery.

Chronic bronchitis is a disorder characterized by excessive mucus secretion in the bronchial tubes. It is manifested by a chronic or recurrent productive cough. In Great Britain, chronic bronchitis is the second most common cause of death in men aged 40 to 55, and is the leading cause of disability in this age group.* Studies in Britain have shown chronic bronchitis mortality to be associated with the level of air pollution. Chronic bronchitis morbidity in postmen in Great Britain has been shown to be related to the amount of pollution to which they are exposed.

Emphysema is the breakdown and destruction of the alveolar walls in the lung. The destruction of these alveolar sacs leads to great difficulty in

---

* The high incidence of chronic bronchitis in Great Britain may in part be due to the more frequent diagnosis of chronic bronchitis as a disease. In the United States physicians tend to diagnose emphysema and other respiratory diseases more frequently. The distinction between many of these diseases is rather fuzzy.

breathing. Available evidence suggests a relationship between emphysema and polluted air. In the United States, for example, the disease is more common in urban than in nonurban areas and is a leading cause of death, especially for cigarette smokers.

In addition to the chronic diseases, air pollution can have a serious influence on acute diseases such as the common cold and pneumonia. Statistical evidence has shown that both of these illnesses can be aggravated by breathing dirty air.

EFFECTS ON VEGETATION

Vegetation is injured by air pollutants in three ways: (1) necrosis (collapse of the leaf tissue), (2) chlorosis (bleaching or other color changes), and (3) alterations in growth. The types of injury caused by various pollutants differ markedly (Figure 17-10).

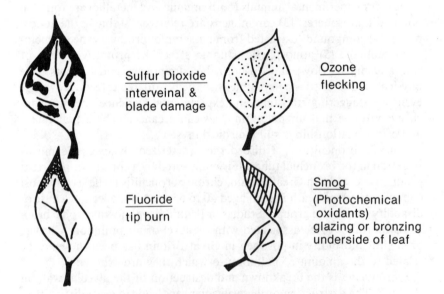

Figure 17-10   Typical air pollutant injury to vegetation.

Sulfur dioxide produces marginal or interveinal blotches which are white to straw in color on broad-leafed plants (Figure 17-11). Grasses injured by sulfur dioxide show a streaking (light tan to white) on either side of the midvein. Brown necrosis occurs on the tips of conifer needles with adjacent chlorotic areas. Alfalfa, barley, cotton, wheat and apple are among the plants most sensitive to sulfur dioxide. Sensitive species are injured at concentrations of 780 $\mu g/m^3$ for 8 hours.

In conifers and grasses, fluoride exposure produces an injury known as tip burn. In broad-leafed plants the fluoride injury is a necrosis at the periphery of the leaf. Among the plants most sensitive to fluorides are gladiolus, Chinese apricot, Italian prune and pine.

At sufficient concentrations ozone produces tissue collapse and markings of the upper surface of the leaf. These markings are known as stipple (pigmented red-brown) and flecking (bleached straw to white). One to two hours of exposure at air levels of about 300 $\mu g/m^3$ produces injury in sensitive species. Sensitive varieties include tomato, tobacco, bean, spinach and potato.

Peroxyacyl nitrates (PAN) are present in photochemical smog and

Figure 17-11   Sulfur dioxide injury to oak.

produce typical smog injury. The smog-produced injury is a bronzing on the underside of the leaves of vegetables. In grasses the collapsed tissue shows up as bands bleached tan to yellow. Smog exposure has also been shown to produce early maturity or senescence in plants. Even low PAN concentrations will injure sensitive species. Among the most sensitive species are petunia, romain lettuce, pinto bean and annual bluegrass.

Acute devastation to vegetation has occurred around plants emitting pollutants, such as the SO_2 damage around the copper smelter shown in Figure 17-12.

EFFECTS ON DOMESTIC ANIMALS

As might be expected, air pollutants affect animals other than people. During the Donora episode, 20% of the canaries and 15% of the dogs were affected.

At Poza Rica an unknown number of canaries, chickens, cattle, pigs, geese, ducks and dogs became ill or died as a result of the hydrogen sulfide exposure. All canaries exposed to the H_2S died. (The canary has long been used in British mines as a bioassay method for toxic gases.)

In the 1952 London episode, of 351 cattle on the ground floor at the Smithfield Cattle Show, 52 became seriously ill, 5 died, and 9 others were slaughtered later.

Chronic poisoning usually results from ingesting forage contaminated by the pollutant. Pollutants important in this connection are the heavy metals arsenic, lead and molybdenum. At Anaconda, Montana, in 1902, extensive poisoning of cattle, horses and sheep occurred in the vicinity of a copper smelter. Of a flock of 3500, about 620 sheep died after feeding on vegetation some 15 miles from the smelter. Horses remote from the smelter but fed hay from the smelter area also died. The grass and moss ingested by the sheep were found to contain 52 and 405 $\mu g/g$ respectively, of arsenic trioxide. The hay ingested by the horses contained 285 $\mu g/g$ arsenic trioxide.

Lead poisoning of cattle and horses was reported from Germany as recently as 1955. Cattle and horses grazing within a radius of 5 km of each of two lead and zinc foundries were affected, and some had to be slaughtered. Dust samples collected in the vicinity of the sources showed lead concentrations ranging from 17 to 45% and zinc 5 to 23%.

In 1954 cattle grazing at a distance of 0.1 km from a steel plant in Sweden were poisoned by molybdenum. An analysis of the pasture vegetation showed a molybdenum concentration of 230 $\mu g/g$ of dry matter.

Perhaps the best known of the pollutants affecting livestock is fluor-

Figure 17-12   Barren landscape around a copper smelter due to acute SO_2 damage.

ide. The problem of *fluorosis* of livestock is an old one. Farm animals, particularly cattle, sheep and swine, are susceptible to fluorine poisoning. Horses and poultry on the other hand seem to have a high level of resistance. Fluorosis is characterized by mottled teeth and a condition of the joints known as exostosis leading to lameness and ultimately death.

EFFECTS ON MATERIALS

Perhaps the most familiar effect of air pollution on materials is soiling of building surfaces, clothing and other articles. Soiling results from the deposition of smoke (fine particles of approximately 0.3 μ diameter) on surfaces. Over a period of time this deposition becomes noticeable as soiling, a discoloring or darkening of the surface. Damage to the surface, of course, results from the cleaning operation. In the case of exterior building materials, sandblasting is often required to clean the surface— and part of the surface is removed in the process of cleaning.

Another effect of air pollution is that of accelerating the corrosion of metals. For example, it has been observed that in the presence of sulfur dioxide many materials corrode much faster than they would otherwise.

One of the early noted effects of the Los Angeles smog was rubber cracking. Indeed, the effect of ozone, a principal ingredient of smog, on rubber is so specific that rubber cracking can be used to measure ozone concentrations (Chapter 19). The economic significance of rubber cracking is apparent.

Fabrics are also affected by air pollutants. A few years ago women in Toronto, Canada, and Jacksonville, Florida, noted that the nylon hose they were wearing seemed to disintegrate. This effect was due to a fine sulfuric acid mist present in the air of those cities. Other effects of pollutants on fabrics include bleaching and discoloration, both of which lead to an unacceptable deterioration of products made of fabrics.

The action of hydrogen sulfide on lead base paints is well known. Hydrogen sulfide, in the presence of moisture, reacts with lead dioxide in paint to form lead sulfide, producing a brown to black discoloration. As might be imagined, this is unattractive on a white house.

EFFECTS ON ATMOSPHERE

The ability of air pollutants, especially particulates, to reduce visibility is well known to passengers on airplanes approaching nearly any large urban area in the United States. The visibility reduction results from light scattering rather than obscuration of light. The particles primarily responsible for this effect are quite small—in the range of 0.3 to 0.6 μ in diameter.

Although carbon dioxide is usually not considered a pollutant, CO_2 concentrations in the atmosphere have been increasing in recent years, attributable to increased combustion of fossil fuels. Since CO_2 absorbs strongly at about 15 μ (heat energy) it retards the radiative cooling of the earth. The CO_2 concentrations will increase from the 1968 level of 0.032% to about 0.038% by the year 2000. This would theoretically result in an increase of 0.5°C, and is often referred to as the "greenhouse effect."

In contrast to the CO_2 effect is the observation that the temperature near the earth's surface has declined in recent years. Some investigators attribute this decline to an increase in the earth's albedo (reflectivity) as a result of an increased atmospheric aerosol from pollution. Measurements of atmospheric turbidity in Washington, DC, and Davos, Switzerland, indicate increases of particulates of up to 57 and 70%, respectively, during the first half of this century. These observations support the suggestion of increased albedo or reflectivity. Is it possible that the aerosol effect is keeping pace with the CO_2 effect?

CONCLUSION

It is important to realize that while the effect of air pollutants on animals, vegetation and materials is easy to determine, the health effects on humans can only be estimated on the basis of epidemiological evidence. Due to moral and ethical considerations which preclude deliberate exposure of human subjects to concentrations that might result in disease, this is the only avenue of study open. Epidemiological studies have been criticized because the results are expressed as statistical associations. It should be pointed out that it is probably the best evidence we will ever have. In addition, these studies are with people in the real world, exposed to real pollution, not to a synthetic atmosphere concocted in the laboratory.

One thing is certain—all of the evidence, both experimental and epidemiological, suggests that air pollution is indeed a serious threat to our health and well-being.

PROBLEMS

17.1 If you drive a 1974 car an average of 1000 miles/month, how much CO and HC would be emitted during the year? (Note: The EPA

1974 standards are 3.4 g/mi for HC and 39 g/mi of CO.) How long would it take to achieve lethal concentration of CO in a common double-car garage?

17.2 A 2.5% level of CO in hemoglobin (COHb) has been shown to cause impairment in time-interval discrimination. The level of CO on crowded city streets sometimes hits 100 ppm CO. An approximate relationship between CO and COHb (after prolonged exposure) is

$$COHb(\%) = 0.5 + 0.16 \times CO(ppm)$$

What level of COHb would a traffic cop be subjected to during a working day directing traffic on a city street?

17.3 Draw a graph showing the concentration of NO, NO_2, HC and O_3 in the Los Angeles area during a sunny, smoggy day. Superimpose on this graph, in another color, the curves as they appear on a cloudy day.

17.4 Give three examples of synergism in air pollution.

17.5 If SO_2 is so soluble in water, how can it get to the deeper reaches of the lung without first dissolving in the mucus?

17.6 Using data found in the most recent edition of *Environmental Quality,* published by the Council on Environmental Quality, trace the concentration of SO_2 and other pollutants in your community. What has occurred to change these concentrations?

17.7 Match the pollutants with the appropriate description.

A. Hydrogen fluoride (HF)   ____ a secondary pollutant in photo-chemical smog

B. Hydrogen sulfide (H_2S)   ____ cause of "weather fleck"
C. Carbon monoxide (CO)   ____ moved by cilia
D. Carbon dioxide (CO_2)   ____ one major source is power production

E. Sulfur dioxide (SO_2)   ____ combines with hemoglobin
F. Particulates   ____ a pungent, toxic gas often found in sewers

G. Oxides of nitrogen (NOx)   ____ result of phosphate fertilizer production

H. Peroxyacyl nitrate (PAN)   ____ produced naturally by decaying matter

I. Ozone (O_3)   ____ produced only by high compression combustion

17.8 How does synergism make it difficult to establish cause/effect relationships between air pollutants and disease? Give at least one example.

SUGGESTED READING

Air Pollution Control Association. *Recognition of Air Pollution Injury to Vegetation: A Pictorial Atlas* (1970).

Brodine, V. *Air Pollution* (New York: Harcourt Brace Jovanovich, 1971).

Lynn, D. A. *Air Pollution, Threat and Response* (Reading, MA: Addison-Wesley, 1976).

American Lung Association. *Air Pollution Primer* (1971).

American Lung Association. *Introduction to Respiratory Disease* (1969).

Perkins, H. C. *Air Pollution* (New York: McGraw-Hill Book Company, 1974).

Wark, K., and C. F. Warner. *Air Pollution, Its Origins and Control* (New York: Dun-Donnelly Publishers, 1976).

Williamson, S. J. *Fundamentals of Air Pollution* (Reading, MA: Addison-Wesley Publishing Co., 1973).

Wise, W. *Killer Smog* (Chicago: Rand-McNally & Co., 1969).

Chapter 18

Meteorology and Air Quality

One would think that with the earth's atmosphere being about 100 miles deep, we should be able to dilute effectively all the garbage thrown into it. But actually, about 95% of the total air mass is within only 12 miles of the surface. This layer, called the *troposphere,* is where we have our weather and air pollution problems.

The science of meteorology has great bearing on air pollution. An air pollution problem involves three parts: the source, the movement of the pollutant, and the recipient (Figure 18-1). Whereas Chapter 17 covers sources and effects, this chapter covers the transport mechanism—how the pollutants travel through the atmosphere.

Figure 18-1   Meteorology of air pollutants.

BASIC METEOROLOGY

Weather changes in the form of fronts, which can be warm or cold. Warm fronts usually are associated with steady rain and drizzle; cold fronts bring heavy local rain.

Another way of picturing weather is in terms of barometric pressure. Low pressure systems are associated with both hot and cold fronts. The air movement around low pressure systems is counterclockwise (in the northern hemisphere) and vertical winds are upward, where condensation and precipitation take place. High pressure systems bring sunny and calm weather, with the wind spiraling clockwise and downward. There are no fronts associated with high pressure systems, which represent stable atmospheric conditions. The low and high pressure systems, commonly called *cyclones* and *anticyclones* respectively, are illustrated in Figure 18-2. Because of the high stability in anticyclones, these are usually precursors to air pollution episodes.

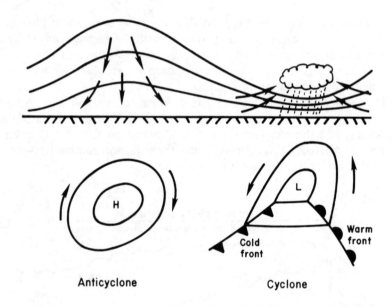

Anticyclone   Cyclone

Figure 18-2   Anticyclone and cyclone.

Figure 18-3 shows a weather map for the eastern United States for 18 January 1968. Note the high pressure system over the entire area, with clear skies and little or no wind. The air pollutants emitted into the atmosphere on this day would not be dispersed (no wind) and since there is no rain, would not be removed. Such stable conditions commonly result in a buildup of pollutants. In this case, had the cold front in the middle of the United States not come through, this weather condition might have produced a serious air pollution episode.

A primary objective of air quality management is to attain low air pollutant concentrations in community air. This can be accomplished of

DAILY WEATHER MAP

U.S. DEPARTMENT OF COMMERCE
ENVIRONMENTAL SCIENCE SERVICES ADMINISTRATION
WEATHER BUREAU

THURSDAY, JANUARY 18, 1968

Figure 18-3   Weather map for 18 January 1968.

course by reducing the amount of pollutants emitted. It is also possible to reduce concentrations by attaining adequate dispersal and dilution in the atmosphere. Such dispersal is controlled by atmospheric conditions, and can be vertical as well as horizontal. The principles governing both vertical and horizontal movement of air are therefore important in air quality management.

HORIZONTAL DISPERSION OF POLLUTANTS

The earth acts like a wave converter, accepting the sun's light energy (high-frequency waves) and converting this to heat energy (low-frequency waves), which is then radiated back to space. The heat transfer from the earth to space is by radiation, conduction and convection. Radiation involves the transfer of heat by energy waves, a minor effect on the atmosphere; conduction is the transfer of heat by physical contact; convection is the process of heating by the movement of air masses.

If the earth did not rotate, the air near the equator would be heated, then it would rise and move toward the poles, where it would sink (Figure 18-4). But the earth does rotate and thus always presents new areas for the sun to shine on and to warm. Accordingly, a pattern of winds is set up around the world, some seasonal (e.g., hurricanes) and some permanent. Local conditions and cloud cover further complicate the picture.

For example, land masses heat and cool faster than water, thus shoreline winds blow out to sea during the night and inland during the day. Valley and slope winds are caused by the cooling of air on mountain slopes. In cities, brick and stone absorb and hold heat, creating a *heat island* about the city during the night (Figure 18-5). This effect sets up a self-contained circulation pattern called a *haze hood* from which the pollutants cannot escape.

The horizontal motion of winds is measured as wind velocity. These wind velocity data are plotted as a *wind rose,* a graphic picture of the direction and velocity *from which the wind came.* The wind rose in Figure 18-6 shows that the prevailing winds were from the southwest.

Air pollution engineers often use a variation of the wind rose, called a *pollution rose,* to determine the source of a pollutant. Instead of plotting all winds on a radial graph, only those days during which the concentration of a pollutant is above a certain minimum are used. Figure 18-7 is an actual plotting of such pollution roses. Only winds carrying SO_2 levels greater than 250 $\mu g/m^3$ were plotted. Note how the fingers of the roses point to Plant 3. Pollution roses can be plotted for other pollutants as

A. If the earth did not turn, the air would circulate in a fixed pattern

B. The earth turns, creating variable wind patterns

Figure 18-4   Global wind patterns (courtesy American Lung Association).

well and are useful for pinpointing sources of atmospheric contamination.

VERTICAL DISPERSION OF POLLUTANTS

As a parcel of air rises in the earth's atmosphere, it experiences lower and lower pressure from surrounding air molecules, and thus it expands.

Figure 18-5   Heat island formed over a city.

Figure 18-6   Typical wind rose.

Figure 18-7   Pollution roses, with SO_2 concentrations greater than 250 $\mu g/m^3$. The major suspected sources are the four chemical plants, but the data indicate that Plant 3 is the primary culprit.

This expansion lowers the temperature of the air parcel. Ideally a parcel of air cools at about 1°C/100 m or 5.4°F/1000 ft (or warms at 1°C/100 m if it is coming down). This warming or cooling is termed the *dry adiabatic* lapse rate,* and is independent of prevailing atmospheric temperatures. The 1°C/100 m *always* holds (for dry air), regardless of what the actual temperature at various elevations might be.

The actual temperature-elevation measurements are called *prevailing lapse rates* and can be classified as shown in Figure 18-8.

A *superadiabatic lapse rate,* also called a *strong lapse rate,* occurs when the atmospheric temperature drops more than 1°C/100 m. A *subadiabatic lapse rate,* also called a *weak lapse rate,* is characterized by a

---

*Adiabatic is a term denoting no heat transfer (e.g., between the air parcel and the surrounding air).

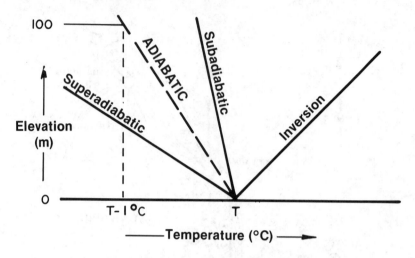

Figure 18-8   Prevailing lapse rates, and the dry adiabatic lapse rate.

drop of less than 1°C/100 m. A special case of the weak lapse rate is the *inversion*, a condition which has warmer air above colder air.

During a superadiabatic lapse rate the atmospheric conditions are unstable; a subadiabatic and especially an inversion characterizes a stable atmosphere. This can be demonstrated by depicting a parcel of air at 500 m (see Figure 18-9A). If the temperature at 500 m is 20°C, during a superadiabatic condition the temperature at ground level might be 30° and at 1000 m it might be 10° (Note: a change of more than 1°C/100 m).

If the parcel of air at 500 m is moved upward to 1000 m, what would be its temperature? Remember that assuming adiabatic condition, the parcel would cool 1°C/100 m. The temperature of the parcel at 1000 m is thus 5° less than 20° or 15°C.

The prevailing temperature, however, is 10°C, and the air parcel finds itself surrounded by cooler air. Will it rise or fall? Obviously, it will rise, since warm air rises. Once the parcel of air under superadiabatic conditions is displaced upward, it keeps right on going.

Similarly, if a parcel is displaced downward, say to ground level, the air parcel is 20° + (500 m × [1°/100 m]) = 25°C. It finds the air around it a warm 30° and thus the cooler air parcel would just as soon keep going down if it could. Superadiabatic conditions are thus unstable, characterized by a great deal of vertical air movement and turbulence.

The subadiabatic condition shown in Figure 18-9B is by contrast a very stable system. Consider again a parcel of air at 500 m and at 20°C. A typical subadiabatic system has a ground level temperature of 21°C and 19°C at 1000 m. If the parcel is displaced to 1000 m, it will cool by

Figure 18-9   Stability and vertical air movement.

5-15°. But, finding the air around it a warmer 19°, it will fall right back to its point of origin. Similarly, if the air parcel were brought to ground level, it would be at 25°, and finding itself surrounded by 21° air, it would rise back to 500 m. Thus the subadiabatic system would tend to dampen out vertical movement and is characterized by a very limited vertical mixing.

An inversion is an extreme subadiabatic condition, and thus the vertical air movement within an inversion is almost nil.

Figure 18-10 is an actual temperature sounding for Los Angeles. Note the beginning of the inversion at about 1000 ft. This puts an effective cap on the city and holds in the air pollution. The L.A. inversion is called a *subsidence inversion,* caused by a large warm air mass subsiding over the city.

Figure 18-10   Actual temperature sounding for Los Angeles, October 1962, 4 PM. The dotted lines show the dry adiabatic lapse rate, 5.4°F/1000 ft.

A more common type of inversion is the *radiation inversion,* caused by the radiation of heat to the atmosphere from the earth. During the night, as the earth cools, the air close to the ground loses heat, thus causing an inversion (Figure 18-11). The pollution emitted during the night is caught under this lid and does not escape until the earth warms sufficiently to break the inversion.

Atmospheric stability can often be recognized by the shapes of plumes emitted from smokestacks (Figures 18-12 and 18-13). One potentially serious condition is called *fumigation,* where the pollutants are caught under an inversion and are mixed due to a strong lapse rate. A looping plume also can be dangerous due to a very high ground level concentration of pollutants as the plume touches ground.

Figure 18-11   Typical prevailing lapse rates during a sunny day and clear night.

If we assume perfect adiabatic conditions in a plume, we can estimate how far it will rise (or sink) and what type of plume it will be during any given atmospheric temperature condition. This is illustrated in Example 18.1.

Example 18.1
A stack 100 m tall emits a plume at 20°C. The prevailing lapse rates are shown in Figure 18-14. How high will the plume rise (assuming perfect adiabatic conditions) and what type of plume will it be?

Note that the prevailing lapse rate is subadiabatic to 200 m and an inversion exists above 200 m. The smoke at 20°C finds itself surrounded by colder (18.5°C) air, and thus rises. As it rises, it cools, so that at 200 m it is 19°C. At about 220 m, the surrounding air is at the same temperature as the smoke (about 18.7°C) and the smoke ceases to rise.

Below 220 m, the plume would have been stable and slightly coning. It would not, however, have penetrated 220 m and thus there would have been a cap on the plume.

Effect of Water in the Atmosphere

Thus far we have assumed that there was no water in the atmosphere, hence we speak of the *dry* adiabatic lapse rate. Water will of course condense and evaporate, and in so doing emit and absorb heat, making the calculations of stability quite a bit more complicated.

Water also affects air quality in other ways. Fogs are formed when moist air cools and the moisture condenses. Aerosols provide the condensation nuclei, and fogs thus tend to occur more readily in urban areas.

Figure 18-12   Plume shapes and atmospheric stability.

In addition to inversions, serious air pollution episodes are almost always accompanied by fogs. These tiny droplets of water are detrimental in two ways. In the first place, fog makes it possible to convert SO_3 to H_2SO_4. Secondly, fog sits in valleys and prevents the sun from warming the valley floor and breaking inversions, thus often prolonging air pollution episodes.

ATMOSPHERIC DISPERSION

Dispersion is the process of spreading out the emission over a large area and thus reducing the concentration of the specific pollutants. The

Figure 18-13   Iron oxide dust looping plume from a steel mill.

Figure 18-14   Atmospheric conditions for an emission at 20°C (Example 18-1).

plume spread or dispersion is in two dimensions: horizontally and vertically. We assume that the greatest concentration in the pollutants is in the plume centerline, that is, in the direction of the prevailing wind. The further we get from the centerline, the lower the concentration. If we assume that the spread of a plume in both directions is approximated by a Gaussian probability curve, we can calculate the concentration of a pollutant at any distance x downwind from the source by

$$X_{(x, y, z)} = \frac{Q}{2\pi\bar{u}\sigma_y\sigma_z} \exp(-\tfrac{1}{2}[(y/\sigma_y)^2 + (z/\sigma_z)^2])$$

where

X = concentration at some point in the x, y, z coordinate space, kg/m^3

Q = emissions, kg/sec

$\bar{u}$ = average wind speed, m/sec

σ_y and σ_z = standard deviation of the dispersion in the y and z directions

The coordinates are shown in Figure 18-15. Note that z is in the vertical direction, y is horizontal crosswind, and x is downwind.

Figure 18-15 Gaussian dispersion model.

The standard deviations are measures of how much the plume spreads. If y and z are large, the spread is great, and the concentration is of course low. The opposite is true if the spread is small.

The dispersion is dependent on both atmospheric stability (as we saw earlier) and the distance from the source. Figure 18-16 is one approximation for the dispersion coefficients.

Atmospheric stability is denoted in Figure 18-6 by letters ranging from A to F. Table 18-1 is a key for selecting the proper stability condition.

Suppose a stack has an effective height (stack height plus plume rise) of H meters, as shown in Figure 18-15. The elevation of the plume centerline is $z = H$, and the diffusion equation is thus

$$X_{(x, y, z)} = \frac{Q}{2\pi \bar{u} \sigma_y \sigma_z} \exp(-\tfrac{1}{2}[(y/\sigma_y)^2 + ((z-H)/\sigma_z)^2])$$

Note that at the plume centerline elevation $z = H$ and the last exponential term drops out, yielding the equation

$$X_{(x, y, z)} = \frac{Q}{2\pi \bar{u} \sigma_y \sigma_z} \exp(-\tfrac{1}{2}[(y/\sigma_y)^2])$$

These equations hold as long as the ground does not influence the diffusion. This is usually not a good assumption, since the ground is not a

Figure 18-16   Dispersion coefficients.

Table 18-1.   Atmospheric Stability Key for Figure 18-16

Surface Wind Speed (at 10 m) (m/sec)	Day[a] Incoming Solar Radiation (Sunshine)			Night[a] Thinly Overcast or 4/8 Low Cloud	3/8 Cloud
	Strong	Moderate	Slight		
< 2	A	A–B	B	–	–
2–3	A–B	B	C	E	F
3–5	B	B–C	C	D*	E
5–6	C	C–D	D	D	D
>6	C	D	D	D	D

[a]The neutral category, D, should be assumed for overcast conditions during day or night.

100% efficient sink for the pollutants, and the levels must be higher at ground level due to inability of the plume to disperse into the ground. This effect can be taken into account if we think of an imaginary mirror image source at elevation $z - H$, as shown in Figure 18-17. Taking this into account, we can write

$$\chi_{(x,y,z)} = \frac{Q}{2\pi \bar{u}\sigma_y\sigma_z} \left(\exp(-\tfrac{1}{2}[(y/\sigma_y)^2])\right)$$
$$\times \left(\exp(-\tfrac{1}{2}[(z-H)^2/\sigma_z^2]) + \exp(-\tfrac{1}{2}[(z+H)^2/\sigma_z^2])\right)$$

Example 18.2

Given a sunny summer afternoon with average wind, $\bar{u} = 4$ m/sec, emission $Q = 0.01$ kg/sec, and the effective stack height 20 m, find the ground level concentration at 200 meters from the stack.

Using the above equation, and from Figure 18-16 finding that at 200 meters, $\sigma_y = 36$ and $\sigma_z = 20$ for an unstable superadiabatic strong solar radiation (Table 18-1), the atmospheric conditions are (Type B), and noting that maximum concentrations occur on the plume centerline, at $y = 0$

$$\chi = \frac{Q}{2\pi \bar{u}\sigma_y\sigma_z} \left(\exp(-\tfrac{1}{2}[(y/\sigma_y)^2])\right)$$
$$\times \left(\exp(-\tfrac{1}{2}[(z-H)^2/\sigma_z^2]) + \exp(-\tfrac{1}{2}[(z-H)^2/\sigma_z^2])\right)$$

$$\chi = \frac{0.01}{2(3.14)(4)(36)(20)} \left(\exp(-\tfrac{1}{2}[(0/36)^2])\right)$$
$$\times \left(\exp(-\tfrac{1}{2}[(0-20)^2/20^2]) + \exp(-\tfrac{1}{2}[(0+20)^2])\right)$$

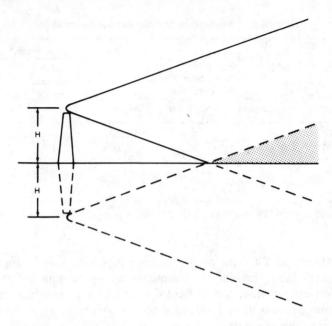

Figure 18-17   Source and image source. In the shaded area, the concentration is doubled due to the image source.

$$X = (5.53 \times 10^{-7})(1)(e^{-1/2} + e^{-1/2})$$

$$X = (5.53 \times 10^{-7})(0.6 + 0.6) = 6.64 \times 10^{-7} \text{ kg/m}^3$$

$$X = 664 \ \mu\text{g/m}^3$$

Among the many assumptions made in this model is the accurate estimation of a uniform, steady and unidirectional wind. None of these are realistic of course. Wind direction shifts, and wind speed varies with time (unsteady) as well as with elevation (nonuniform). The latter can be approximated by assuming a parabolic wind velocity profile, and calculating the velocity $\bar{u}_1$ at any given elevation h_1 as a function of a measured velocity $\bar{u}_2$ at another elevation, h_2 as

$$\bar{u}_1 = \bar{u}_2 \left(\frac{h_1}{h_2} \right)^n$$

where $n = 0.25$ for unstable and $n = 0.5$ for stable atmospheric conditions.

The effective height of the stack was defined as the elevation at which the plume levels off. The analysis assuming perfect adiabatic conditions discussed earlier is not realisitc, and plume rise is usually estimated by empirical models, all based on the assumption that plumes rise higher if

- they are hot
- they have a high exit velocity
- the wind velocity is small.

Thus, for example, the following equation has been shown to be in good agreement with observed data.

$$\Delta h = -0.029 \left(\frac{vd}{\bar{u}} \right) + 2.62 \left(\frac{\sqrt{Q_h}}{\bar{u}} \right)$$

where Δh = plume rise in meters (see Figure 18-15)
        v = stack exit velocity, m/sec
        d = stack exit diameter, m
        $\bar{u}$ = wind speed at stack exit elevation, m/sec
        Q_h = heat emission rate from the stack kJ/sec

The last term can be approximated as

$$Q_h = m \, C_p \, (T_s - T_a)$$

where m = stack gas mass flow rate, kg/sec
        C_p = constant pressure specific heat of the stack gas. (Note: Air at 25°C has a specific heat $C_p = 1$ kJ/kg-°K or 0.24 Btu/lb-°F.)
        T_s = stack gas temperature, °K
        T_a = atmospheric temperature, °K

Example 18.3
A power plant has a 2-m-diam stack and emits gases with a velocity of 15 m/sec and a heat emission rate of 4800 kJ/sec. The wind speed is 5 m/sec. Estimate the plume rise.

$$\Delta h = -0.029 \, \frac{vd}{\bar{u}} + 2.62 \, \frac{\sqrt{Q_h}}{\bar{u}}$$

$$= -0.029 \left(\frac{15 \cdot 2}{5} \right) + 2.65 \, \frac{(4800)^{1/2}}{5}$$

$$= 36.2 \text{ m}$$

Finally, it should be pointed out that the accuracy of this plume rise and dispersion analysis is very poor. Air pollution modelers are usually pleased to find their models predicting concentrations to within an order of magnitude! We thus should refrain from placing undue validity on these models.

CLEANSING THE ATMOSPHERE

Since we obviously have not yet all suffocated, and yet prodigious amounts of pollutants have been thrown into the air over the millions of years, there must exist a series of processes by which air is cleansed. These include the effect of gravity, contact with the earth's surface, and removal by precipitation.

Gravity

Particulates, if of sufficient size, are removed by simple settling out under the influence of gravity. Unfortunately, the settling velocities of common particulates are very small. For example, a 1-μm particle will have a settling velocity of about 1 cm/sec, in ideal quiescent conditions. Practically, due to turbulence in the atmosphere, particles smaller than 20 μm will seldom settle out by gravity. Gases are removed by gravity only if they are adsorbed onto particulates. Sulfur dioxide, for example, is readily adsorbed and is thus partially removed by gravity.

Surface Sink

Many of the gases are adsorbed by the earth's surface, including stone, vegetation and other materials. Some gases such as SO_2 are readily dissolved in surface waters.

Precipitation

The third major removal mechanism is by precipitation. Two types of removal occur, the first being an "in-cloud" process called *rainout,* where submicron particles become nuclei for the formation of rain droplets which will grow and eventually fall as precipitation. The second mechanism is called *washout* and is a "below-cloud" process, where the rain falls through the air pollutants, the pollutants are impinged or dissolved in the droplets, and then carried to earth.

The relative importance of these removal mechanisms was illustrated by a study of SO_2 emissions in Great Britain, where the surface sink accounted for 60% of the SO_2, 15% was removed by precipitation and 25% left Great Britain (heading you-know-where).

CONCLUSION

Air pollution episodes are the results of high emissions and a combination of meteorological factors. Some of these factors are:

1. little horizontal wind movement
2. stable atmospheric conditions, resulting in very limited vertical air movement
3. fog, which promotes the formation of secondary pollutants and hinders the sun from warming the ground and breaking inversions
4. high pressure areas resulting in downward vertical air movement and absence of rain for washing the atmosphere

It would seem reasonable therefore that episodes can be predicted on the basis of meteorological data, provided the potential exists. The EPA, in cooperation with the Weather Bureau, has indeed established procedures for evaluating meteorological data so as to provide early warning for impending episode conditions and has developed emergency plans (including shutting down industries) should the conditions warrant.

PROBLEMS

18.1 Given the following temperature soundings:

Elevation (m)	Temperature (°C)
0	20
50	15
100	10
150	15
200	20
250	15
300	20

what type of plume would you expect if the exit temperature of the plume were 10°C and the smoke stack were

    a. 50 m tall?
    b. 150 m tall?
    c. 250 m tall?

18.2 Consider a prevailing lapse rate which has these temperatures: ground $= 21°C$, 500 m $= 20°C$, 600 m $= 19°C$, 1000 m $= 20°C$. If we released a parcel of air at 500 m and at 20°C, would it tend to sink, rise or remain where it was? If a stack is 500 m tall, what type of plume would you expect to see?

18.3 Draw a map with X and Y coordinates (X horizontal, Y vertical) and place on the map the following:

    Industrial Plant "A" at $X = 3$, $Y = 3$
    Industrial Plant "B" at $X = 8$, $Y = 1$
    Industrial Plant "C" at $X = 8$, $Y = 8$
    Air Sampling Station at $X = 5$, $Y = 5$

The data at the air sampling station are:

Day	Wind Direction	Particulates ($\mu g/m^3$)	SO_2 ($\mu g/m^3$)
1	N	80	80
2	NE	120	20
3	NW	30	30
4	N	90	40
5	NE	130	20
6	SW	20	180
7	S	30	100
8	SW	40	200
9	E	100	60
10	W	10	100

Draw pollution roses to show which plant is guilty of the air pollution.

18.4 If your job were to continuously analyze meteorological data and watch for conditions which might lead to the occurrence of an "episode," what specific conditions would you be looking for? (That is, what are the meteorological criteria for the formation of an episode?)

18.5 Take a photograph of a smoke plume and describe: (a) the time of day, (b) the climatological conditions, (c) the atmospheric stability and (d) the type of plume. You may submit a photograph or a 2 × 2 slide. Put your name on the picture and stable or tape it to a sheet of paper.

18.6 A power plant burns 1000 tons of coal/day, 2% of which is sulfur, and all of this is emitted from the 100-m stack. For a wind speed of 10 m/sec, calculate: (a) the maximum ground level concentration of SO_2, 10 km downwind from the plant, (b) the maximum ground level concentration and the point at which this occurs for stability categories A, C and F. (Do part *b* only if you have a computer at your disposal.)

18.7 A power plant emits 20 metric tons of particulates per hour out of a 100-m (effective height) stack. At 50 m downwind at a wind velocity of 2 m/sec, at plume centerline, what type of atmospheric stability would yield the *highest* ground level concentration? (Don't try to use Figure 18-15. Use common sense.)

18.8 Given a wind velocity of 2 m/sec, calculate the maximum expected SO_2 concentration at ground level in a town 10 km downwind from a power plant which burns 1000 metric tons of coal per day (sulfur content of 1%). The stack is 100 m high.

18.9 Consider the following atmospheric temperature soundings:

Elevation (ft)	Temperature (°F)
0	70.0
200	68.0
400	66.0
600	72.0
800	70.0
1000	68.0

a. Indicate below the type of lapse rate involved
   - 0 to   400 ft _____
   - 400 to   600 ft _____
   - 600 to 1000 ft _____
b. Indicate below the plume type if the stack were
   - 300 ft tall _____
   - 500 ft tall _____
   - 700 ft tall _____
c. What would be the ground level concentration of the pollutant (negligible, moderate, high) if the stack were
   - 300 ft tall _____
   - 500 ft tall _____
   - 700 ft tall _____

18.10 The odor threshold of H_2S is about 0.7 $\mu g/m^3$. If an industry emits 0.08 g/sec of H_2O out of a 40-m stack during an overcast night with a wind speed of 3 m/sec, estimate the area (in terms of x and y coordinates) where H_2O would be detected.

18.11 If the maximum ground level concentration of SO_2 is to be limited to 50 $\mu g/m^3$ on a clear day with a wind speed of 3 m/sec and an emission rate of 100 g/sec, what is the required effective stack height?

18.12 A copper smelter processes 1000 metric tons per day of ore, principally $CuFeS_2$, producing SO_2 as a pollutant. The stack height is 280 m with an exit velocity of 15 m/sec through a 2-m-diameter stack and temperature of 250°C. The wind speed at a height of 5 m is 4 m/sec. During an overcast night (10°C):
   a. Determine the maximum ground level concentration.
   b. Estimate the percent SO_2 removal necessary to reduce that maximum to below EPA primary ambient air quality standards, 24-hr exposure (Chapter 21).

18.13 The concentration of H_2S at a location 200 m downward from an abandoned oil well is 3.2 $\mu g/m^3$. What is the emission rate, on a partially overcast afternoon?

18.14 A power plant has a 500-m stack with an outlet diameter of 10 m. The plant burns 15,500 metric tons coal/day, with a sulfur content of 3.5%. The flue gas exit temperature is 200°C and the weather is overcast on a winter afternoon, 0°C. The plant uses 8 kg of air per kg of coal. Wind measurements at 3 m from the ground show an average velocity of 4 m/sec, from due north. Plot a graph of ground level SO_2 with distance downwind. At the 10-km point, plot the SO_2 at a right angle to the wind, using perspective (see Figure 18-13) in both y and z directions.

18.15 What is the expected SO_2 concentration (Problem 18.14) if the sampling site is SE of the plant at a distance of 8 km?

18.16 A coal-fired power plant emits SO_2 at rate of 1500 g/sec. What is the centerline concentration of SO_2 3 km downwind on an overcast summer afternoon if the wind speed is 4 m/sec. The stack height is 110 m, its diameter is 1.3 m, exit velocity is 9 m/sec and exit temperature is 330°C.

18.17 A power plant burns coal with a 2% sulfur content and has a stack height of 70 m and a diameter of 1.5 m. The exit velocity and temperature are 15 m/sec and 340°C respectively. The atmospheric conditions are 28°C, barometric pressure 98 kPa and the weather is a clear summer afternoon with a wind speed of 2.8 m/sec. At a point 2.5 km downwind and 0.3 km off plume centerline, the ground level concentration of particulates was found to be 1430 μg/m³. How much coal does the plant burn?

LIST OF SYMBOLS

C_p = specific heat, kJ/kg-°K
d = stack exit diameter, m
H = effective height of the stack, m
h = stack elevation, m
m = mass flow rate, kg/sec
Q = emission rate, kg/sec
Q_h = heat emission rate, kJ/sec
T = temperature
$\bar{u}$ = average wind speed, m/sec
v = stack exit velocity, m/sec
σ_y = standard deviation, y direction, m
σ_z = standard deviation, z direction, m
χ = concentration of pollutant, kg/m³

SUGGESTED READING

Bach, W. *Atmospheric Pollution* (New York: McGraw-Hill Book Company, 1972).

Environmental Protection Agency. *Guide for Air Pollution Episode Avoidance* (1971).

Perkins, H. *Air Pollution* (New York: McGraw-Hill Book Company, 1974).

Seinfeld, J. H. *Air Pollution* (New York: McGraw-Hill Book Company, 1975).

Stern, A. C., Ed. *Proceedings, Symposium on Multi-Source Urban Diffusion Models,* U.S. Environmental Protection Agency, Research Triangle Park, NC (1972).

Williamston, S. J. *Fundamentals of Air Pollution* (Reading, MA: Addison-Wesley Publishing Co., 1972).

Chapter 19

Measurement of Air Quality

It was previously observed that air quality measurement, like water quality measurement, is complicated by the lack of knowledge as to what is "clean" and by the difficulty in defining "quality."

"Pure air" was ordinarily defined as containing only the naturally occurring gases, but "pure air" doesn't exist in nature. Pollen, dust, fog, etc., are all contaminants, although they occur without any assistance from industrialized man. No attempt is usually made to differentiate between natural and man-made pollutants and air quality measurements are designed to measure all types of contaminants.

The measurements of air quality generally fall into three classes:

1. *Measurement of emissions.* This is called *stack sampling* when a stationary source is analyzed. A hole is punched into the stack and samples drawn out for on-the-spot analyses. Sampling moving sources is more difficult.
2. *Meteorological measurements.* The measurement of meteorological factors is necessary if we are to know how and why the pollutants travel from the source to the recipient. Some of these are discussed in the previous chapter.
3. *Ambient air quality.* The quality of ambient air is of course of major concern. Almost all evidence of health effects is based on these measurements. Monitoring air quality can also provide data for recognizing episodes and thus can provide some warning of impending health problems.

The development of air quality instrumentation can be divided into three distinct phases or generations. The original, or first-generation, devices were developed by the air pollution control engineers and scientists and were needed for obtaining quantitative information on air quality. There was little precedent for measuring minute quantities of various gases in the atmosphere, nor was there much money for development of the necessary instrumentation. Accordingly, first-generation devices were simple, inexpensive and required no external power for operation. They were also inconvenient, very slow and of questionable accuracy.

Second-generation measurements evolved when more accurate and rapid data were required. These devices use a power source (usually electrical power to drive an air pump) and thus can sample more air in a shorter time. The techniques for gas measurement generally involved wet chemistry, in that the gas was either dissolved into or reacted with a collecting fluid.

Third-generation devices differ from their predecessors in that they provide a continuous readout. It is thus possible to get a continuous graph of levels of various pollutants with the measurement occurring almost instantaneously.

Examples of all three types of instrumentation are discussed below, for both particulates and gaseous pollutants.

MEASUREMENT OF PARTICULATES

First-generation devices for measuring particulates involved the determination of how much dust settles to the earth. Such *dustfall* measurements are by far the simplest means of evaluating air quality. Dust is collected either in open buckets (Figure 19-1) or on sticky tapes wrapped around jars. The sampling period for dustfall buckets is generally 30 days; sticky tapes are usually read in 7 days and give a qualitative indication of the direction of the particulate pollution. The dustfall jars are dried to remove moisture and weighed to determine the amount of dust in the jar, which is reported commonly as tons (2000 lb) of dust settled per square mile in 30 days.*

Example 19.1
A dustfall jar, 6 in. in diameter, weighs 100.0 g. After one month, the jar and collected dust weigh 100.5 g. What is the dustfall?

---

* A more realistic method of reporting dustfall data might have been as g dust/cm² - 30 days. The *tons* is however much more impressive, and this consideration no doubt lead to the use of the more awkward tons/mi²-month.

Figure 19-1   Dustfall jar.

Weight of dust = 100.5 − 100.0 = 0.5 g

$$= 0.5 \text{ g} \times 2.2 \times 10^{-3} \text{ lb/g} \times 5 \times 10^{-4} \text{ ton/lb}$$

$$= 5.5 \times 10^{-7} \text{ tons of dust}$$

$$\text{Area of jar} = \frac{\pi D^2}{4} = \frac{3.14 \times (6 \text{ in.} \times [1 \text{ ft}/12 \text{ in.}])^2}{4} \times 3.6 \times 10^{-8} \text{ mi}^2/\text{ft}^2$$

$$= 7.1 \times 10^{-9} \text{ mi}^2$$

$$\text{Dustfall} = \frac{5.5 \times 10^{-7} \text{ tons}}{7.1 \times 10^{-9} \text{ mi}^2}$$

$$= 77 \text{ tons/mi}^2\text{-month}$$

(Note: This represents a very heavy dustfall. Urban dustfall rarely exceeds 50 ton/mi²-month).

The measurement of particulates by dustfall jars is subject to many problems (not the least of which is uncooperative pigeons) and only one data point per month is obtained. The second-generation particulate sampling device, the *high-volume sampler* (or hi-vol) is a substantial improvement since its sampling time is normally only 24 hours.

The high-volume sampler, the workhorse of particulate sampling (Figure 19-2), operates much like a vacuum cleaner by simply forcing over 2000 cubic meters (70,000 ft³) of air through the filter in 24 hours. The analysis is gravimetric; the filter is weighed before and after, and the difference is the particulates collected.

Filter

Blower

Figure 19-2   Hi-vol sampler.

The air flow is measured by a small flow meter, usually calibrated in cubic feet of air per minute. Because the filter gets dirty during the 24 hours of operation, less air goes through the filter during the latter part of the test than in the beginning and the air flow must therefore be measured at both the start and end of the test period and the values averaged.

Example 19.2
    A clean filter is found to weigh 10.00 grams. After 24 hours in a hi-vol, the filter plus dust weighs 10.10 grams. The air flow at the start and end

of the test was 60 and 40 ft^3/minute, respectively. What is the particulate concentration?

$$\text{Weight of the particulates (dust)} = (10.10 - 10.00)\,\text{g} \times 10^6\ \mu\text{g/g}$$

$$= 0.1 \times 10^6\ \mu\text{g}$$

$$\text{Average air flow} = \frac{60 + 40}{2} = 50\ \text{ft}^3/\text{min}$$

$$\text{Total air through the filter} = 50\ \text{ft}^3/\text{min} \times 60\ \text{min/hr} \times 24\ \text{hr/day} \times 1\ \text{day}$$

$$= 72{,}000\ \text{ft}^3$$

$$= 72{,}000\ \text{ft}^3 \times 28.3 \times 10^{-3}\ \text{m}^3/\text{ft}^3$$

$$= 2038\ \text{m}^3$$

$$\text{Total suspended particulates} = \frac{0.1 \times 10^6\ \mu\text{g}}{2038\ \text{m}^3}$$

$$= 49\ \mu\text{g/m}^3$$

The particulate concentration thus measured is often referred to as *total suspended particulates* (TSP) to differentiate it from other measurements of particulates.

Another widely used measure of particulates in the environmental health area is of *respirable particulates,* or those particulates that would be respired into lungs. These are generally defined as being less than 0.3 μ in size, and the measurements are done with stacked filters. The first filter removes only particulates >0.3 μ, and the second filter, having smaller spaces, removes the small respirable particulates.

Another device for collecting airborne particulates is the *cascade impactor* (Figure 19-3). The impactor consists of four tubes, each with a progressively smaller opening, thus forcing higher and higher velocities. A particle entering the device may be small enough to follow the streamline of flow without hitting the microscope slide. At the next nozzle, however, the velocity may be sufficiently high to prevent the particle from quite negotiating the turn, and it will impinge on the slide.

Interestingly, no third-generation particulate measuring devices have yet been developed and accepted. The difficulty, of course, is that the measurement must be gravimetric, and it is difficult to construct a device which continuously weighs minute quantities of dust. Some indirect devices are used for estimating particulates, the most notable being the

Figure 19-3   Cascade impactor.

nepthelometer, which actually measures light scatter, the assumption being that an atmosphere which contains particulates also scatters light. Unfortunately, different sizes of particulates scatter light differently and atmospheric moisture (fog) also interferes with the passage of light, but this should not be measured as particulates.

MEASUREMENT OF GASES

As is the case with particulates, the first-generation methods for measuring the concentration of gaseous pollutants are simple yet ingenious gadgets, using no power, but providing a quantitative measure of the atmospheric concentration of specific gases. Two of the most interesting (and important) gases for which first-generation devices were developed are ozone (O_3) and sulfur dioxide (SO_2).

The original ozone measuring technique takes advantage of the fact that ozone attacks and cracks rubber. In this test, specially prepared and weighed strips of rubber are hung outside, and the cracks formed in the rubber strip are measured and related to ozone levels in the atmosphere.

Sulfur dioxide is measured by impregnating filter papers with chemicals which react with SO_2 and change color. For example, lead peroxide reacts as

$$PbO_2 + SO_2 \rightarrow PbSO_4$$

forming a dark lead sulfate. The extent of this reaction is estimated by the dark areas on the filter paper.

These devices are obviously slow, inaccurate and subject to many interferences. Second-generation techniques are by contrast much faster (hours instead of days) and more accurate. These techniques almost all involve the use of a *bubbler,* shown in Figure 19-4. The gas is literally bubbled through the liquid, which either reacts chemically with the gas of

Figure 19-4   A typical bubbler used for the measurement of gaseous air pollutants.

interest, or into which the gas is dissolved. Wet chemical techniques are then used to measure the concentration of the gas.

A simple (but now seldom used) bubbler technique for measuring SO_2 is to bubble air through hydrogen peroxide, so that the following reaction occurs:

$$SO_2 + H_2O_2 \rightarrow H_2SO_4$$

The amount of sulfuric acid formed can be determined by titrating the solution with a base of known strength.

One of the better methods of measuring SO_2 is the colorimetric pararosaniline method, in which SO_2 is bubbled into a liquid containing tetrachloromercurate (TCM). The SO_2 and TCM combine to form a stable complex. Pararosaniline is then added to this complex with which it forms a colored solution. The amount of color is proportional to the SO_2 in the solution, and the color is measured with a spectrophotometer at a wavelength of 560 mm. (See Chapter 3 for ammonia measurement—another example of a colorimetric technique.) Figure 19-5 graphically illustrates the pararosaniline method for SO_2 measurement.

Most bubblers are not 100% efficient, for not all of the gas bubbled in will be absorbed by the liquid, and some will escape. Obviously, this creates problems since it is usually necessary to have quantitative determinations. Most bubblers are thus tested to establish their efficiencies, and these values are used as contrasts by which the measured values are multiplied to obtain the actual concentrations.

Hundreds of different techniques have been used in third-generation gaseous measurement. One widely used device is the *nondispersive infrared analyzer,* used for carbon monoxide measurement. This technique relies on the fact that CO absorbs infrared radiation of a specific frequency. Schematically, as shown in Figure 19-6, the detector consists of two chambers, with the air sample pumped into one and the other containing an inert reference gas (e.g., nitrogen). The two infrared lamps shine through the cells and the CO absorbs a part of that energy, directly proportional to the amount of CO in the chamber. After passing through the two cells, the radiant energy is absorbed by the gas in two detector cells, each one containing CO. This absorption causes the gas in the detector cells to be heated, but the detector under the reference cell which receives more energy is heated more. The hotter gas expands, moving a diaphragm separating the two detector cells. This movement is detected electrically and the signal is read out on a recorder, thus giving a continuous trace of CO in the sample.

Figure 19-5   Schematic of the pararosaniline method for measuring SO_2.

Figure 19-6   Nondispersive infrared spectrometry for CO measurement.

REFERENCE METHODS

Because of the great number of methods available for measuring air pollutants, the U.S. Environmental Protection Agency has chosen a series of reference methods (Table 19-1). These are not necessarily absolutely accurate, but they are judged the best available, and results from other methods are to be compared to these. The existence of reference methods is especially important when compliance with air quality standards is in question.

GRAB SAMPLES

Often it is necessary to obtain a sample of a gas for future analysis in the laboratory. Whereas in water pollution work the collection of

Table 19-1.   Standard EPA Reference Methods for Air
Quality Measurements

Pollutant	Reference Method	Comments
Particulates (TSP)	High-volume sampler	Note that this is a second-generation device, giving 24-hr readings.
Sulfur dioxide	Pararosaniline	
Carbon monoxide	Nondispersive infrared spectrometry	
Nitrogen dioxide	Chemiluminescence	This method was adopted after the original reference method was judged unreliable (and years of NO_2 monitoring data became worthless).
Photochemical oxidants (O_3)	Chemiluminescence	This method measures only O_3.
Hydrocarbons (nonmethane)	Flame ionization	Methane is not measured because it is a nonreactive hydrocarbon, naturally occurring.

samples does not present a serious problem, in air quality measurements it is a real challenge.

One method is the use of evacuated containers, drawing a vacuum in a glass jar and releasing it when the air is to be sampled. Unfortunately, it is not possible to evacuate the jar 100%, and contamination is always a possibility.

Bubbling gas into a container with water in it and displacing the water is an effective method only if the gases of interest are not at all soluble in water.

Plastic and aluminum bags have been used extensively. Usually the gas is pumped in and allowed to escape through a hole. By thus displacing two or three volumes of the bag, contamination problems are avoided.

STACK SAMPLING

Stack sampling is an art worthy of individual attention. As the name implies, this involves the sampling of gas from a smokestack, and is necessary for evaluating compliance with emission standards and determining efficiencies of air pollution control equipment.

The most serious problem with stack sampling is the risk of obtaining an unrepresentative sample. Accordingly, a thorough survey is usually made of the flow, temperature and pollutant concentration in a stack. It is also usually desirable to obtain samples from a series of locations within a stack to be assured of accurate measurements. A train of instruments (Figure 19-7) is often utilized for stack sampling so that a number of measurements can be determined at each positioning of the intake nozzle.

SMOKE

Air pollution has historically been associated with smoke—the darker the smoke, the more pollution. We now of course know that this isn't necessarily true, but many regulations (e.g., for municipal incinerators) are still written on the basis of smoke density.

The density of smoke has for many years been measured on the Ringelmann scale, running from 0 for white or transparent smoke to 5 for totally black opaque smoke. The test is conducted by holding a card such as those shown in Figure 19-8 and comparing the blackness of the card to the smoke. A fairly dark smoke is thus said to be "Ringelmann 4."

Figure 19-7   A stack sampling train.

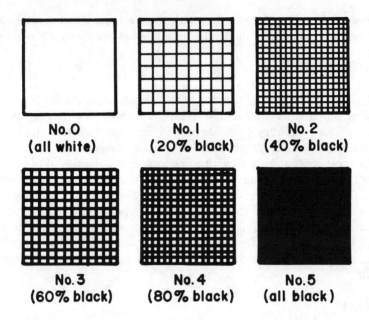

No. 0
(all white)

No. 1
(20% black)

No. 2
(40% black)

No. 3
(60% black)

No. 4
(80% black)

No. 5
(all black)

Figure 19-8   Ringlemann scale for measuring the opacity of smoke.

CONCLUSION

As with water pollution, the analytical tests of air quality can be only as good as the samples or sampling techniques used. In addition, the prevailing analytical techniques leave a great deal to be desired in both precision and accuracy. It is important to remember, therefore, that most measurements of environmental quality, and especially air quality, are at best reasonable estimates and should not be believed to the fourth decimal point.

PROBLEMS

19.1 An empty 6-in.-diameter dustfall jar weighed 1560 g. After sitting out a prescribed amount of time, it weighed 1570 g. Report the dustfall in the usual manner.

19.2 A hi-vol clean filter weighs 20.0 g and the dirty filter weighs 20.5 g. The initial and final air flows were 70 and 50 ft³/min. What volume of air went through the filter in 24 hours? What was the level of particulates in the atmosphere? How does this compare to the National Air Quality Standards? (See Chapter 21.)

19.3 A high-volume sampler draws air in at an average rate of 70 ft^3/min. If the particulate reading is 200 μg/m^3, what was the weight of the dust on the filter?

SUGGESTED READING

American Public Health Association. *Methods of Air Sampling and Analysis* (1973).

Leithe, W. *The Analysis of Air Pollutants* (Ann Arbor, MI: Ann Arbor Science Publishers, Inc., 1970).

Methods of Air Sampling and Analysis (Washington, DC: American Public Health Association, 1972).

Painter, D. E. *Air Pollution Technology* (Reston, VA: Reston Publishing Co., Inc., 1974).

Wark, K., and C. F. Warner. *Air Pollution* (New York: IEP, 1976).

Chapter 20

Air Pollution Control

It is not difficult to see how air has become the ubiquitous waste-basket. For ages man has been dumping wastes into the atmosphere, and these pollutants have "disappeared" with the wind. It is a difficult transition, therefore, to suddenly force people to limit emissions, and almost impossible to get them to pay for it. Nevertheless, industrial man must now prevent the emission of what would a few years ago have been wasted into the atmosphere.

The control of emissions can be realized in a number of ways. Five separate possibilities for control are pictured in Figure 20-1. Dispersion is covered in Chapter 18, and the four remaining control points are discussed individually in this chapter.

SOURCE CORRECTION

Often the easiest solution to an air pollution problem is to stop or change the guilty process. Once the decision has been made that a product or process is necessary, the engineer must consider the possibility of controlling emissions by changing the process. For example, if automobiles are blamed for high lead levels in urban air, the most reasonable solution is simply to eliminate the lead in the gasoline. The source has been corrected and the problem solved.

In addition to a change of raw material, a modification of the process might also be used to achieve a desired result. For example, municipal refuse incinerators have been known to stink. The odors can often be readily controlled if the incinerators are operated at a high enough temperature to completely oxidize the organics that cause the odor. Provided this higher-temperature operation is possible, it is a very reasonable process change to obtain the desired air pollution control.

Figure 20-1   Points of possible air pollution control.

Strictly speaking, such measures as process change, raw material conversion or equipment modification to meet emission standards are known as *controls*. In contrast, *abatement* is the term used for all devices and methods for decreasing the quantity of pollutant reaching the atmosphere, once it has already been emitted from the source. For the sake of simplicity, however, we refer to all of the procedures as controls.

COLLECTION OF POLLUTANTS

Often the most serious problem in air pollution control is collection of the pollutants so as to provide treatment. Automobiles are notorious polluters, but only because their emissions cannot be readily collected. If we could channel the exhausts from automobiles to some central facilities, their treatment would be much more reasonable than controlling each individual car.

One success in collecting pollutants has been the recycling of blow-by gases in the internal combustion engine (see page 496). By reigniting these gases and emitting them through the car's exhaust system, the necessity of installing a separate treatment device for the car has been eliminated.

Air pollution control engineers have their toughest problems when the pollutants from an industry are not collected but are emitted from windows, doors and cracks in the walls. It is not, therefore, always possible to solve a problem by simply installing some piece of control equipment. Often a complete overhaul of the air flow in the entire plant is required.

COOLING

The exhaust gases to be treated are sometimes too hot for the control equipment and the gases must first be cooled. This can be done in three general ways: dilution, quenching or heat exchange coils (Figure 20-2). Dilution is acceptable only if the total amount of hot exhaust is small. Quenching has the added advantage of scrubbing out some of these gases and particulates, but this method may result in a dirty and hot liquid which must be disposed of. The cooling coils are probably the most widely used, and are especially appropriate when heat can be conserved.

TREATMENT

Selection of the correct treatment device requires matching characteristics of the pollutant with features of the control device. It is important to realize that the sizes of air pollutants range many orders of magnitude, and it is therefore not reasonable to expect one device to be effective and efficient for all pollutants. In addition, the types of chemicals in emissions often will dictate the use of some devices. For example, a gas containing a high concentration of SO_3 could be cleaned by water sprays, but the resulting sulfuric acid might present serious corrosion problems.

The various air pollution control devices are conveniently divided into those applicable for controlling particulates and those used for controlling gaseous pollutants. The reason, of course, is the difference in the size. Gas molecules have diameters of about 0.0001 microns, particulates range from 0.1 microns and up. In any case, however, the efficiency, η, of a collection device is evaluated according to the following relationship (see Figure 20-3):

Figure 20-2   Cooling hot waste gases.

$$\eta = \frac{M_c}{M_a} \times 100$$

where M_c = amount of pollutant collected, kg/sec
M_a = amount of pollutant that was collectable, kg/sec

In other words, M_a is the percent of the pollutant that could have been collected were the devices perfect. The terms M_c and M_a are in terms of mass per unit time.

If M_b is the pollutant that *escapes* capture, then

$$M_a = M_b + M_c$$

and efficiency could also be expressed as

$$\eta = \frac{M_c}{M_b + M_c} \times 100$$

If the concentration of the pollutants and the rate of air flow are known, the efficiency is of course expressed as

$$\eta = \frac{Q_c C_c \times 100}{Q_a C_a}$$

where Q = air flow rate, m^3/sec
C = concentration of pollutant, kg/m^3

and the subscripts refer to inflow and captured pollutant (Figure 20-3). In situations where the air flow rates are not known and only concentrations in the three streams are measured, the efficiency can be shown to be equal to

$$\eta = \frac{C_c(C_a - C_b) \times 100}{C_a(C_c - C_b)}$$

Figure 20-3   Definition sketch of a collecting device.

Settling Chambers

The simplest devices for controlling particulates are settling chambers consisting of nothing more than wide places in the exhaust flue where larger particles can settle out, usually with a baffle to slow the gas stream. Obviously, only very large particulates ($>100\ \mu$) can be efficiently removed in settling chambers.

Cyclones

Possibly the most popular, economical and effective means of controlling particulates is the cyclone. Figure 20-4 shows a simple schematic and Figure 20-5 is a more detailed picture of a cyclone. The dirty air is blasted into a conical cylinder, but off the centerline. This creates a violent swirl within the cone, and the heavy solids migrate to the wall of the cylinder where they slow down due to friction and exit at the bottom of the cone. The clean air is in the middle of the cylinder and exits out the top.

Figure 20-4   Four methods of controlling particulates from stationary sources.

CLEAN GAS OUT

DIRTY GAS IN

HOPPER

PARTICULATE DISCHARGE

Figure 20-5   Cyclone (courtesy American Lung Association).

Cyclones are widely used as precleaners, to remove heavy material before further treatment.

Cyclones are sized on the basis of a *separation factor,* which is developed as follows: Consider a particle "settling" radially in a cyclone (from the centerline to the outside wall). The velocity of this movement, if we assume laminar flow (not a good assumption) and spherical particles (likewise) is

$$v_R = \frac{(\rho_s - \rho)\, r\omega^2\, d^2}{18\,\mu}$$

where

v_R = radial velocity, m/sec
ρ_s = density of particle, kg/m^3
ρ = density of air, kg/m^3
r = radial distance, m
ω = rotational velocity, rad/sec
d = diameter of particle, m
μ = viscosity of air, kg/m-sec

Since $r\omega^2$, the centrifugal acceleration, equals v_{tan}^2/r, the tangential velocity squared divided by the radius,

$$v_R = \left[\frac{(\rho_s - \rho)\, d^2}{18\,\mu}\right] \cdot \frac{v_{tan}^2}{r}$$

The term in brackets is the gravitational velocity with the gravitational acceleration, g, missing. Hence

$$v_R = \left(\frac{v}{g}\right)\left(\frac{v_{tan}^2}{r}\right)$$

The separation factor, S, is defined as

$$S = \frac{v_R}{v} = \frac{v_{tan}^2}{rg}$$

A large separation factor requires high tangential velocities, thus small diameters and high velocities, leading to high pressure drops. A large-diameter cyclone, on the other hand, would have a low separation factor. The value of S can range from 5 to as high as 2500. (Note: S is dimensionless.)

The pressure drop through a cyclone can be calculated by an empirical relationship:

$$\Delta P = \frac{3950\, K\, Q^2\, P\, \rho}{T}$$

where ΔP = pressure drop, m of water
        Q = m^3/sec of gas
        P = absolute pressure, atm
        ρ = gas density, kg/m^3
        T = temperature, °K
        K = factor, function of cyclone diameter. Approximate values are shown in Table 20-1.

Table 20-1.   Values of K for Calculating Pressure Drop in Cyclones

Cyclone diameter (in.)	K
29	10^{-4}
16	10^{-3}
8	10^{-2}
4	10^{-1}

Typical pressure drops in cyclones are from 1 to 8 inches of water.

The efficiency of a cyclone can be estimated by using the concept of a *cut diameter,* or that particle diameter at which 50% of the particles are removed by the cyclone. This is defined as

$$d_c = \left[\frac{9 \, \mu \, b}{2 \pi N v_i \, (\rho_s - \rho)} \right]^{1/2}$$

where μ = gas viscosity, kg/m-hr
        b = cyclone inlet width, m
        N = effective number of outer turns in the cyclone (normally about 4)
        v_i = inlet gas velocity, m/sec
        ρ_s = particle density, kg/m^3
        ρ = gas density, kg/m^3 (usually estimated as zero)

This cut diameter can be used to establish the collection efficiency for any other diameter particle, d, as shown in Figure 20-6. The number of turns (N) can also be approximated as

$$N = \frac{\pi}{H}(2L_1 + L_2)$$

where H = height of the inlet, m

L_1 = length of cylinder, m
L_2 = length of cone, m

Example 20.1

A cyclone has an inlet width of 10 cm and four effective turns. The gas has a temperature of 350°K, an inlet velocity of 10 m/sec and the particles have an average size of 8 μ and a density of 1.5 g/cm³. What is the efficiency of collection?

$$d_c = \left[\frac{9 \mu b}{2 \pi N v_i (\rho_s - \rho)} \right]^{1/2}$$

The viscosity of air at 350°K is about 0.0748 kg/m-hr. We can assume $\rho = 0$ and $N = 4$. All parameters are in m-hr-kg.

$$d_c = \left[\frac{(9)(0.0748)(0.1)}{2(3.14)(4)(36,000)(1.5 \times 10^{-3} - 1 \times 10^{-6})} \right]^{1/2}$$

$$= 7.12 \times 10^{-6} \text{ m} = 7.12 \mu$$

Then, $d/d_c = 8/7.12 = 1.12$. From Figure 20-6, the expected removal efficiency is about 55%.

Bag (or Fabric) Filters

The filters used for controlling particulates (Figures 20-4 and 20-7) operate like the common vacuum cleaner. Fabric bags are used to collect

Figure 20-6 Cyclone efficiency.

the dust which must be periodically shaken out of the bags. The fabric will remove nearly all particulates, including submicron sizes. Bag filters are widely used in many industrial applications, but are sensitive to high temperatures and humidity.

CLEAN GAS OUT

FILTER BAGS

DIRTY GAS IN

HOPPER

DISCHARGE

Figure 20-7   Fabric filters (courtesy American Lung Association).

The basic mechanism of dust removal in fabric filters is thought to be similar to the action of sand filters in water quality management (see Chapter 5). The dust particles adhere to the fabric due to entrapment and surface forces. They are brought into contact by impingment and/or Brownian diffusion. Since fabric filters commonly have an air space-to-fiber ratio of 1:1, the removal mechanism cannot be simple sieving.

As the particles adhere to the fabric, the removal efficiency increases, but so does the pressure drop. The latter is thus a sum of the pressure drop due to the fabric and the drop due to the caked or adhered particles. This is expressed as

$$\Delta P = v\mu\left(\frac{x_f}{K_f} + \frac{x_p}{K_p}\right)$$

where ΔP = overall pressure drop, m of water
          v = superficial gas velocity through the fabric, m/sec

μ = gas viscosity, poise
x = thickness of filter (f) and particle layer (p), m
K = filter (f) and particle layer (p) permeability.

The values of K must be obtained by experiment for each type of fabric and type of dust to be collected.

Wet Collectors

The simple spray tower, pictured in Figures 20-4 and 20-8, is an effective method for removing large particulates. More efficient scrubbers promote the contact between air and water by violent action in a narrow throat section into which the water is introduced. Generally, the more violent the encounter, and hence the smaller the gas bubbles or water droplets, the more effective the scrubbing.

A commonly used high-energy wet collector is a *venturi scrubber* (see Figure 20-9) where the gas flow is constricted through a venturi throat section and water is introduced at high-pressure streams perpendicular to the gas flow. The venturi scrubber is essentially 100% efficient in removing particles $>5 \mu$. The pressure drop is estimated by an empirical equation as

$$\Delta P = v_g^2 L \times 10^{-6}$$

where ΔP = pressure drop across the venturi, cm of water
v_g = gas velocity in the throat, cm/sec
L = water-to-gas volume ratio, liter/m^3

Wet scrubbers are efficient devices, but have two major drawbacks:

1. They produce a visible plume, albeit only water vapor. The lay public seldom differentiates, and hence public relations often dictate no visible plume.
2. The waste is now in liquid form and some manner of treatment is necessary.

Electrostatic Precipitators

Today, electrostatic precipitators are widely used in power plants, mainly because the power is inexpensive and readily available. The particulate matter is removed by first being charged by electrons jumping

CLEAN GAS OUT

SPRAY RISER

WATER JETS

TANGENTIAL INLET
FOR DIRTY GAS

WATER IN

FLUSHING JETS,
DIRECTED
DOWNWARD

WATER AND
POLLUTING
PARTICLES OUT

Figure 20-8   Scrubber (courtesy American Lung Association).

from one high-voltage electrode to the other, and then migrating to the positively charged collecting electrode (electrons are negatively charged). The type of electrostatic precipitator shown in Figure 20-4 consists of a pipe with a wire hanging down the middle. The particulates collect on the pipe and must be removed by banging the pipes with hammers. Figure 20-10 shows a plate-type precipitator, with flat plates acting as the collection electrodes. Electrostatic precipitators have no moving parts, require only electricity to operate, and are extremely effective in removing sub-micron particulates. They are also expensive.

The efficiency of an electrostatic precipitator is commonly estimated by an empirical equation

DIRTY AIR

WATER

CLEAN AIR +
DIRTY WATER

Figure 20-9   Venturi scrubber.

$$\eta = 1 - \exp\left(\frac{-A\,v_d}{Q}\right)$$

where   A = area of the collection electrodes (in the case of a pipe type unit, the inside area of the pipes), m^2
Q = flow rate of gas through the pipe, m^3/sec
v_d = the drift velocity, m/sec

The drift velocity is the velocity of the particles toward the collecting electrode. It can be calculated theoretically by equating the electrostatic force on the charged particle in an electrical field, and a drag force as the

NEGATIVE ELECTRODE CONNECTED
TO ELECTRICAL POWER SOURCE

NEGATIVELY CHARGED WIRE

GROUNDED COLLECTING PLATE
WITH POSITIVE CHARGE

DIRTY GAS IN

HOPPER TO DISCHARGE

CLEAN GAS OUT

Figure 20-10   Electrostatic precipitator (courtesy American Lung Association).

particle moves through the gas. This is perfectly analogous to the terminal settling velocity in a wastewater settling tank, except that in the latter case the force is due to gravity. Practically, the drift velocity can be estimated by simply relating it to particle size as

$$v_d = 0.5 \ d$$

where d = particle size, μ
          v_d = drift velocity, m/sec

Commonly, drift velocities are on the order of 0.03 to 0.2 m/sec.

A major problem with electrostatic precipitators is that as the dust layer builds up inside the pipe, the resistivity increases, thus decreasing the drift velocity. Ironically, one means of reducing the insulating effect of the dust is to inject sulfur (!) as SO_2 or sulfuric acid into the gas.

Figure 20-11 shows the effectiveness of an electrostatic precipitator in controlling emission from a power plant. The large white boxes in the foreground are the electrostatic precipitators. The one on the right, leading to the two stacks on the right, has been turned off to show the effectiveness by comparison with the almost undetectable emission from the stacks on the left.

Figure 20-11   Effectiveness of electrostatic precipitators on a coal-fired power plant. The electrostatic precipitator on the right has been turned off.

Comparison of Particulate Control Devices

The efficiencies and costs of the various control devices obviously vary widely. Figure 20-12 shows approximate collection efficiency curves, as a function of particle size, for the various devices discussed.

Figure 20-12 Comparison of removal efficiencies. A = baffled settling chamber, B = simple cyclone, C = high-efficiency cyclone, D = electrostatic precipitator, E = spray tower wet scrubber, F = venturi scrubber, G = bag filter.

CONTROL OF GASEOUS POLLUTANTS

The control of gases involves the removal of the pollutant from the gaseous emissions, a chemical change in the pollutant, or a change in the process producing the pollutant.

Wet scrubbers, such as already discussed, can remove gaseous pollutants by simply dissolving them in the water. Alternatively, a chemical may be injected into the scrubber water which then reacts with the pollutants. This is the basis for most SO_2 removal techniques, as discussed below. *Adsorption* is a useful method when it is possible to bring the pollutant into contact with an efficient adsorber like activated carbon. This method is effective for many organic pollutants (Figure 20-13).

Incineration (Figure 20-14) or flaring is used when an organic pollutant can be oxidized to CO_2 and water. A variation of incineration is *catalytic combustion* where the temperature of the reaction is lowered by the use of a catalyst which mediates the reaction. Figure 20-15 is a comparison of simple incineration and catalytic combustion. The catalytic "reactor" is used in many cars to reduce HC and CO emissions by oxidizing them to CO_2 and H_2O (see below).

DIRTY GAS IN

CARBON BED

CLEAN GAS OUT

Figure 20-13   Adsorber for control of gaseous pollutants (courtesy American Lung Association).

Control of Sulfur Oxides

We noted earlier that sulfur oxides (SO_2 and SO_3) are serious and yet ubiquitous air pollutants. The major source of sulfur oxides (or SO_x as they are often referred to in shorthand) is coal-fired power plants. The increasingly strict standards for SO_x control have prompted the development of a number of options and techniques for reducing emissions of sulfur oxides. Among these options are:

1. *Change to Low–Sulfur Fuel.* Natural gas and oil are considerably lower in sulfur than coal. However, uncertain and expensive supplies make this option risky.

CLEAN GAS OUT

BURNER PORTS

BURNER PIPING
DIRTY GAS INLET

Figure 20-14   Incinerator for controlling gaseous pollutants (courtesy American Lung Association).

2. *Desulfurize the Coal.* Sulfur in coal is either organic or inorganic. The inorganic form is iron pyrite (FeS_2) which, since it occurs in discrete particles, can be removed by washing. The removal of the organic sulfur (generally about 60% of the total) requires chemical reactions, and is most economically accomplished if the coal is gasified (changed into a gas resembling natural gas).

3. *Tall Stacks.* A short-sighted method, albeit locally economical, of SO_2 control is to build incredibly tall smokestacks and disperse the SO_x. This option has been employed in Great Britain and is in part responsible for the acid rain problem plaguing Scandinavia.

4. *Flue–Gas Desulfurization.* The last option is to reduce the SO_x

Incineration

Catalytic Combustion

Figure 20-15 Catalytic combustion compared to simple incineration.

emitted by cleaning the gases coming from the combustion, the so-called flue-gases. Many systems have been employed, and a great deal of research is presently under way to make these processes more efficient.

The most widely used method of SO_2 removal is to contact the sulfur with lime. The reaction is

$$SO_2 + CaO \rightarrow CaSO_3$$

or if limestone is used,

$$SO_2 + CaCO_3 \rightarrow CaSO_4 + CO_2$$

Both calcium sulfite and sulfate (gypsum) are solids which have low solubilities and thus can be separated in gravity settling tanks. The three commonly employed options, illustrated in Figure 20-16, are injecting limestone into the scrubber, calcining the limestone first and injecting lime into the scrubber, and adding limestone to the boiler. In the last two cases the limestone reacts to form quicklime

$$CaCO_3 \overset{\Delta}{\rightarrow} CaO + CO_2$$

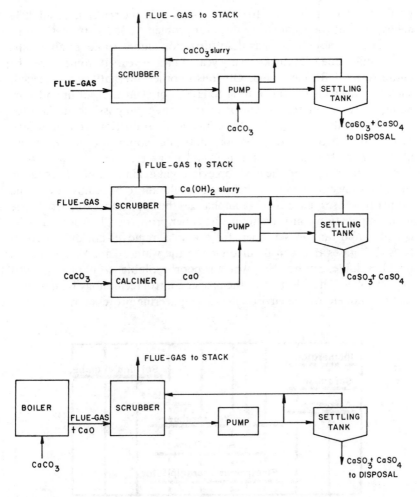

Figure 20-16   Three options for desulfurization of flue-gases using limestone.

In the one case the quicklime so produced is hydrated (to $Ca(OH)_2$) which is then injected into the scrubber. In the last system the quicklime is produced in the boiler without the need of a separate calciner and the lime is carried to the scrubber with the exit gases.

The calcium salts thus formed represent a staggering disposal problem. If, according to U.S. Environmental Protection Agency (EPA) estimates, the total emission of SO_2 from all sources in the United States is about 40×10^6 metric tons per year, and if 90% of this is removed, 35×10^6 metric tons of SO_2 or 76×10^6 metric tons of $CaSO_4$ would be produced. This is in slurry form after settling in the tank, at about 20% solids, so that the total slurry disposal problem is 380×10^6 metric tons per year, or enough flue-gas desulfurization sludge to cover about 400 square miles one foot thick—every year! In three years it would cover the entire state of Rhode Island. Obviously, other alternatives are needed, such as converting the sulfur to H_2S, H_2SO_4 or elemental sulfur and marketing these raw materials. Unfortunately, the total possible markets for these chemicals is far less than the anticipated production from desulfurization. Much work needs to be done in the development of more reasonable methods for removing sulfur from flue-gases.

Before moving on to the next topic, it might again be useful to reiterate the importance of matching the type of pollutant to be removed with the proper process. Figure 20-17 reemphasizes the importance of particle size in the application control equipment. Other properties, however, may be equally important. A scrubber, for example, removes not only particulates, but gases that can be dissolved in the water. Thus SO_2, which is readily soluble, but not NO, which is poorly soluble, would be removed in a scrubber. The selection of the proper control technology is an important component of the environmental engineering profession.

Figure 20-17   Comparison of control options by particle size.

CONTROL OF MOVING SOURCES

Although many of the above control techniques can apply to moving sources as well as stationary ones, one very special moving source—the automobile—deserves special attention. As noted in Chapter 18, the automobile has many potential sources of pollution. Realistically, however, there are only a few important points requiring control (Figure 20-18):

A. evaporation of hydrocarbons (HC) from the fuel tank
B. evaporation of HC from the carburetor
C. emissions of unburnt gasoline and partially oxidized HC from the crankcase
D. the NO_x, HC and CO from the exhaust.

Figure 20-18   Internal combustion engine showing four major emission points.

The evaporative losses from the gas tank and carburetor have been eliminated by storing any vapors emitted in an activated carbon canister. This would normally occur when the engine is turned off and the hot gasoline in the carburetor vaporizes. The vapors can later be purged by air and burned in the engine, as shown schematically in Figure 20-19.

The third source of pollution, the crankcase vent, can be eliminated by closing off the vent to the atmosphere and recycling the blowby gases into the intake manifold. The positive crankcase ventilation (PCV) valve is a small check valve that prevents the buildup of pressure in the crankcase.

The most difficult control problem is the exhaust, which accounts for about 60% of the HC and almost all of the NO_x, CO and lead. One immediate problem is how to measure these emissions. It is not as simple

Figure 20-19   Internal combustion engine showing methods of controlling emissions.

as sticking a sampler up the tailpipe, since the quantity of pollutants emitted changes with the mode of operation. The effect of operation on emissions is illustrated as Table 20-2. Note that when the car is accelerating the combustion is efficient (low CO and HC) and the high compression produces a lot of NO_x. On the other hand, decelerating results in low NO_x and very high HC due to partially burned fuel.

Because of these difficulties the EPA has instituted a standard test for measuring emissions. This test procedure includes a cold start, acceleration and cruising on a dynamometer to simulate a load on the wheels, and a hot start.

Table 20-2.   Effect of Engine Operation on Automotive Exhaust Characteristics, Shown as Fraction of Idling Emissions

| | Component | | |
	CO	HC	NO_x
Idling	1.0	1.0	1.0
Accelerating	0.6	0.4	100
Cruising	0.6	0.3	66
Decelerating	0.6	11.4	1.0

Emission control techniques for the internal combustion automobile engine include tune-ups, catalytic reactors, and engine modifications. A tune-up can have a significant effect on emission components. For example, a high air/fuel ratio (a lean mixture) will reduce both CO and HC, but will increase NO_x. A well tuned car is nevertheless the first line of defense for controlling automobile emissions.

The second control strategy, now extensively used in all cars sold in the United States, is the catalytic reactor which oxidizes the CO and HC to CO_2 and H_2O. The most popular catalyst seems to be platinum, which must be periodically regenerated.

Catalytic reactors have two serious drawbacks, however. First, they are easily fouled by lead. In fact, the move to nonlead gasoline has been prompted by this consideration and not by the concern for lead levels in the atmosphere. The second problem with catalytic reactors is that the sulfur compounds in gasoline are oxidized to particulate SO_3, and thus increase the sulfur levels in urban environments. It is unlikely that catalytic reactors will be standard equipment for cars of the future.

The obvious solution, and the third control technique, is engine modification. The stratified charge engine, for example, has achieved rigid standards without using catalytic reactors. In these engines, the cylinders have two compartments, with one compartment receiving a rich mixture, igniting, and then providing a broad flame for an efficient burn in the main cylinder compartment. Other modifications are now under investigation.

Even with extensive investment in engineering, it will be difficult to manufacture a totally clean internal combustion engine. A great deal of research is presently underway to develop an alternative that can meet strict air pollution requirements. Electric cars are clean but can only store limited power, thus their range is limited. In addition, the electricity used to power such vehicles must be generated, thereby creating more pollution.

The diesel engines used in trucks and buses also are important sources of pollution. There are, however, fewer of these vehicles in operation than gasoline-powered cars, and the emissions may not be as harmful due to the nature of diesel engines. The main problems associated with diesel engines are the visible smoke plume and odors, two characteristics that have led to considerable public irritation with diesel-powered vehicles. But from a public health viewpoint, diesel exhaust does not constitute the problem that the exhaust from gasoline-powered cars does. Diesel engines in passenger cars can, in fact, readily meet strict emission control standards.

CONCLUSION

This chapter is devoted mainly to the description of air pollution control alternatives by "bolt-on" devices. It should be reemphasized that this is usually the most expensive method of control. A general environmental engineering truism is that the least expensive and most effective control point is always the farthest up the process line. It is at the beginning of the process or, better yet, consideration of alternatives to the process, where the most effective control is achieved. This is obvious in the case of flue-gas desulfurization. We should not seek to bury Rhode Island, but rather ask if all the electricity is really necessary. Are there other power sources? Not only is this good engineering technology and economics, but it is also sensitive and enlightened analysis of our lifestyle and its impact on the environment.

PROBLEMS

20.1 Taking into account cost, ease of operation, and ultimate disposal of residuals, what type of control device would you suggest for the following emissions?
   a.   A dust with particle range of 5–10 μ?
   b.   A gas containing 20% SO_2 and 80% N_2?
   c.   A gas containing 90% HC and 10% O_2?

20.2 A stack emission has the following characteristics: 90% SO_2, 10% N_2, no particulates. What treatment device would you suggest and why?

20.3 How big is 10,000 microns in inches? How big is 1 micron in inches?

20.4 An industrial emission has the following characteristics: N_2—80%, O_2—15%, CO_2—5%. You are called in as a consultant to advise on the type of air pollution control equipment required. What would be your recommendation?

20.5 A whiskey distillery has hired you as a consultant to design the air pollution equipment for their new plant, to be built in a residential area. What problems would you encounter, and what would be your control strategy?

20.6 A dust has a particle size analysis as follows:

Mean Size		Percent of Particles by Weight
0.05	----------------	10
0.1	----------------	20
0.5	----------------	25
1.0	----------------	35
5.0	----------------	10

The particle specific gravity is 1.2.

A 1.51-m-diameter cyclone with an inlet width of 15 cm and an inlet length of 25 cm is operating at 5 effective turns and 20 m^3/sec air flow.
   a. What is the efficiency of removal?
   b. What is the pressure drop (assume normal temperature and pressure of the air)?
   c. What is the separation factor? Is this a high- or low-efficiency cyclone?

20.7 A plate-type electrostatic precipitator is to remove a particulate with a diameter of 0.5 μ. The 40 plates are 5 cm apart and 3 m high. How deep must the plates be to achieve removal (ideally)?

20.8 A dust has particles with a drift velocity of about 0.15 m/sec. For a total air flow of 60 m^3/sec, what must be the number of 10-m × 10-m collecting plates in order to achieve 90% removal?

20.9 Design a start/stop/drive test for measuring the emission of pollutants from an automobile. Justify your choice.

20.10 Calculate the removal efficiencies for the dust described in Problem 20.6 for a standard cyclone, an electrostatic precipitator and a bag filter, using the curves in Figure 20-12.

20.11 Determine the efficiency of a cyclone of diameter 0.5 m, at a flow rate of 4 m^3/sec and a gas temperature of 25°C. The inlet width is 0.13 m^2 and the area of the entrance is 0.04 m^2. The particulates have a diameter of 10 μ.

20.12 An electrostatic precipitator has the following specifications:
   height: 7.5 m
   length: 5 m
   number of passages: 3
   plate spacing: 0.3 m

The flow rate is 18 m³/sec, the particle size is 154 μ, and the particles have a drift velocity of 0.162 m/sec. Find the expected efficiency.

   20.13 For the problem above, what would be the increase in efficiency by
   a. doubling the length
   b. doubling the number of plates and reducing the plate spacing by one-half?

LIST OF SYMBOLS

A = area, m²
b = cyclone inlet width, m
C = concentration, kg/m³
d = particle diameter, m
d_c = cut diameter, m
H = height of cyclone inlet, m
L = water-to-gas volume ratio, liter/m³
L_1 = length of cylinder in a cyclone, m
L_2 = length of cone in a cyclone, m
M_a = mass of pollutant that could have been collected, kg
M_b = mass of pollutant that escaped capture, kg
M_c = mass of pollutant collected, kg
N = effective number of turns in a cyclone
P = pressure, atm
ΔP = pressure drop, m water
Q = flow rate, m³/sec
r = radial distance, m
v_d = drift velocity, m/sec
v_i = inlet gas velocity, m/sec
v_R = radial velocity, m/sec
x = thickness, m
η = collection efficiency, %
μ = gas viscosity, poise
ρ = gas density, kg/m³
ρ_s = particle density, kg/m³
ω = rotational speed, rad/sec

SUGGESTED READING

Control Technology: Gases and Odors. APCA Reprint, Pittsburgh (1973).

Control Technology: Particulates. APCA Reprint, Pittsburgh (1973).

Crawford, M. *Air Pollution Control Theory* (New York: McGraw-Hill Book Company, 1976).

Hesketh, H. E. *Air Pollution Control* (Ann Arbor, MI: Ann Arbor Science Publishers, Inc., 1979).

Nonhebel, G., Ed. *Processes for Air Pollution Control* (Cleveland, OH: CRC Press, 1964).

Ward, K., and C. F. Warner. *Air Pollution* (New York: IEP, 1976).

Chapter 21

Air Pollution Law

As with water pollution, a complex system of laws and regulations governs the use of air pollution abatement technologies. In this chapter, the evolution of air pollution law is described, from its roots in common law through the passage of federal statutory and administrative initiatives. Problems encountered by regulatory agencies and polluters are addressed with particular emphasis on the impacts the system may or may not have on future economic development. Figure 21-1 offers a roadmap to be followed through this maze.

AIR QUALITY AND COMMON LAW

When dealing with common law, an individual or groups of individuals injured by a source of air pollution may cite general principles in two branches of that law which have developed over the years and may apply to their particular damages:

- tort law
- property law

The harmed party, the plaintiff, could enter a courtroom and seek remedies from the defendant for damaged personal well-being or damaged property.

Tort Law

A tort is an injury incurred by one or more individuals. Careless accidents, defamation of character, and exposure to harmful airborne chem-

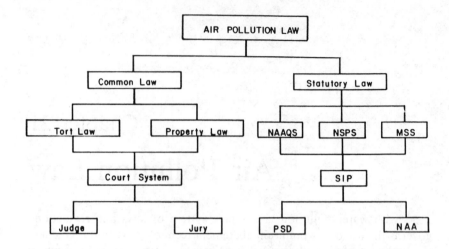

Figure 21-1   Air pollution law in the United States.

icals are the types of wrongs included under this branch of common law. A polluter could be held responsible for the damage to human health under three broad categories of tort liability: intentional liability, negligence and strict liability.

Intentional liability requires proof that somebody did a wrong to another party *on purpose.* This proof is especially complicated in the case of damages from air pollution. The fact that a "wrong" actually occurred must first be established, a process which may rely on direct statistical evidence or strong inference, such as the results of lab tests on rats. Additionally, intent to do the "wrong" must be established, which involves producing evidence in the form of written documents or direct testimony from the accused individual or group of individuals. Such evidence is not easily obtained. If intentional liability can be proven to the satisfaction of the courts, actual damages as well as punitive (punishment) damages can be awarded to the injured plaintiff.

Negligence may involve mere inattention by the air polluter which allowed the injury to occur. Proof in the courtroom focuses on the lack of reasonable care taken on the defendant's part. Examples of such neglect in air pollution include failure to inspect the operation and maintenance (O & M) of electrostatic precipitators or the failure to design and size an adequate abatement technology. Again, damages can be awarded to the plaintiff.

Strict liability does not consider the *fault* or state of mind of the defendant. Under certain extreme cases, a court of common law has held that some acts are *abnormally dangerous* and individuals conducting those acts are strictly liable if injury occurs. The court does tend to balance the danger of an act against the public utility associated with the act.

Again, if personal damage is caused by air pollution from a known source, the damaged party may enter a court of common law and argue for monetary damages to be paid by the defendant and/or an injunction to stop the polluter from polluting. Sufficient proof and precedent is often difficult if not impossible to muster, and in many cases, tort law has been found inadequate in controlling air pollution and awarding damages.

Property Law

Property law, on the other hand, focuses on the theories of nuisance and property rights; nuisance is based on the interference with the use or enjoyment of property and property rights is based on actual invasion of the property. Property law is founded on ancient actions between land owners, and involves such considerations as property damage and trespassing. A plaintiff basing a case on property law rolls the dice and hopes the court will rule favorably as it balances social utility against individual property rights.

Nuisance is the most widely used form of common law action concerning the environment. Public nuisance involves unreasonable interference with a right, such as the "right to clean air," common to the general public. A public official must bring the case to the courtroom and represent the public that is harmed by the air pollution. Private nuisance, on the other hand, is based on unreasonable interference with the use and enjoyment of private land. The key to a nuisance action is how the courts define "unreasonable" interference. Based on precedents and the arguments of the parties involved, the common law court balances the equities, hardships and injuries in the particular case and rules in favor of either the plaintiff or the defendant.

Trespass is closely related to the theory of nuisance. The major difference is that some physical invasion, no matter how minor, is technically a trespass. Recall that nuisance theory demands an "unreasonable" inter-ference with land and the outcome of a particular case depends on how a court defines "unreasonable." Trespass is relatively cut and dried. Examples of trespass include physical walk-ons, vibrations from nearby surface or subsurface strata, and possibly gases and microscopic particles flowing from an individual smoke stack.

In conclusion, common law has generally proven inadequate in dealing with problems of air pollution. The strict burdens of proof required in the courtroom often result in decisions that favor the defendant and lead to smoke stacks that continue to pollute the atmosphere. Additionally, the technicality and complexity of individual cases often limit the ability of a court to act; complicated tests and hard-to-find experts often leave a court and a plaintiff with their hands tied. Furthermore, the absence of standing in a common law courtroom often prevents private individuals from bringing a case before the judge and jury unless the individual actu-ally suffered material or bodily harm from the air pollution.

One key aspect of these common law principles is their degree of varia-tion. Each state has its own body of common law, and individuals relying on the court system are generally confined to using the common laws of the applicable state.

Given these shortcomings inherent in common law, Congress adopted a federal Clean Air Act, with the objective of plugging some of the pollu-tion holes in common law.

STATUTORY LAW

The federal statutory law dealing with air pollution has developed over the past several decades. The initial legislative effort was the Clean Air Act of 1963. Although this Act outlined the broad goals and provided money for research, it and the subsequent Air Quality Act of 1967 gener-ally did not lead to cleaner air. Not until late in 1970 did Congress pass amendments to the 1967 Act that put some teeth into the federal regu-latory effort.

In these 1970 amendments, the U.S. Environmental Protection Agency (EPA) was empowered to develop air quality criteria, define National Ambient Air Quality Standards (NAAQS) and set maximum emission standards for moving as well as stationary sources. The air quality cri-teria are massive collections of data and information about the effects of various pollutants on public health and private property. These criteria, or background documents, were developed by the EPA for seven classes of pollutants: particulates, sulfur oxides, nitrogen oxides, hydrocarbons, oxidants (ozone), carbon monoxide and lead. Based on the data in these documents, the EPA set the NAAQS air quality levels to be attained in all parts of the country.

These NAAQS focus on two levels of controls: primary and secondary. The primary standards apply exclusively to the protection of human health; the secondary standards, which are more stringent, apply to property damage. These standards are tabulated in Table 21-1. Air quality standards, including ambient limitations and moving and stationary source emission levels, become part of an individual state's clean air program. Plants emitting asbestos, beryllium, mercury, or vinyl chloride are subject to special regulations under the National Emission Standards for Hazardous Air Pollutants (NESHAPS).

Table 21-1. Selected National Ambient Air Quality Standards (NAAQS)

Pollutant	Primary		Secondary	
	(ppm)	(μg/m^3)	(ppm)	(μg/m^3)
Particulate Matter (μg/m^3)				
Annual geometric mean	–	75	–	60
Max 24-hr concentration	–	260	–	150
Sulfur Oxides				
Annual arithmetic mean	0.03	80	0.02	60
Max 24-hr concentration	0.14	365	0.1	260
Max 3-hr concentration	–		0.5	1,300
Carbon Monoxide				
Max 8-hr concentration	9	10,000	same	
Max 1-hr concentration	35	40,000	as primary	
Photochemical Oxidants				
Max 1-hr concentration	0.10	200	same as primary	
Hydrocarbons				
Max 3-hr concentration	0.24	160	same as primary	
Nitrogen Oxides				
Annual arithmetic mean	0.05	100	same as primary	
Lead				
Avg of 3 months		1.5	same as primary	

State Implementation Plans (SIP) developed by individual states and approved by the EPA contain the actual abatement requirements devised to reduce air pollution as necessary to achieve compliance with NAAQS within that state. The respective responsibilities which state agencies have for developing an emissions inventory, formulating SIP, and issuing construction and operating permits vary from state to state. The SIP must include ambient and emission standards that are at least as strict as those mandated by the EPA. With these standards in place, the states have the power to initiate abatement orders to achieve acceptable air quality. Additionally, under powers granted by the federal Act, private citizens are explicitly given the right to sue the state and/or government if the agency is lax in its enforcement of the law.

In theory, the Clean Air Act and Amendments should result in cleaner air. States develop SIP, model air masses, and impose emissions standards to meet the NAAQS. The regulatory process ideally consists of four basic smooth-running operations:

- *prepermitting:* the identification of discharges, determination of which ones to regulate, and the establishment of emission limits.
- *permitting:* preparation of detailed source-specific documents which set emission limits.
- *construction:* installation of control equipment and/or modification of production processes at the site.
- *continuing compliance:* operation and maintenance of the abatement facility and equipment to maintain the specified emission limitations.

In reality, there are several flies in this ointment, and regulatory agencies face several significant problems in the drive to clean up the air.

PROBLEMS OF IMPLEMENTATION

Both regulatory agencies and the sources of pollution have encountered difficulties in the implementation of air pollution abatement measures. Because most major stationary sources of air pollution now have installed abatement equipment in response to federal pollution control requirements, regulatory agencies must develop programs which promote continuing compliance. However, the track records of the agencies indicate that current programs are not always effective.

Flaws in design and construction of control equipment contribute to significant noncompliance. Problems include the use of improper materials in constructing controls, undersizing of controls, inadequate instrumentation of the control equipment, and inaccessibility of control components for proper operation and maintenance. Numerous design flaws such as these cast doubt on how effectively the air pollution control agencies evaluate the permit applications for these sources. Additionally, many permit reviewers working in government lack the necessary practical experience to fully evaluate proposed controls and tend to rely too heavily on the inadequate technical manuals available to them.

A source permit, as it is typically designed within the regulatory framework, could ensure that an emitter plans to install necessary control equipment. It could also require the emitter to perform recordkeeping functions which facilitate proper operation and maintenance. However, agencies do not always use the permit program to accomplish these objectives. For example, in some states, a majority of the sources fail to keep any operating records at all that would enable an independent assessment of past compliance. In other states, although many sources tend to keep good operating records, they are not required to do so by the control agency.

An agency's ability to detect violations of emission requirements can be critical to the success of its regulatory effort. However, many agencies rely on surveillance by sight and smell as opposed to stack testing or monitoring. For example, air pollution control agencies often issue notices of violation for such problems as odor, dust and excessive visible emissions, even though it is recognized that this approach neglects other perhaps more detrimental pollutants.

Many state and local agency programs rely on the voluntary approach to achieving compliance by sources. Although a majority of sources appear to respond to this strategy, in some instances such programs are ineffective. In one state, for example, eight major emitters who agreed to voluntary controls in 1978 were later referred for legal action. This indicates a failure of good faith efforts to comply with their commitments. In addition, cases referred for civil or criminal action rarely result in significant penalties. In the same state six sources received finds in 1977, one was for $2000 and five ranged from $200 to $500. This compares to a legal maximum of $10,000 per day of violation.

The emitters of air pollution also face problems in complying with regulations. Although most large stationary sources of pollution have installed control equipment, severe emission problems can be caused by design or upsets either with the control equipment or in the production process. Poor design of the control equipment is probably the primary cause of excess emissions in the greatest number of sources, and process upsets and routine component failure appear to be the second major cause of excess emissions. Other causes of excess emissions include improper maintenance, lack of spare parts, improper construction materials and lack of instrumentation. Additionally, wide variations in the frequency and duration of excess emissions are generally experienced by different industrial categories.

Industrial Growth

Industrial growth is a concern of regulatory agencies in all states where ambient air quality standards are not met (nonattainment areas) and in areas wishing to prevent significant deterioration (PSD) of relatively high levels of existing air quality. The problems that agencies face in implementing air pollution control programs and the negative responses of many industrial sources toward regulations greatly slow the progress toward achieving the NAAQS and have the potential to slow industrial development. By 1975, with the passage of the initial deadline for achieving air quality standards, it became clear that continuing NAAQS violations would be widespread. This presented the dilemma for the EPA of how to handle the proposed construction of a new facility in an area that has not attained air quality standards, since any new facility emissions would further delay attainment of the NAAQS.

The EPA made an attempt to reconcile the conflicting national concerns for clean air and the need for economic growth in issuing its *Offsets Policy* as an Interpretative Ruling in the *Federal Register* on December 21, 1976. The ruling stated that new plants could be constructed in nonattainment areas, but only under certain conditions. These conditions required that emissions must be controlled to the greatest degree possible and that more than equivalent offsetting emission reduction must be obtained from other sources to ensure progress toward achievement of the standards.

The formulation of the EPA policy occurred simultaneously with the Congressional review of the Clean Air Act and the subsequent passage of the Clean Air Act Amendments of 1977. Congress provided that the interpretative ruling should remain in effect until July 1, 1979, unless modified by another EPA rule. Following that date, revised state implementation plans are intended to provide the basic framework for the review of new sources in nonattainment areas.

On January 16, 1979, the EPA issued new regulations revising the original interpretative ruling. These revisions incorporate a number of significant changes prompted by the 1977 amendments but preserve the basic framework of the earlier ruling. Under this current policy, a new source wishing to locate in any nonattainment area must satisfy three stringent conditions.

First, the facility must obtain emission reductions from other sources greater than the emissions expected from the proposed source. These emission "offsets" should be made on a pound per hour basis when facilities in the emission calculation are operating at maximum production levels. There must be more than a one for one reduction of emissions from existing sources in order to demonstrate reasonable progress toward attainment of ambient standards. The premise of the offset policy is that even in areas where abatement measures have been taken, opportunities may exist for further emission reduction. The offset must represent emission reductions that otherwise would not be required. Types of actions which could generate offsets include:

1. tighter controls on existing operations at the same site
2. a binding agreement with another facility to reduce emissions
3. the purchase of another facility in order to reduce emissions by installation of control equipment or by closing the facility down

Secondly, the new source must meet an emission limitation that specifies the lowest achievable emission rate (LAER) for such a source. In setting such emission limitations, the state must consider the lowest rate achieved in practice and the most stringent limitations contained in any state plan for such a source. The state must then establish the more stringent limitations, unless the applicant can demonstrate that the most stringent limitation is not achievable. If numerical emission standards are found to be infeasible, a design, operational or equipment standard may be prescribed. In no event can the emission rate exceed the applicable new source performance standard.

Finally, the source must certify that all existing major sources owned, operated or controlled by the applicant in the same state are in compliance with the applicable state implementation plan requirements and standards under the Act or are on an approved timetable for compliance.

In addition to the considerations made in nonattainment areas, a new source of pollution must comply with the PSD regulatory program requiring preconstruction approval of new plants with significant potential emissions to be built in "clean air" areas. There are three components of this program. First, air classification systems were developed whereby all areas of the country meeting air quality standards were classified as Class I, II or III, with varying limitations on growth in each class. Class I includes pristine areas subject to the tightest air pollution controls, Class II covers areas of moderate growth, and Class III areas of major industrialization. All areas of the nation were originally labeled as Class II under regulations promulgated by the EPA in 1974, with individual states authorized to reclassify areas at later dates.

Secondly, additional increments of air quality—defined as the quantity of additional pollution permitted through the combined impacts of all future growth within a given area—were defined for each area in the nation. This increment theoretically defines the additional amount of emissions each area could withstand and still be in compliance with the ambient air quality standard. Picture a child inflating a balloon. The balloon will hold just so much air before it pops. This increment is the additional pollution permitted in an area before the area pops through the NAAQS.

Finally, the new facility must gain preconstruction approval from the regulating agency. Each new plant must develop and submit data and analyses supporting the area's pollution control program. Included in the plan must be provisions for construction and operation of the best available control technology onsite at the plant.

CONCLUSION

Air pollution law is a complex web of common and statutory law. Although common law has offered and continues to offer checks and balances between polluters and economic development, shortcomings do exist. Federal statutory law has attempted to fill the voids and, to a certain extent, has been successful in cleaning the air. Engineers must be aware of the requirements placed on industry by this system of laws. Particular attention must be paid to the siting of new plants in different sections of the nation.

PROBLEMS

21.1 Acid rain is a mounting problem, particularly in the northeastern states. Discuss how this problem can be controlled under the system of common law. Compare this approach with the remedy under federal statutory law.

21.2 Pollution from tobacco smoke can significantly degrade the quality of certain air masses. Develop sample town ordinances to improve the quality of air: (1) inside city buses and (2) over the spectators at both indoor and outdoor sporting events. Rely on both structural and nonstructural alternatives; i.e., solutions that mechanically, chemically or electrically clean the air, and solutions that prevent all or part of the pollution in the first place.

21.3 Would you favor an international law that permits open burning of hazardous waste on the high seas? Open burning refers to combustion without emission controls. Would you favor such a law if emissions were controlled? Discuss your answers in detail; would you require permits which specify time of day of burning, or distance from shore of burning, or banning certain wastes from incineration?

21.4 Emissions from a nuclear power generating facility pose a unique set of problems for local health officials. Such facilities typically have very low emissions, but each has the potential for a catastrophic discharge. What precautions would you take if you were charged with protecting the air quality of nearby residents. How would your set of rules consider: (1) direction of prevailing winds, (2) age of residents, and (3) siting new schools?

21.5 Assume you live in a small town that has two stationary sources of air pollution: a laundry/dry cleaning establishment and a regional hospital. Automobiles and front porches in the town have a habit of turning black literally overnight. What air laws apply in this situation and, given that they presumably are not being enforced, how would you: (1) determine if, in fact, ambient and/or emissions standards were being violated and, if either was, (2) advise the residents to respond to see that the laws are enforced?

LIST OF SYMBOLS

EPA = U.S. Environmental Protection Agency
LAER = lowest achievable emission rate
NAAQS = National Ambient Air Quality Standards
NESHAPS = National Emission Standards for Hazardous Air Pollutants
PSD = prevention of significant deterioration
SIP = State (Air Pollution Control) Implementation Plan

SUGGESTED READING

Crocker, T. D. "The Structuring of Atmospheric Pollution Control Systems," in *The Economics of Air Pollution*, Harold Wolozin, Ed. (New York: W. W. Norton Press, 1966).

Dale, J. H. *Pollution, Property and Prices* (Toronto: Toronto University Press, 1968).

DeLucia, R. J. *Evaluation of Marketable Effluent Permit Systems,* Office of Research and Development, U.S. Environmental Protection Agency, (1974).

Edelman, S. *The Law of Air Pollution Control,* Environmental Sciences Services (1970).

Landan, N. J. and P. D. Rheingold. *The Environmental Law Handbook* (New York: Ballantine Books, 1971).

Mackintosh, D. K. *The Economics of Airborne Emissions: The Case of an Air Rights Market* (New York: Praeger Publishers, 1973).

The Conservation Foundation. *A Citizens' Guide to Clean Air* (Washington, DC: the Conservation Foundation, 1976).

U.S. Congress, *Clean Air Act Amendments: Public Law 93–319,* U.S. Government Printing Office (1970); and *Clean Air Act Amendments: Public Law 95–95,* U.S. Government Printing Office, (1977).

Chapter 22

Noise Pollution

The ability to make and detect sound provides humans with the facility to communicate among each other as well as to receive useful information from the environment. Sound can provide warning (the first alarm), useful information (a whistling tea kettle), and enjoyment (music).

In addition to such useful and pleasurable sounds there is noise, often defined as unwanted or extraneous sound. What is and is not noise is often in dispute, especially in the area of music.*

We generally think of noise as an unwanted by-product of our civilization, and classify such sources as trucks, airplanes, industrial machinery, air conditioners and similar sound producers as noise.

Urban noise is not, surprisingly, a modern phenomenon. Legend has it, for example, that Julius Caesar forbade the driving of chariots on Rome cobblestone streets during nightime so he could sleep.[1] Before the widespread use of automobiles, the 1890 noise pollution levels in London were described by an anonymous contributor to the *Scientific American* as:[2]

> The noise surged like a mighty heart-beat in the central districts of London's life. It was a thing beyond all imaginings. The streets of workaday London were uniformly paved in 'granite' sets . . . and the hammering of a multitude of iron-shod hairy heels, the deafening

---

*George Bernard Shaw entered a posh London restaurant, took a seat, and was confronted by the waiter. "While you are eating sir, the orchestra will play anything you like. What would you like them to play?" Shaw's replay? "Dominoes."

[1] Baron, R. A. *The Tyranny of Noise* (New York: St. Martins Press, 1970).

[2] Still, H. *In Quest of Quiet* (Harrisburg, PA: Stackpole Books, 1970).

side-drum tattoo or tyred wheels jarring from the apex of one set (of cobblestones) to the next, like sticks dragging along a fence; the creaking and groaning and chirping and rattling of vehicles, light and heavy, thus maltreated; the jangling of chain harness, augmented by the shrieking and bellowings called for from those of God's creatures who desired to impart information or proffer a request vocally— raised a din that is beyond conception. It was not any such paltry thing as noise.

Noise, as we will see, can adversely affect humans both physiologically as well as psychologically. It is an insidious pollutant in that damage is usually long range and permanent. And yet it is certainly the pollutant of least public concern, and (except for radioactivity) the least understood.

THE CONCEPT OF SOUND

The man-on-the-street has little if any concept of what sound is or what it can do. This is perhaps best exemplified by a well-meaning industrial plant manager who decided to decrease the noise level in his factory by placing microphones in the plant and channelling the noise through loudspeakers to the outside.[3]

Sound is in effect a transfer of energy. For example, rocks thrown at you would certainly get your attention, but this would require the transfer of mass (rocks). Alternatively, your attention can be gained by poking you with a stick, in which case the stick is not lost, but energy is transferred from the poker to the pokee.[3] In the same way, sound travels through a medium such as air without a transfer of mass. Just as the stick had to move back and forth, so must air molecules oscillate in waves in order to transfer energy.

The small displacement of air molecules which creates pressure waves in the atmosphere is illustrated in Figure 22-1. As the piston is forced to the right in the tube, the air molecules next to it are reluctant to move and instead pile up on the face of the piston (Newton's First Law). These compressed molecules now act as a spring, and release the pressure by jumping forward, creating a wave of compressed air molecules that move through the tube. The potential energy has been converted to kinetic energy.

These pressure waves move down the tube at a velocity of 344 m/sec (at 20°C). If the piston oscillates at a frequency of say 10 cycles/sec,

[3]Taylor, R. *Noise* (New York: Penguin Books, 1970).

Figure 22-1   A piston creates pressure waves which are transmitted through air.

there would be a series of pressure waves in the tube each 34.4 m apart. This relationship is expressed as

$$\lambda = \frac{c}{f}$$

where λ = wavelength, m
        c = velocity of the sound in a given medium, m/sec
        f = frequency, cycles/sec

Sound travels at different speeds in different materials, depending on the material's elasticity.

Example 22.1

In cast iron, sound waves travel at about 3440 m/sec. What would be the wavelength of a sound from a train if it rumbles at 50 cycles/sec and one listens to it placing an ear on the track?

$$\lambda = \frac{c}{f} = \frac{3440}{50} \cong 69\,m$$

In acoustics, the frequency as cycles per second is denoted by the name Hertz, and written Hz.* The common audible range for humans is between 20 and 20,000 Hz. The middle A on the piano, for example, is 440 Hz.

*In honor of German Physicist Heinrich Hertz (1857–1894).

The frequency is one of the two basic parameters which describe sound.*

If the amplitude of a pressure wave of a pure sound (a sound with only one frequency) is plotted against time, the wave is seen to produce a sinusoidal trace, (Figure 22-2). All other nonrandom (dirty) sounds are made up of a number of suitable sinusoidal waves, as demonstrated originally by Fourier.

The amplitude of a sound wave (pure or dirty) is measured in units of pressure, such as N/m^2. The pressure waves which our ears can detect (and thus we classify as audible sound) range from 0.00002 N/m^2 to 200,000 N/m^2, an obviously inconveniently large range. As a result, sound pressures are expressed in *decibels* (dB)** a unit of measurement that requires some explanation.

Actually, a decibel isn't a unit of anything—it is a ratio. As such, anything can be measured in decibels. For example, it is possible to express distance from Washington, DC to San Francisco as 20 times as long as from Washington to Philadelphia. The latter distance is a *reference* dis-

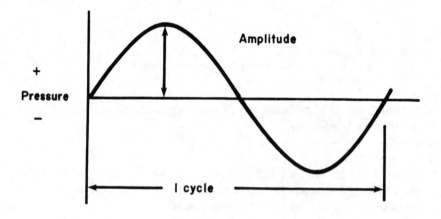

Figure 22-2   A representation of a sound wave.

---

*Incidentally, the frequency of sound detected by the ear is also dependent on the velocity of the source relative to the listener. This effect was first noted by a distinguished Austrian physicist, Christian Doppler, who in 1845 placed trumpet players on a railroad car and ran the cars past trained musicians who noted the change in pitch (frequency) as the train and trumpeters sped past. The "Doppler Effect" explains the apparent change in frequency of sound from an approaching and receding source.

**After Alexander Graham Bell.

tance to which other distances are compared. Similarly, the amplitude or loudness of sound is defined by a ratio of two pressures, one being a reference pressure. Amplitude is expressed as sound pressure level (SPL) and defined as

$$SPL \text{ (as dB)} = 20 \log_{10} \frac{P}{P_{ref}}$$

where   SPL (dB) = sound pressure level in dB
P = pressure of sound wave
P_{ref} = some reference pressure, generally chosen as the threshold of hearing, 0.00002 N/m^2

This equation is derived in more detail in the next chapter. The idea of using a ratio to describe any measurement is developed further in the example below.

Example 22.2

Suppose a pizza emporium sells three pizzas, 10, 100, and 1000 cm in diameter (there's no need for realism in examples, of course) and since there isn't enough space on the sign for all those zeros, it has been decided to express the size of the pizzas in ratios. We can first define a new term

$$\text{Pizza} (\pi) = \frac{D}{D_{ref}}$$

where D = diameter of any pizza
D_{ref} = reference diameter

and choose $D_{ref} = 1$ cm. Hence the three sizes of pizzas would be

$$\pi_1 = \frac{10}{1} = 10$$

$$\pi_2 = \frac{100}{1} = 100$$

$$\pi_3 = \frac{1000}{1} = 1000$$

But this, of course, doesn't help much, so we redefine

$$\text{Pizza} (\pi) = \log_{10} \frac{D}{D_{ref}}$$

and the three sizes become

$$\pi_1 = \log_{10} \frac{10}{1} = 1$$

$$\pi_2 = \log_{10} \frac{100}{1} = 2$$

$$\pi_3 = \log_{10} \frac{1000}{1} = 3$$

To carry this example further, we could again redefine a pizza as

$$\text{Pizza (deci-}\pi) = 10 \log_{10} \frac{D}{D_{ref}}$$

Thus the three sizes are deci-$\pi = 10$, 20 and 30.

Note that the above example, as deci-π increases by 10, the diameter increases 10-fold.

Likewise in sound measurement, as the sound pressure level (SPL) increases by 10, the sound pressure increases by an order of magnitude (10-fold). The reference pressure (P_{ref}) is selected as the threshold of hearing, which for a normal healthy ear is about 2×10^{-5} N/m^2.*

Since dB are logarithmic ratios of two pressures, the sound pressure levels in dB emitted from two sources cannot be added arithmetically. For example a source emitting 80 dB and another emitting 60 dB do not, if put in the same location, emit 140 dB. On the contrary, the 80 dB noise would mask the 60 dB, and the combined effect would be only a little greater than 80 dB.

The mathematics of adding decibels is a bit complicated. The procedure can be simplified a great deal by using the graph shown as Figure 22-3. As a rule of thumb, adding two equal sounds increases the SPL by 3 dB, and if one sound is more than 10 dB louder than a second sound, the contribution of the latter is negligible.

Background noise (or ambient noise) must also be subtracted from any measured noise. Using the above rule of thumb, if the sound level is more than 10 dB greater than the ambient level, the contribution can be ignored.

---

*It should be obvious then that a person with exceptionally keen ears can hear *negative* decibels.

Figure 22-3   Chart for combining different sound pressure levels. For example: Combine 80 and 75 dB. The difference is 5 dB. The 5-dB line intersects the curved line at 1.2 dB, thus the total value is 81.2 dB. (Courtesy Gen Rad.)

The covering of a sound with a louder one is known as masking. Speech can be masked by industrial noise, for example, as shown in Table 22-1. These data show that an 80-dB SPL in a factory will effectively prevent conversation. Telephone conversations are similarly affected, with a 65-dB background making communication difficult and 80-dB making it impossible. In some cases, it has been found advisable to use *white noise,* a broad frequency hum, to mask other more annoying noises.

THE ACOUSTIC ENVIRONMENT

The types of sounds around us vary from a Rachmaninoff concerto to the roar of a jet plane. Noise, the subject of this chapter, is generally considered to be unwanted sound, or the sound incidental to our civilization which we would just as soon not have to put up with. The intensities of some typical environmental noises are shown in Figure 22-4.

Laws against noise abound. Most local jurisdictions have ordinances against "loud and unnecessary noise." The problem is that precious few

Table 22-1.   Sound Levels for Speech Masking

Distance (ft)	Speech Interference Level (dB)	
	Normal	Shouting
3	60	78
6	54	72
12	48	66

of these ordinances are enforced. Only recently have some cities begun to enforce the laws already on the books. Memphis, for example, gave 10,000 traffic tickets one year for hornblowing. Recently the churchbells of a Chicago church were muffled because they were beamed directly into nearby apartments. Traffic cops in Illinois are handing out citations to truck drivers for overly loud trucks.

"But officer! I wasn't going too fast."

"No, but you were going too loud."

In the industrial environment, noise is regulated by federal legislation called the Occupational Safety and Health Act (OSHA), which sets limits for noise in the working place. Table 22-2 lists these limits.

There is some disagreement as to the level of noise that should be allowed for an 8-hour working day. Some researchers and health agencies insist that 85 dB(A)* should be the limit. This is not, as might seem on the surface, a minor quibble, since the jump from 85 to 90 dB is actually an increase of about 4 times the sound pressure!

HEALTH EFFECTS OF NOISE

In the Bronx borough of New York City, one spring evening, four boys were at play, shouting and racing in and out of an apartment building. Suddenly, from a second-floor window, came the crack of a pistol. One of the boys sprawled dead on the pavement. The victim happened to be thirteen years old, son of a prominent public leader, but there was no political implication in the tragedy. The killer confessed to police that he was a nightworker who had lost control of himself because the noise from the boys prevented him from sleeping.

We have only recently become aware of the devastating psychological

---

*The designation dB(A) is explained in the next chapter.

dB

- 140 — ← Threshold of pain
- 130 — 50 HP Siren at 100 ft

  Pneumatic riveter
- 120 —
- Jetliner 500 ft overhead
- 110 —
- 100 — Air hammer

  Air compressor
- 90 — Heavy city traffic

  ← Beginning of hearing damage
        if prolonged
- 80 —
- 70 — Average traffic
- 60 — Conversational speech, 3 ft
- 50 — Business office

  Average residence
- 40 —

  Soft whisper, 5 ft
- 30 — Room in a quiet house at night
- 20 — Motion picture studio
- 10 —
- 0 — ← Threshold of hearing

Figure 22-4 Environmental noise.

effects of noise. The effect of excessive noise on our ability to hear, on the other hand, has been known for a long time.

The human ear is an incredible instrument. Imagine having to design and construct a scale for weighing just as accurately a flea or an elephant. Yet this is the range of performance to which we are accustomed from our ears.

A schematic of the human auditory system is shown in Figure 22-5.

Table 22-2.   OSHA Maximum Permissible Industrial
Noise Levels

Sound Level, dB(A)	Maximum Duration During Any Working Day (hr)
90	8
92	6
95	4
100	2
105	1
110	½
115	¼

Sound pressure waves caused by vibrations set the eardrum (*tympanic membrane*) in motion. This activates the three bones in the middle ear. The *hammer, anvil* and *stirrup* physically amplify the motion received from the eardrum and transmit it to the inner ear. This fluid-filled cavity contains the *cochlea,* a snail-like structure in which the physical motion is transmitted to tiny hair cells. These hair cells deflect, much like seaweed swaying in the current, and certain cells are responsive only to certain frequencies. The mechanical motion of these hair cells is transformed to bioelectrical signals and transmitted to the brain by the auditory nerves. Acute damage can occur to the eardrum, but this occurs only with

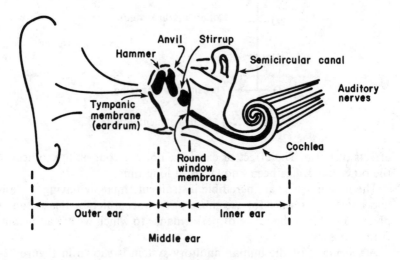

Figure 22-5   Cut-away drawing of the human ear.

very loud sudden noises. More serious is the chronic damage to the tiny hair cells in the inner ear. Prolonged exposure to noise of a certain frequency pattern can cause either temporary hearing loss, which disappears in a few hours or days, or permanent loss. The former is called *temporary threshold shift,* and the latter is known as *permanent threshold shift.* Literally, your threshold of hearing changes, so you are not able to hear some sounds.

Temporary threshold shift is generally not damaging to your ear unless it is prolonged. People who work in noisy environments commonly find that they hear less well at the end of the day. Performers in rock bands are subjected to very loud noises (substantially above the allowable OSHA levels), and commonly are victims of temporary threshold shift. In one study, the results of which are shown in Figure 22-6, the players suffered as much as 15 dB temporary threshold shift after a concert.

Repeated noise over a long time leads to permanent threshold shift. This is especially true in industrial applications where people are subjected to noises of a certain frequency. Figure 22-7 shows data from a study performed on workers at a textile mill. Note that the people who worked in the spinning and weaving parts of the mill, where noise levels

Figure 22-6   Temporary threshold shift for rock band performers. (Source: Data by the U.S. Public Health Service, in Ref. 4.)

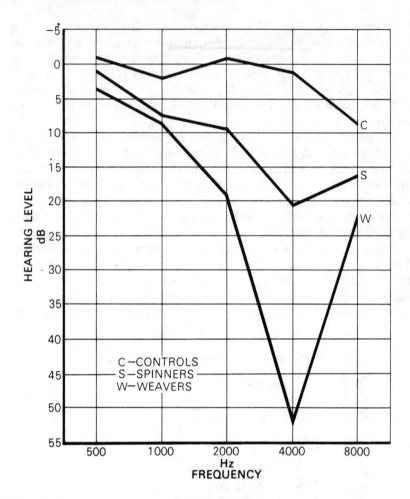

Figure 22-7   Permanent threshold shift in textile workers. [Source: Burns, W., *et al.* "An Exploratory Study of Hearing and Noise Exposure in Textile Workers," *Ann. Occup. Hyg.* 7, 323 (1958).

are highest, suffered the most severe loss in hearing, especially at around 4000 Hz, the frequency of noise emitted by the machines.

As people get older, hearing becomes less acute simply as one of the effects of aging. This loss of hearing, called *presbycusis*, is illustrated in Figure 22-8. Note that the greatest loss occurs at the higher frequencies. Speech frequency is about 1000–2000 Hz, and thus older people commonly accuse others of "whispering."

In addition to presbycusis, however, there is a serious loss of hearing

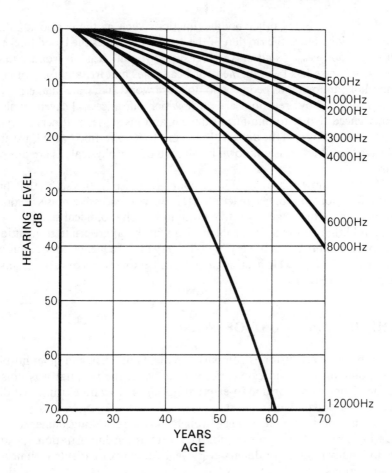

Figure 22-8 Hearing loss with age. [Source: Hinchcliffe, R. "The Pattern of the Threshold of Perception of Hearing and other Special Senses as a Function of Age," *Gerontologica* 2: 311 (1958).]

due to environmental noise. In one study[4] 11% of ninth graders, 13% of twelfth graders and 35% of college freshmen had a greater than 15 dB loss of hearing at 2000 Hz. The study concluded that this severe loss resulted from exposure to loud noises such as motorcycles and rock music, and that as a result "the hearing of many of these students had already deteriorated to a level of the average 65-year-old person."

---

[4]Lipscomb, D. M., reported in White, F. A., *Our Acoustic Environment* (New York: John Wiley & Sons, 1975).

Noise also affects other bodily functions such as the cardiovascular system. Noise alters the rhythm of the heartbeat, makes the blood thicker, dilates blood vessels and makes focusing difficult. It is no wonder that excessive noise has been blamed for headaches and irritability. Noise is especially annoying to people who do close work, like watchmakers.

All of the above reactions are those which our ancestral caveman also experienced. Noise to him meant danger and his senses and nerves were "up," ready to repel the danger. In the modern noise-filled world, we are always "up," and it is unknown how much of our physical ills are due to this.

We also know that man cannot adapt to noise in the sense that his body functions no longer react a certain way to excessive noise. People do not, therefore, "get used to" noise in the physiological sense.

In addition to the noise problem, it might be appropriate to mention the potential problems of very high- or very low-frequency sound, out of our usual 20–20,000 Hz hearing range. The health effects of these, if any, remain to be studied.

THE DOLLAR COST OF NOISE

Numerous case histories comparing patients in noisy and quiet hospitals point to increased convalescent time when the hospital was noisy (either from within or due to external noise). This can be translated directly to a dollar figure.

Recent court cases have been won by workers seeking damages for hearing loss suffered on the job. The Veterans Administration spends many, many millions of dollars every year for care of patients with hearing disorders.

Other costs, such as sleeping pills, lost time in industry, and apartment soundproofing are difficult to quantify. The John F. Kennedy Cultural Center in Washington spent $5 million for soundproofing, necessitated by the jets using the nearby National Airport.

DEGRADATION OF ENVIRONMENTAL QUALITY

It is even more difficult to measure the effect noise has had on the quality of life. How much is noise to blame for irate husbands and grumpy wives, for grouchy taxi drivers and surly clerks?

Children reared in noisy neighborhoods must be taught to listen. They cannot focus their auditory senses on one sound, such as the voice of a teacher.

Within the 80- to 90-dB circle around the Kennedy airport in New York are 22 schools. Every time a plane passes the teacher must stop talking and try to reestablish attention.

The effect of these insults cannot be measured. Yet we intuitively feel that the effect *must* be negative.

CONCLUSION

Noise is a real and dangerous form of environmental pollution. Since people cannot adapt to it physiologically, we are perhaps adapting psychologically instead. Noise can keep our senses "on edge" and prevent us from relaxing. Our mental powers must therefore control this insult to our bodies. Since noise, in the context of human evolution, is a very recent development, we have not yet adapted to it, and must thus be living on our buffer capacity. One wonders how plentiful this is.

PROBLEMS

22.1 If an office has a noise level of 70 dB and a new machine emitting 68 dB is added to the din, what is the combined sound level?

22.2 The OSHA standard for 8-hr exposure to noise is 90 dB(A). The EPA suggests that this should be 85 dB(A). Show that the OSHA level is almost 400% louder in terms of sound level than the EPA suggestion.

22.3 A machine with an overall noise level of 90 dB is placed in a room with another machine putting out 95 dB.
    a. What will be the sound level in the room?
    b. Based on OSHA criteria how long should workers be in the room during one working day?

22.4 The advertisement in Figure 22-9 shows a white noise generator.
    a. What is "white noise?"
    b. Discuss the effects of such white noise on health.

22.5 What sound pressure level results from the combination of three sources: 68 dB, 78 dB and 72 dB?

LIST OF SYMBOLS

C = velocity of sound wave, m/sec
dB = decibels

Figure 22-9   Advertisement for a white noise generator (courtesy Edmund Scientific).

f = frequency of sound wave, cycle/sec
P = pressure, N/m²
P_{ref} = reference pressure, generally stated as 0.00002 N/m²
SPL = sound pressure level
λ = wavelength, m

SUGGESTED READING

Burns, W. *Noise and Man* (Philadelphia: J. B. Lippincott Co., 1968).
"Motor Vehicle Noise Control Special Report 152," Transportation, Research Board, National Research Council, Washington, DC (1975).
"Noise Abatement and Control," Highway Research Record No. 448, Highway Research Board, Washington, DC (1973).
Rettinger, M. *Acoustic Design and Noise Control* (New York: Chemical Publishing Company, Inc., 1973).
Taylor, R. *Noise* (New York: Penguin Books, 1970).
White, F. A. *Our Acoustic Environment* (New York: John Wiley & Sons, 1975).

Chapter 23

Noise Measurement and Control

Although the human ear is a remarkable instrument, able to detect sound pressures over 7 orders of magnitude, it is not a perfect receptor of acoustic energy. In the measurement and control of noise, it is therefore important to know not only what a sound pressure is, but also to have some notion of how loud a sound *seems* to be. Before we address that topic, however, we must review some basics of sound.

SOUND PRESSURE LEVEL, FREQUENCY AND PROPAGATION

In 1825, E. H. Weber found that humans can perceive differences in small weights, but if a person is already holding a substantial weight, that same increment is not detectable. The same idea is true with sound. For example, a 2 N/m² increase from an initial sound pressure of 2 N/m² is readily perceived, whereas the same 2 N/m² difference is not noticed if it is added to a background of 200 N/m² sound pressure. Mathematically, this can be expressed as

$$ds = K \frac{dW}{W}$$

where   ds = is the minimum perceptible increase in a sensation
             (e.g., hearing)
      W = the load, (e.g., background sound pressure level)
   dW = the change in the load
      K = a constant

The integrated form of this concept is known as the Weber-Fechner Law

$$s = K \log_{10} W$$

This idea is used in the definition of a *decibel*:

$$dB = 10 \log_{10} \frac{W}{W_{ref}}$$

where the power level W is divided by a constant which is a reference value, both measured in watts.

Since the pressure waves in air are half positive pressure and half negative pressure (illustrated in the preceding chapter, Figure 22-2), adding these would result in zero. Accordingly, sound pressures are measured as root-mean-square (rms) values, which are related to the energy in the wave. It can also be shown that the power associated with a sound wave is proportional to the square of the rms pressure. We can therefore write

$$dB = 10 \log_{10} \frac{W}{W_{ref}} = 10 \log_{10} \frac{P^2}{P^2_{ref}}$$

$$= 10 \log_{10} \left(\frac{P}{P_{ref}} \right)^2 = 20 \log_{10} \frac{P}{P_{ref}}$$

where P and P_{ref} are the pressure and reference pressure. If we define $P_{ref} = 0.00002 \ N/m^2$,* the threshold of hearing, we can define the *sound pressure level* (SPL) as

$$SPL \ (dB) = 20 \log_{10} \frac{P}{P_{ref}}$$

Although it is common to see the sound pressure measured over a full range of frequencies, it is sometimes necessary to describe a noise by the amount of sound pressure present at a specific range of frequencies. Such a *frequency analysis,* shown in Figure 23-1, can be used for solving industrial problems or evaluating the danger of a certain sound to the human ear.

In addition to amplitude and frequency, sound has two more characteristics of importance. Both can be visualized by imagining the ripples

---

*$0.00002 \ N/m^2 = 20 \ \mu N/m^2 = 0.0002$ microbars.

Figure 23-1   Typical frequency analysis of a machine noise with background noise.

created by dropping a pebble into a large, still pond. The ripples, analogous to sound pressure waves, are reflected outward from the source, and the magnitude of the ripples is dissipated as they get farther away from the source. Similarly, sound levels decrease as the distance between the receptor and source is increased.

In summary, then, the four important characteristics of sound waves are:

1. Sound pressure is the magnitude or amplitude of sound.
2. The pitch is determined by the frequency of the pressure fluctuations.
3. Sound waves propagate away from the source.
4. Sound pressure decreases with increasing distance from the source.

But these characteristics ignore the human ear. We know that the ear is an amazingly sensitive receptor, but is it equally sensitive at all frequencies? Can we hear low and high sounds equally well? The answers to these questions lead us to the concept of *sound level.*

SOUND LEVEL

Suppose you were put into a very quiet room and subjected to a pure tone at 1000 Hz at 40-dB SPL. If, in turn, this sound was turned off and a pure sound at 100 Hz was piped in and adjusted in loudness until you judged it to be "equally loud" to the 40-dB 1000-Hz tone you had just heard a moment ago, you would, surprisingly enough, judge the 100-Hz tone to be equally loud when it was at about 55 dB. In other words, more energy is needed at the lower frequency in order to hear a tone at the same loudness, indicating that the human ear is rather inefficient for low tones.

It is possible to conduct such experiments for many sounds and with many people, and draw average equal loudness contours (Figure 23-2). These contours are in terms of *phons* which correspond to the sound pressure level in dB of the 1000-Hz reference tone.

Using Figure 23-2, a person subjected to a 65-dB SPL tone at 50 Hz would judge this to be equally as loud as a 40-dB 1000-Hz tone. Hence, the 50-Hz 65-dB sound has a *loudness level* of 40 phons.

Such measurements are commonly called *sound levels* and are not

Figure 23-2   Equal loudness contours (courtesy Gen Rad).

based on basic physical phenomena only, but have a "fudge factor" thrown in which corresponds to the inefficiency of the human ear.

Sound level (SL) is measured with a *sound level meter,* consisting of a microphone, amplifier, a frequency weighing circuit (filters) and an output scale. The sound level meter is shown schematically in Figure 23-3, and Figure 23-4 shows a typical hand-held sound level meter. The weighing network filters out specific frequencies to make the response more characteristic of human hearing. Through use, three scales have become internationally standardized (Figure 23-5).

Note that the A scale in Figure 23-5 corresponds closely to an inverted 40-phon contour in Figure 23-2. Similarly, the B scale in response approximates an inverted 70-phon contour. The C scale is an essentially flat response, giving equal weight to all frequencies. It approximates the response of the ear to intense sound pressure levels.

The results of noise measurement using the standard sound level meter are expressed in terms of dB but with the scale designated. If on the A scale the meter reads 45 dB, the measurement is reported as 45 dB(A).

Most noise ordinances and regulations are in terms of dB(A). This is a good approximation of human response for not very loud sounds. For very loud noise the C scale is a better approximation, but because the use of multiple scales complicates matters, scales other than A are seldom used.

In addition to the A, B and C scales, a new D scale has recently been introduced to approximate human response to aircraft noise (exclusively).

This is another complication involved in the measurement of noise. We do not generally respond the same way to different types of noise, even though they might be equal in sound level. For example, a symphony orchestra hitting a concert C at 120 dB(A) and a jet engine at an equal 120 dB(A) will draw very different reactions from people.

In response to this problem, a plethora of parameters have been de-

Figure 23-3   Schematic representation of a sound level meter.

Figure 23-4   A typical sound level meter (courtesy Gen Rad).

Figure 23-5   The A, B and C filtering curves for a sound level meter.

vised, all supposedly "the best" means of quantitatively measuring human response to noise. Among these are:

- Traffic Noise Index (TNI)
- Sones
- Perceived Noise Level (PNdB)
- Noise and Number Index (NNI)
- Effective Perceived Noise Level (EPNdB)
- Speech Interference Level (SIL)

More parameters are suggested all the time.

The reason for this proliferation is that a physical phenomenon (pressure waves) is related to physiological human response (hearing) and then to psychological response (pleasure or irritation). Along the way, the science becomes progressively more subjective.

There is a further complication in measuring some noises, particularly those commonly called "community noise," such as traffic, loud parties, etc. Although we have thus far treated noise as if it were constant in intensity and frequency with time, this obviously is not true for transient noises such as trucks moving past a sound level meter, or loud parties.

MEASURING TRANSIENT NOISE

Transient noise is still measured with a sound level meter, but the results must be reported in statistical terms. The common parameter is the *percent of time a sound level is exceeded,* denoted by the letter L with a subscript. For example, $L_{10} = 70$ dB(A) means that 10% of the time the noise was louder than 70 dB as measured on the A scale.

Transient noise data are gathered by reading the SL at regular intervals. These numbers are then ranked and plotted, and the L values read off the graph.

Example 23.1

Suppose the traffic noise data in Table 23-1 were gathered at 10-second intervals. These numbers are then ranked as indicated in the table and plotted as in Figure 23-6. Note that since 10 readings were taken, the lowest reading (Rank No. 1) corresponds to a SL that is equalled or exceeded 90% of the time.

Hence, 70 dB(A) is plotted versus 90% in Figure 23-6.* Similarly, 71 dB(A) is exceeded 80% of the time.

One widely used parameter for gauging the perceived level of noise from transient sources is the noise pollution level (NPL), which takes

**Table 23-1.   Sample Traffic Noise Data and Calculations
for Example 23.1**

Data		Rank No.	% of time equal to or exceeded	dB(A)
Time (sec)	dB(A)			
10	71	1	90	70
20	75	2	80	71
30	70	3	70	74
40	78	4	60	74
50	80	5	50	75
60	84	6	40	75
70	76	7	30	76
80	74	8	20	78
90	75	9	10	80
100	74	10	0	84

---

*You could also argue that since 70 dB(A) is the lowest value, it is the SL exceeded 100% of the time and, therefore, you could plot 70 dB(A) versus 100%. The second value, 71 dB(A), is exceeded 90% of the time, etc. Either method is correct, and the error diminishes as the number of data points increases.

Figure 23-6. Results from a survey of transient noise (data from Table 23-1). The data are plotted as percent of time the sound level is exceeded, versus the sound level (SL) in dB(A).

into account the irritability of impulse (sudden) noises. The NPL is defined as

$$\text{NPL in dB(A)} = L_{50} + (L_{10} - L_{90}) + \frac{(L_{10} - L_{90})^2}{60}$$

As defined earlier, the symbol L refers to the percentage of time the noise is equal to or greater than some value. The percentage is indicated by the subscript. L_{10} is thus the dB(A) level exceeded 10% of the time.

With reference to Figure 23-6, L_{10}, L_{50} and L_{90} are thus 80, 75 and 70 dB(A) respectively. These can then be substituted into the equation and the NPL calculated.

It is always advisable to take as much data as possible. The ten readings illustrated above are seldom sufficient for a thorough analysis. The percentages must be calculated on the basis of the total number of readings taken. If, for example, 20 readings are recorded, the lowest (Rank No. 1) corresponds to 95%, the second lowest 90%, etc.

NOISE CONTROL

The control of noise is possible at three levels:

1. reducing the sound produced
2. interrupting the path of the sound
3. protecting the recipient

When we consider noise control in industry, in the community or in our home, we should keep in mind that all problems have these three possible solutions.

Industrial Noise Control

Industrial noise control generally involves the replacement of noise-producing machinery or equipment with quiet alternatives. For example, the noise from an air fan can be reduced by increasing the number of blades or the pitch of the blades and decreasing the rotational speed (thus obtaining the same air flow).

The second method of decreasing industrial noise is to interrupt the path, for example, by covering a noisy motor with insulating material.

The third method of noise control, often used in industry, is to protect the recipient by distributing earmuffs to the employees. But the problem is thus not really solved, and often the workers refuse to wear earmuffs, considering them sissy. There seems to be something masculine about being able to "take it." Such misdirected masculinity will only end up in deafness.

The Occupational Safety and Health Act applies industrial noise limits to all jobs involving significant federal funds. These limits are tabulated in Chapter 22 as Table 22-2.

Community Noise Control

The three major sources of community noise are aircraft, highway traffic and construction.

Construction noise must be controlled by local ordinances (unless federal funds are involved). Control usually involves the muffling of air compressors, jack hammers, hand compactors, etc. Since mufflers cost money, contractors will not take it upon themselves to control noise, and outside pressures must be exerted.

Aircraft noise in the United States is the province of the Federal Aviation Administration, which has instituted a two-pronged attack on this problem. First, it has set limits on aircraft engine noise and will not allow aircraft exceeding these limits to use the airports. This has forced manufacturers to design for quiet as well as for thrust.

The second effort has been to divert flight paths away from populated areas and whenever necessary to have pilots use less than maximum power when the takeoff carries them over a noise-sensitive area. Often this approach is not enough to prevent significant noise-induced damage or annoyance, and aircraft noise remains a real problem in urban areas.

Supersonic aircraft present a special problem. Not only are their engines noisy, but the sonic boom can create property damage and mental anguish. The magnitude of this problem will become known only when (and if) supersonic airlines begin regular service over land.

The third major source of community noise is from highways. The car or truck creates noise by a number of means:

1. exhaust noise
2. tire noise
3. engine intake noise
4. gears and transmission
5. aerodynamic (wind) noise

A modern passenger car is so well muffled that its most important contribution, at moderate and high speeds, is tire noise. Sports cars and motorcycles, on the other hand, contribute exhaust, intake and gear noise.

The worst offender on the highways is the heavy truck. Truck noise is generally from all of the above sources. In most cases, the total noise generated by vehicles can be correlated directly to the truck volume. Figure 23-7 is a typical plot showing sound level as a function of traffic volume (measured in number of trucks per hour). Clearly, truck volume is of great importance. Incidentally, this graph is plotted as "sound level exceeded 10% of the time." Peak sound levels could be a great deal higher.

A number of alternatives are available for reducing highway noise. First, the source can be controlled by making quieter vehicles; second, highways could be routed away from populated areas; and third, noise could be baffled with walls or other types of barriers. Other methods include lowering speed limits, designing for nonstop operation, and reducing all highways to less than 8% grade.

Vegetation, surprisingly, makes a very poor noise screen. The most

Figure 23-7   The effect of truck density and distance from a highway.

effective buffers have been to raise or lower the highway, or to build physical barriers beside the road and thus screen the noise. All of these have limitations. For example, noise will bounce off the walls and create little or no noise shadow. In addition, walls hinder highway ventilation, thus contributing to the buildup of dangerous air pollutants.

The Department of Transportation has established design noise levels for various land uses, as shown in Table 23-2.

Table 23-2.   Design Noise Levels Set by the Federal Highway Administration

Land Category	Design Noise Level, L_{10}	Description of Land Use
A	60 dB(A) Exterior	Activities requiring special qualities of serenity and quiet, such as amphitheatres
B	70 dB(A) Exterior	Residences, motels, hospitals, schools, parks, libraries
	55 dB(A) Interior	Residences, motels, hospitals, schools, parks, libraries
C	75 dB(A) Exterior	Developed land not covered in categories A and B
D	no limit	Undeveloped land

Table 23-3.   Some Domestic Noisemakers

Item	Sound Level dB(A)
Vacuum cleaner (10 ft)	75
Inside quiet car (50 mph)	65
Inside sports car (50 mph)	80
Flushing toilet	85
Garbage disposal (3 ft)	80
Window air conditioner (10 ft)	55
Ringing alarm clock (2 ft)	80
Lawn mower (operator's position)	105
Snowmobile (driver's position)	120
Rock band (10 ft)	115

Noise in the Home

Private dwellings are getting noisier because of internally produced noise as well as external community noise. The list of gadgets in a modern American home reads like a list of Halloween noisemakers. Some examples of domestic noise are listed in Table 23-3.

Otherwise similar products of different brands often will vary significantly in noise levels. When shopping for an appliance, it is just as important therefore to ask the clerk "How noisy is it?" as it is to ask him "How much does it cost?" And if he looks at you as if you had two heads, explain to him that he should know the dB(A) at the operator's position for all of his wares. He may actually bother to find out.

CONCLUSION

As long as noise was considered just another annoyance in a polluted world, not much attention was given to it. We now have enough data to show that noise is a definite health hazard, and should be numbered among our more serious pollutants. It is possible, using available technology, to lessen this form of pollution. However, the solution costs money and private enterprise cannot afford to give noise a great deal of consideration until forced to by either the government or the consuming public.

PROBLEMS

23.1 Given the following noise data, calculate the L_{10} and L_{50}.

Time (sec)	dB(A)	Time (sec)	dB(A)
10	70	60	65
20	50	70	60
30	65	80	55
40	60	90	70
50	55	100	50

23.2 In addition to the data listed in Example 23.1 (Table 23.1), the following SL measurements were taken:

Time (sec)	dB(A)	Time (sec)	dB(A)
110	80	160	95
120	82	170	98
130	78	180	82
140	87	190	88
150	92	200	75

Calculate the L_{50}, L_{10} and NPL using all 20 data points.

23.3 Suppose your dormitory is 200 yards from a highway. What truck traffic volume would be "allowable" in order to stay within the Federal Highway Administration guidelines?

23.4 If the SL were 80 dB(C) and 60 dB(A), would you suspect that most of the noise was of high, medium or low frequency? Why?

23.5 If the pressure wave was 0.3 N/m², what is the SPL in dB (re 20×10^{-6} N/m²)? If this is all the information about the sound you have, what can you say about the SL in dB(A)? What data would you need to make a more accurate estimate of the SL in dB(A)?

23.6 Many animals hear better than humans. Dogs, for example, can hear sounds at pressures close to 2×10^{-6} N/m². What is this in decibels?

23.7 If you sing at a level 10,000 times greater than the power of the faintest audible sound, at which dB level are you singing?

23.8 How many times more powerful is a 120-decibel sound than a zero-decibel sound?

23.9 On a graph of dB versus Hz (10 to 50,000), show a possible frequency analysis for: (a) a passing freight train, (b) a dog whistle, (c) "white noise."

23.10 Carry a sound level meter with you for one entire day. Measure and record the sound levels as dB(A) in classes, in your room, during sports events, in the dining halls, or wherever you go during the day.

23.11 Seek out and measure the three most obnoxious noises you can think of. Compare these to the noises in Table 23-3.

23.13 In your room measure and plot the sound level in dB(A) of an alarm clock versus distance. At what distance will it still wake you if it requires 70 dB(A) to get you up? Draw the same curve outside. What is the effect of your room on the sound level?

23.14 Construct a sound level frequency curve for a basketball game. Calculate the noise pollution level.

23.15 A noise is found to give the following responses on a sound level meter: 82 dB(A), 83 dB(B) and 84 dB(C). Is the noise of a high or low frequency?

23.16 A machine produces 80 dB(A) at 100 Hz (almost pure sound).
   a. Would a person who has suffered a noise-induced threshold shift of 40 dB at that frequency be able to hear this sound? Explain.
   b. What would this noise measure on the C scale of a sound level meter?

LIST OF SYMBOLS

dB = decibel
Hz = Hertz, cycles/sec
L_x = x percent of the time the stated sound level (L) was exceeded, %
NPL = noise pollution level
P = pressure, N/m^2
P_{ref} = reference pressure, N/m^2
s = sensation (hearing, touch, etc.)
SLM = sound level meter
SPL = sound pressure level
W = power level, watts

SUGGESTED READING

Bell, L. H. *Fundamentals of Industrial Noise Control* (Harmony Publications, 1973).

Bragdon, C. R. *Noise Pollution* (Philadelphia: University of Pennsylvania Press, 1970).

Gatley, W. S., and E. E. Frye. *Regulation of Noise in Urban Areas,* U.S. Environmental Protection Agency (1971).

National Research Council. "Noise Abatement and Control," Highway Research Record No. 448, Highway Research Board, Washington, DC (1973).

Taylor, R. *Noise* (New York: Penguin Books, 1970).

White, F. A. *Our Acoustic Environment* (New York: John Wiley & Sons, 1975).

Chapter 24

Environmental Impact and Economic Assessment

Ideally, engineers are expected to respond to a given problem in a rational manner. Decision-making in the public sector should follow a definite sequence: (1) problem definition, (2) generation of alternative solutions, (3) evaluation of alternatives, (4) implementation of a selected solution and (5) review and appropriate revision of the implemented solution. This process is even mandated by law. At the federal level the National Environmental Policy Act (NEPA) requires that alternative solutions to a problem be considered by federal agencies. This is also true at the state level where, for example, a similar law in Wisconsin (WEPA) requires state agencies to make a conscious effort to develop alternatives before a solution to a problem is implemented. In some states agencies also operate under the planning, programming and budgeting system (PPBS), which again stresses the need to consider alternative solutions to problems they address.

In reality, this rational planning sequence is seldom followed. Charles Lindbloom argues that *disjointed incrementalism* ("muddling through") is the more accurate description of the decision-making process, be it at the federal, state or local level.[1] Nevertheless, more recent national and state policies, such as NEPA, WEPA, and PPBS, are law, and their intent has been to integrate the muddling through with some degree of more rational planning. The mandate to generate and consider all feasible solutions is an example.

The result, for better or worse, has been a hybrid strategy of "mud-

---

[1] Lindbloom, C. E. "The Science of Muddling Through," *Public Administration* (Spring 1959).

dling through'' in which limited groups of alternative solutions are proposed and studied successively, with a choice being made *de facto* when all relevant parties cease to disagree. In decision-making theory, this particular process has been called *mixed scanning*.[2] Frequently used in administrative agencies by design or default, it has proven practical, with the majority of day-by-day middle-level decisions going unchallenged and the credibility of most agencies remaining stable. However, mixed scanning falls short of achieving the ideal of rational planning, in which a broad range of feasible alternative solutions are reviewed simultaneously rather than successively. Recent trends in court action have resulted in delaying or stopping the implementation of many solutions because alternatives were not adequately developed.

It is granted that administrative constraints of time, limited information, expertise, expense, and often conflicting preordained priorities make it very hard to realize the ideal of rational planning. What ought not be allowed, however, is the lack of knowledge of how to perform the steps of rational planning. It is one thing to require that alternatives be reviewed simultaneously; it is another to know how to generate and evaluate truly alternative solutions for any given problem. This chapter discusses how alternative solutions to environmental engineering are analyzed in terms of their projected environmental and economic impacts.

ENVIRONMENTAL IMPACT

On January 1, 1970, then-President Nixon signed into law the National Environmental Policy Act, which declared a national policy to encourage productive and enjoyable harmony between people and their environment. This law established the Council on Environmental Quality (CEQ), which monitors the environmental effects of federal activities and assists the President in evaluating environmental problems and determining the best solutions to these problems. But few people realized the NEPA contained a real sleeper article: Section 102(2)(C), which requires federal agencies to evaluate the consequences of any proposed action on the environment:

> The Congress authorizes and directs that, to the fullest extent possible: (1) the policies, regulations, and public laws of the United

---

[2]Etzioni, A. "Mixed Scanning—A Third Approach to Decision Making," *Public Administration Review* (December 1967).

States shall be interpreted and administered in accordance with the policies set forth in this chapter, and (2) all agencies of the Federal Government shall include in every recommendation or report on proposals for legislation and other major Federal actions significantly affecting the quality of the human environment, a detailed statement by the responsible official on—

(i) the environmental impact of the proposed action,

(ii) any adverse environmental effects which cannot be avoided should the proposal be implemented,

(iii) alternatives to the proposed action,

(iv) the relationship between local short-term uses of man's environment and the maintenance and enhancement of long-term productivity, and

(v) any irreversible and irretrievable commitments of resources which would be involved in the proposed action should it be implemented.

In other words, each project funded by the federal government must be accompanied with an "Environmental Impact Statement" (EIS). Such a published statement must assess in detail the potential environmental impacts of a proposed action and alternative actions. All federal agencies are required to prepare statements for projects and programs (a programmatic" EIS) under their jurisdiction. Additionally, the agencies must generally follow a detailed and often lengthy public review of each EIS before proceeding with their projects and programs.

The original idea of the EIS is to introduce environmental factors into the decision-making machinery. The purpose of the EIS is not to provide justification for a construction project, but rather to introduce environmental concerns and have them discussed in public before the decision on a project is made. However, this objective is difficult to apply in practice. Historically, interest groups in and out of government articulated plans to their liking, the sum of which provided the engineer with a set of alternatives to be evaluated. In many other instances, the engineer is left to create his own alternatives, or participate in a group decision-making process like the Delphi technique. In either case, there are normally one or two plans which, from the outset, seem eminently more feasible and reasonable, and these can be legitimatized by juggling time scales or standards of enforcement patterns just slightly and calling them alternatives (as they are in a limited sense). As a result, non-decisions are made,[3] i.e., wholly different ways of perceiving the prob-

---

[3]Bachrach, J., and M. S. Baratz. "Two Faces of Power," *American Political Science Review* (1962), pp. 947–52.

lem and conceiving of solutions have been overlooked, and the primary objective of an EIS has been circumvented. Over the past few years, court decisions and guidelines by various agencies have, in fact, helped to mold this procedure for the development of environmental impact statements.

Ideally, an EIS must be thorough, interdisciplinary, and as quantitative as possible. The writing of an EIS involves three distinct phases: *inventory, assessment* and *evaluation*. The first is a cataloging of environmentally susceptible areas, the second is the process of estimating the impact of the alternatives, and the last is the interpretation of these findings.

Environmental Inventories

The first step in evaluating the environmental impact of a project or project's alternatives is to inventory factors that may be affected by the proposed action. In this step no effort is made to assess the importance of a variable. Any number and many kinds of variables may be included, such as:

1. the "ologies": hydrology, geology, climatology and archeology
2. environmental quality: land, surface and subsurface water, air and sound
3. plant and animal life

This step involves counting, measuring and describing existing conditions.

Environmental Assessment

The process of calculating projected effects that a proposed action or construction project will have on environmental quality is called environmental assessment. It is necessary to develop a methodical, reproducible and reasonable method of evaluating both the effect of the proposed project and the effects of alternatives which may achieve the same ends but which may have different environmental impacts. A number of semiquantitative approaches have been used, among them the checklist, the interaction matrix, and the checklist with weighted rankings.

Checklists are listings of potential environmental impacts, both primary and secondary. Primary effects occur as a direct result of the proposed project, such as the effect of a dam on aquatic life. Secondary effects occur as an indirect result of the action. For example, an interchange for a highway will not directly affect a land area, but indirectly it will draw such establishments as service stations and quick food stores, thus changing land use patterns.

The checklist for a highway project could be divided into three phases: planning, construction and operation. During planning, consideration is given to environmental effects of the highway route and the acquisition and condemnation of property. The construction phase checklist will include displacement of people, noise, soil erosion, water pollution and energy use. Finally, the operation phase will list direct impacts due to noise, air pollution, water pollution due to runoff, energy use, etc., and indirect impacts due to regional development, housing, lifestyle, and economic development.

The checklist technique thus simply lists all of the pertinent factors; then the magnitude and importance of the impacts are estimated. The estimation of impact is quantified by establishing an arbitrary scale, such as:

0 = no impact
1 = minimal impact
2 = small impact
3 = moderate impact
4 = significant impact
5 = severe impact

This scale can be used to estimate both the magnitude and the importance of a given item on the checklist. The numbers can then be combined, and a quantitative measurement of the severity of environmental impact for any given alternative estimated.

Example 24.1

A landfill is to be placed in a floodplain of a river. Estimate the impact using the checklist technique.

First the items impacted are listed, then a judgment concerning both importance and magnitude of the impact is made. In this example, the items are only a sample of the impacts one would normally consider. The numbers in this example are then multiplied and the sum obtained. Thus:

Potential Impact	*Importance × Magnitude*
Groundwater contamination	5 × 5 = 25
Surface water contamination	4 × 3 = 12

Odor	$1 \times 1 =$	1
Noise	$1 \times 2 =$	2
Total		40

This total of 40 can then be compared to totals calculated for alternative courses of action.

In the checklist technique most variables must be subjectively judged. Further, it is difficult to predict future conditions such as land-use pattern changes or changes in lifestyle. Even with these drawbacks, however, this method is often used by engineers in governmental agencies and consulting firms, mainly due to its simplicity.

The *interaction matrix* technique is a two-dimensional listing of existing characteristics and conditions of the environment and detailed proposed actions which may impact the environment. This technique is illustrated in Example 24.2. For example, the characteristics of water might be defined as:

- surface
- ocean
- underground
- quantity
- temperature
- groundwater rechange
- snow, ice and permafrost

Similar characteristics must also be defined for air, land, and other important considerations.

Opposite these listings in the matrix are lists of possible actions. In our example, one such action is labeled *resource extraction,* which could include the following actions:

- blasting and drilling
- surface extraction
- subsurface extraction
- well drilling
- dredging
- timbering
- commercial fishing and hunting

The interactions, as in the checklist technique, are measured in terms of magnitude and the importance. The magnitudes are represented by the extent of the interaction between the environmental characteristics and the proposed actions, and typically can be measured. The importance of the interaction, on the other hand, is often a judgment call on the part of the engineer.

If an interaction is present, for example between underground water and well drilling, a diagonal line is placed in the block. Numbers can then be assigned to the interaction, with 1 being a small and 5 being a large magnitude or importance, and these placed in the blocks with magnitude above and importance below. Appropriate blocks are filled in, using a great deal of judgment and personal bias, and then are summed over a line, thus giving a numerical grade for either the proposed action or environmental characteristics.

Example 24.2

Lignite (brown) coal is to be surface mined in the Appalachian Mountains. Construct an interaction matrix for the water resources (environmental characteristics) versus resource extraction (proposed actions).

Proposed Action

Environmental Characteristics	Blasting + drilling	Surface excavation	Subsurface excavation	Well drilling	Dredging	Timbering	Commercial fishing	Total
Surface water	3/2	5/5						8/7
Ocean water								
Underground water		3/3						3/3
Quantity								
Temperature		1/2						1/2
Recharge								
Snow, ice								
Total	3/2	9/10						

We see that the proposed action would have a significant effect on surface water quality, and that the surface excavation phase will have a large impact. The value of the technique is seen when the matrix is applied to alternative solutions and individual elements in the matrix, as well as row and column totals, are reviewed.

This trivial example cannot fully illustrate the advantage of the interaction technique. With large projects having many phases and diverse impacts, it is relatively easy to pick out especially damaging aspects of the project as well as the environmental characteristics that will be most severely affected.

The search for a comprehensive, systematic, interdisciplinary and quantitative method for evaluating environmental impact has led to the *checklist-with-weighted-rankings* technique. The intent here is to use a checklist as before in order to ensure that all aspects of the environment are covered, as well as to give these items a numerical rating in common units.

The first step is to construct a list of items that could be impacted by the proposed alternative, grouping them into logical sets. One grouping might be:

- Ecology
  - Species and Populations
  - Habitats and Communities
  - Ecosystems
- Aesthetics
  - Land
  - Air
  - Water
  - Biota
  - Man-Made Objects
- Environmental Pollution
  - Water
  - Air
  - Land
  - Noise
- Human Interest
  - Educational and Scientific
  - Cultural
  - Mood/Atmosphere
  - Life Patterns

Each title might have several specific topics under it, for example under Aesthetics, "Air" may list: (1) odor, (2) sound and (3) visual as items in the checklist.

We must now assign ratings to these items, in common units. One procedure is to first estimate the ideal, or natural levels of environmental quality (without man-made pollution) and take a ratio of the expected condition to the ideal. For example, if the ideal dissolved oxygen in the stream is 9 mg/l, and the effect of the proposed action is to lower the dissolved oxygen to 3 mg/l, the ratio would be 0.33. This is sometimes called the environmental quality index (EQI). Another option to this would be to make the relationship nonlinear, as shown in Figure 24-1. Lowering the dissolved oxygen by a few mg/l will not affect the EQI nearly as much as lowering it, for example, below 4 mg/l, since a dissolved oxygen below 4 mg/l definitely has a severe adverse effect on the fish population.

EQIs are calculated for all checklist items, and the values tabulated. Next the weights are attached to the items, usually by distributing 1000 parameter importance units (PIU) among the items. The product of EQI

Figure 24-1   Projected environmental quality index curve for dissolved oxygen.

and PIU, called the environmental impact unit (EIU), is thus the magnitude of the impact multipled by the importance.

This method has several advantages. We can calculate the sum of EIUs and evaluate the "worth" of many alternatives, including the do-nothing alternative. We can also detect points of severe impact, where the EIU after the project may be much lower than before, indicating severe degradation in environmental quality. Its major advantage, however, is that it makes it possible to input data and evaluate the impact on a much less qualitative and a much more objective basis.

Example 24.3

Evaluate the effect of a proposed lignite strip mine on a local stream. Use 10 PIU and linear functions for EQI.

The first step is to list the areas of potential environmental impact. These may be:

- appearance of water
- suspended solids
- odor and floating materials
- aquatic life
- dissolved oxygen

Many other factors could be listed, but these will suffice for this example.

Next we need to assign EQIs to the factors. If we assume a linear relationship, we can calculate them as follows:

Item	Condition Before Project	Condition After Project	EQI
Appearance of water	10	3	0.30
Suspended solids	20 mg/l	1000 mg/l	0.02
Odor	10	5	0.50
Aquatic life	10	2	0.20
Dissolved oxygen	9 mg/l	8 mg/l	0.88

Note that we had to put in subjective quantities for three of the items: "appearance of water," "odor," and "aquatic life" based on an arbitrary scale of decreasing quality from 10 to 1. The actual magnitude is not important since a ratio is calculated. Also note that the sediment ratio had to be inverted to make the EQI indicate improvement, i.e., $EQI < 1$.

Finally, the EQI indices are weighed by the 10 available PIU and the environmental impact units (EIU) are calculated.

Item	Project PIU	After Project $EQI \times PIU = EIU$
Appearance of water	1	$0.3 \times 1 = 0.3$
Suspended solids	2	$0.02 \times 2 = 0.04$
Odor	1	$0.5 \times 1 = 0.5$
Aquatic life	5	$0.2 \times 5 = 1.0$
Dissolved oxygen	1	$0.88 \times 1 = 0.88$
Total	10	2.72

The EIU total of 2.72 for this alternative is then compared to the total EIU for other alternatives.

Evaluation

The final part of the environmental impact assessment is evaluation of the results of the preceding studies. Typically, the evaluation phase is out of the hands of the engineers and scientists responsible for the inventory and assessment phases. Decisions made within the responsible governmental agency ultimately use the EIS to justify past decisions or support new alternatives.

ECONOMIC IMPACT ASSESSMENT

Historically, the President's Council on Environmental Quality has been responsible for overseeing the preparation of EISs and CEQ regulations have listed what should be included in all EISs developed by federal agencies. In the case of the proposed projects discussed earlier in this chapter two basic issues arose: public health dangers and environmental degradation. Under NEPA and the CEQ regulations, both issues must be addressed whenever alternatives are developed and compared.

In many cases, however, consideration of public health and environmental protection alone are not sufficient grounds on which to evaluate a range of alternative programs. Frequently, public acceptability is also a necessary input to an evaluation process. Although an alternative may protect public health and minimize environmental degradation, it may not be publicly acceptable, and thus could prove to be an essentially worthless alternative. Factors which influence public acceptability of a given alternative are generally discussed in terms of economics and broad social concerns. Economics includes the nationwide costs of an alternative, including the state, regional, local and private components, the resulting impacts on user charges and prices, and the ability to finance capital expenditures. Social concerns include public preferences in siting (e.g., no local landfills in wealthy neighborhoods) and public rejection of a particular disposal method (e.g., food-chain landspreading of municipal sludge rejected on "general principle"). Consequently, each alternative that is developed to address the issues of public health and environmental protection must also be analyzed in the context of rigid economic analyses and broad social concerns. Two important considerations that enter an economic impact assessment are discussed below: financing capital requirements and impacts on user charges and prices. Broad social concerns are addressed in Chapter 25.

Financing of Capital Expenditures

A municipality's or industry's inability to finance large capital expenditures will necessarily impact choice among alternatives, and possibly ability to comply with environmental regulations. Traditional economic impact assessments examine the amortized capital and operation and maintenance (O&M) costs of a project and the community's ability to pay, but they typically overlook the problems involved in raising the initial capital funds required for implementation. Financing problems are faced by municipalities and industries of all sizes, but may be particularly troublesome for small communities and small firms which face greater institutional barriers to finance debt. Although the following discussion only examines the issue of financing capability with regard to an ability to comply with water quality regulations, parallel issues can also arise for other types of public and private projects.

For relatively small capital needs, communities may make use of bank borrowing or capital improvement funds financed through operating revenues. However, local shares of wastewater treatment facility capital costs are generally raised by long-term borrowing in the municipal bond market. In the absence of other sources of funds, both the availability of funds through the bond market and the willingness of the community to assume the costs of borrowing may impact the availability of high-cost programs. The availability and cost of funds are, in turn, affected by how the financing is arranged.

Bonds issued by a municipality for the purpose of raising capital for a wastewater treatment are generally general obligation (GO) bonds or revenue bonds. Both have fixed maturities and fixed rates of interest, but they differ in the security pledged by the issuing authority to meet the debt service requirements, i.e., payments for principle plus interest. General obligation bonds are backed by the basic taxing authority of the issuer; revenue bonds are backed solely by the revenue for the service provided by the specific project.

The general obligation bond is generally preferred since the overhead costs of financing GO bonds are lower and their greater security allows them to be offered at a lower rate of interest. Some states are, however, constitutionally prevented from issuing GO bonds, or are limited in the quantity they can issue. In such cases, cities may have to resort to revenue bonds. Nevertheless, most bond issues for wastewater treatment projects are general obligations.

Most municipal bonds carry a credit rating from at least one of the private rating agencies, Standard & Poor's or Moody's Investor Service. Both firms attempt to measure the creditworthiness of borrowers, focusing on the potential for decrease on bond quality by subsequent debt and on the risk of default. Although it is not the sole determinant, the issuer's rating helps to determine the interest costs of borrowing, since individual bond purchasers have little else to guide them, and commercial banks wishing to purchase bonds are constrained by federal regulations to favor investments in the highest rating categories.

Of the rating categories used, only the top ones are considered to be of investment quality. Even among these grades, the difference in interest rates may impose a significantly higher borrowing cost on communities with low ratings. During the 1970s, for example, the interest rate differential between the highest grade (Moody's Aaa) and the lowest investment grade (Moody's Baa) bonds averaged 1.37%. Such a differential implies a substantial variation in financing costs for facilities facing extensive borrowing.

To highlight the importance of financing costs, consider the example of a city planning a $2 million expenditure on an incinerator to serve a publicly owned (wastewater) treatment works (POTW) with a capacity of 15 million gallons per day (mgd). Under one set of assumptions, such a facility would support a population of roughly 75,000, which would bear anywhere from 12.5% to 100% of the total cost, depending on what portion of the capital expense state and federal agencies agreed to pay. Under the more conservative assumption, this would amount to $250,000. Assuming that the capital is raised through one Aaa bond issue amortized over 25 years at an interest rate of 5.18% the interest payments over the entire borrowing period would total $212,500. If the Baa rate were 6.34%, the interest payments would amount to $262,000. Thus, total interest payments would be roughly equal to the principal, and are extremely sensitive to the interest rate.

More important than interest costs, the rating may determine the acceptability of bonds on the market. A city with a low or nonexistent rating might find that credit was simply not available. The fiscal problems of small communities are particularly sensitive to the rating process, which places lower ratings on small towns on the basis of an ill-defined "higher risk of default," despite the lack of supporting evidence. Hence, small towns must often endure higher interest payments. The problem is compounded by the higher average cost of small bond issues due to certain fixed underwriting fees, fees associated with buying and selling the bonds.

Despite the availability of general obligation financing and good credit, municipalities may face difficulties in raising capital due to the state of the market, while some borrowers may be precluded altogether. Two trends are of particular significance. Periods of high rates of inflation are expected to lead to higher interest rates for all municipal borrowers. The impact is significant for borrowers with marginal credit ratings, who will be squeezed out as the difference in interest rates between high- and low-rated bond issues widens. Secondly, the combination of expanding bond supply due to increased municipal borrowing and shrinking demand in periods of slow economic growth is likely to drive interest rates higher, increasing costs for all municipal borrowers.

Clearly, the availability of funding to finance capital expenditures is far from certain. Credit restrictions are not limited to large cities, such as New York and Cleveland, which suffer from severe budget problems. The financial health of a local government can be analyzed. A two-step procedure is useful as engineers consider the impacts of proposed projects:

1.   Define capital requirements of each alternative.
2.   Apply financial classification criteria.

Although rating agency analysts employ a variety of factors in assessing a city's fiscal strength, current outstanding debt is the principal determinant of the bond rating. Debt is generally expressed as a proportion of assets or on a per capita basis in order to make comparisons among cities. Three of these municipal debt ratios are of particular significance in determining market acceptance of bond offerings. These ratios and estimates of market-acceptable thresholds are:

- debt/capita $\geq$ \$300
- debt/capita as % of per capita income $\geq$ 7%
- debt/full property value $\geq$ 4.5%.

Empirical studies have shown that bond offerings by cities which exceed all three debt ratios are not likely to be successful.[4]

Increases in User Charges

A second component of an economic impact assessment is the analysis of projected increases in user charges. For example, if a city constructs a large wastewater treatment facility, what is the likely change in household sewer bills in that city? Based on the number of residential users supporting a given facility, the proportion of wastewater generated by them, and the data on alternatives, a projected cost per household can be computed. This cost can then be compared to the existing charge per household to determine the percent increase attributable to the alternative.

The results of these types of user charge analyses often indicate that engineers may find it difficult to make a distinction among alternatives based solely on increases in user charges. For example, an average difference of only 36¢ per year per household is projected between Alternatives 1 and 2 in Figure 24-2. Respectively, these two alternatives represent a less stringent and a more stringent regulation of effluent water quality. Figure 24-2 summarizes projected impacts on user charges in 350 cities across the nation. By almost any standard, 36¢ per year is insignificant. The findings suggest that the selection process, in this instance between Alternative 1 (little required reduction in pollution) and Alternative 2 (large requirements in pollution reduction), must be based on such other criteria as environmental impact and public health concerns; not on changes in user charges.

CONCLUSIONS

Engineers are required to develop, analyze and compare a range of solutions to any given environmental pollution problem. This range of alternatives must be viewed in terms of their respective environmental impacts and economic assessments. A nagging question exists throughout any such viewing: can individuals really measure, in the strict "sci-

---

[4]Forbes, R. Personal communication; Graduate School of Business Administration, State University of Albany, Albany, New York.

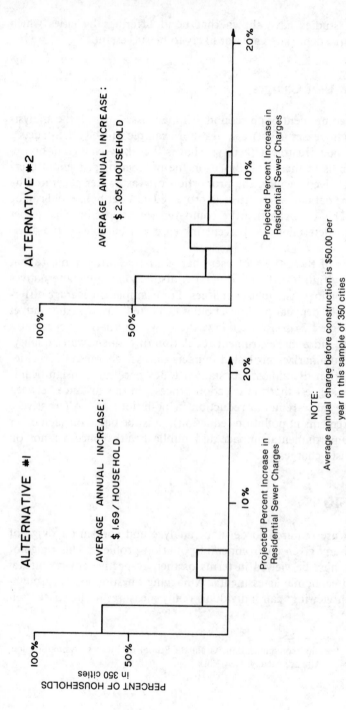

Figure 24-2 Projected increases in sewer use charges paid by a household in a sample community if a new wastewater treatment facility is constructed.

entific" sense, degradation of the environment? Can we place a value on an unspoiled wilderness area? Unfortunately, qualitative judgements are required to assess many impacts of any project. This balancing of values is discussed in the final chapter.

PROBLEMS

24.1 Develop and apply an interaction matrix for the following proposed actions designed to clean municipal wastewater in your home town: (1) construct a large-scale activated sludge facility, (2) require septic tanks for households and small-scale package treatment plants for industries, (3) construct decentralized, small-scale treatment facilities across town, (4) adopt land application technology, (5) continue direct discharge into the river. Draw conclusions from the matrix.

24.2 Discuss the advantages and disadvantages of a benefit-cost ratio in deciding if a town should build a wastewater treatment facility. Focus on the valuation problems associated with analyzing the impacts of such a project.

24.3 What conclusions can be drawn about the abilities of the three towns to finance a proposed water treatment facility:

City	Debt per Capita Current	Debt per Capita Projected	Debt per Capita as % per Capita Income Current	Debt per Capita as % per Capita Income Projected	Debt as % Full Valuation Current	Debt as % Full Valuation Projected
A	$946	$950	22.3	22.3	15.0	15.1
B	$335	$337	9.1	9.1	7.8	7.9
C	$ 6	$411	0.1	10.7	0.1	8.6

Discuss the current financial picture of each town as well as the projected incremental impact of this new proposed construction.

24.4 Determine the average annual *residential* sewer charge for three cities listed below. What is the incremental impact on these charges if the cost of sludge disposal increases as noted:

City	Wastewater Flow (mgd)	% Flow Residential	Annual Budget for Wastewater Treatment	Projected Increase in Sludge Disposal
A	20.0	60	$5,100,000	$1,500,000
B	5.0	90	$3,600,000	$1,000,000
C	0.1	100	$ 50,000	$ 20,000

Discuss your assumptions: gpd per household, % budget pair by residential sector.

24.5 Compare the environmental impact of a coal-fired electricity generating plant with those of a nuclear power plant. In your presentation, look at the flow of fuel from its natural state to the facility. Finalize your comparison with waste disposal considerations.

LIST OF SYMBOLS

CEQ	=	Council on Environmental Quality
EIA	=	economic impact assessment
EIS	=	Environmental Impact Statement
EIU	=	environmental impact unit
EPA	=	Environmental Protection Agency
EQI	=	environmental quality index
NEPA	=	National Environmental Policy Act
PIU	=	parameter importance units
POTW	=	Publicly Owned (Wastewater) Treatment Works
PPBS	=	planning, programming and budgeting system
WEPA	=	Wisconsin Environmental Policy Act

SUGGESTED READING

Burchell, R. W., and D. Listokin. *The Environmental Impact Handbook,* (New Brunswick, NJ: Center for Urban Policy Research, Rutgers, 1976).

Canter, L. *Environmental Impact Assessment* (New York: McGraw-Hill Book Company, 1977).

Dickert, T. G., and K. R. Domeny. *Environmental Impact Assessment: Guidelines and Commentary* (University Extension, University of California, Berkeley, 1974).

Guidelines for Preparation of Environmental Impact Statements, U.S. Environmental Protection Agency, Region 10 (1973).

McHarg, I. *Design With Nature* (Garden City, NY: Doubleday & Co. Inc., 1969).

Rosen, S. J. *Manual for Environmental Impact Evaluation* (Englewood Cliffs, NJ: Prentice-Hall, Inc., 1976).

Chapter 25

The Environmental Ethic

> It probably won't hurt mankind a whole hell of a lot in the long
> run if the whooping crane doesn't quite make it.[1]

This statement, made by a director of air and water resources of a large
papermill, represents an all-too-common approach to our environment.

There is, however, an opposing view which holds that it would be a
major disaster if a bird like the whooping crane became extinct. This
concern for nature, no doubt fueled in part by the uncertain survival of
man on this planet, is slowly developing into a new system of values
which in aggregate might be called the *environment ethic*. More and
more citizens are considering questions not only from the standpoint of
economics, but also from that of environmental quality. Lawyers, economists and politicians are finding it more difficult to answer such questions as "Is growth really necessary?" or "Should the gross national
product always increase?" or "Should we consider the benefits and costs
of public projects solely on the basis of dollars?" A few years ago such
questions would have been brushed aside as foolish.* They weren't
foolish, and they deserve straight answers.

In fact, ecology and economics are on a collision course. Man cannot
continue to grow in numbers and output (and use of resources) without
eventual self-destruction. Present economic systems do not have a feed-
back loop which slows down the processes detrimental to survival.

The conflict between ecologists and economists has been boiled down
to this ditty:

---

[1] Quoted in *The Water Lords,* J. M. Fallows, (New York: Bantam Books, 1971).

* Alfred North Whitehead observed that "The 'silly' question is the first intimation of
some totally new development."

Ecology's uneconomic,
But with another kind of logic,
Economy's unecologic.[2]

Technical economists notwithstanding, the environmental ethic has become a permanent part of our lives. It is in its infancy, a new and bold concept which must be given time to mature and develop. Legal principles, economic calculations and engineering considerations are all evolving rapidly, some perhaps prematurely.

The word "ethic" derives from the Greek "ethos," meaning the character of a person as defined by his actions. This character has been developed during the evolutionary process and has been influenced by the need for adapting to the environment. The "ethic," in short, governs our way of doing things, and this is a direct result of our environment.

When people talk of a "crisis in the environment," they really mean that our way of thinking about and doing things (our ethic) is not adapted to the environment. Humans at one time lived in harmony with the environment, but somehow the species, either as an accident or as a colossal practical joke by some unknown power, changed its lifestyle to the point where it no longer adapted. There is thus no "environmental crisis," but rather a crisis in the recognition that somehow over the years of development we are no longer adapted to our environment.

In the ecological context, such maladaptation results in two options:

1. The organism dies out.
2. The organism evolves to a form and character where it once again is compatible with the environment.

CONCHY   James Childress

---

[2]By Kenneth Boulding, quoted in *Bioethics*, V. R. Potter (Englewood Cliffs, NJ: Prentice-Hall, Inc., 1971).

Assuming that humans choose the latter course, how must this change in character (ethic) occur? It obviously cannot take place by the Darwinian natural selection principle, since many, many generations must pass before individuals with the incompatible characteristics are eliminated. Further, our present social system works in opposition to this selection, in that society rewards the despoilers of the environment more than the conservationists.

This selection must, therefore, occur within only a few generations and the change must be on two levels – the individual and the system. The individual must change his character or ethic, and the social system must change to become compatible with the global ecology.[3]

CAUSES OF THE PROBLEM OF OUR INCOMPATIBILITY WITH NATURE

Why is it that the human being – the most intelligent creature that ever evolved (or was created, depending on your beliefs) – is so incompatible with his environment?

There are three basic lines of argument in attempting to answer that question. One view is that humans created religions whose dogmas held the basic seeds of incompatibility. The second argument is that the social structure created by man makes him inherently unable to attain equilibrium. The third view is that the growth of science and technology is responsible for environmental degradation.

It should be obvious that all three – religion, society and technology – are intertwined and cannot be conveniently separated and individually scrutinized. Nevertheless, for the purposes of our discussion, let us assume that we can pick on each, one at a time.

Religion as the Cause

Although the environment has been severely impacted by man's cultures and civilizations, probably the greatest overall environmental damage has been caused in modern times by Western man. The major religious traditions of Western man are rooted in Judaism and Christianity and these Judeo-Christian traditions have been blamed as the root cause of

---

[3] Kozlovsky, D. *An Ecological and Evolutionary Ethic,* (Englewood Cliffs, NJ: Prentice-Hall, Inc., 1975).

our environmental problems. And there is clearly some justification in this argument. In the first chapter of Genesis, for example, man is commanded by God to subdue nature, to procreate and have dominion over all living things. Such an anthropocentric view of nature runs all through the Judeo-Christian doctrine, a point forcefully made by Lynn White in 1967 in his essay "The Historical Roots of Our Ecologic Crisis."[4]

White argues that the people embracing the Judeo-Christian religions are taught to treat nature as an enemy, that the religious dogmas prescribe that nature and natural resources are to be used only to meet the goals of survival and propagation. The religion of Western man, according to White, is responsible for environmental degradation.[*]

The Christian world reacted predictably to White's essay, with books and papers claiming that Christianity is pro-environment and the environmental ethic can in fact be found in the Bible.[5] This is stretching it a bit, however.

The strongest argument presented to counter Lynn White's indictment of the Judeo-Christian doctrines as a root cause of our environmental problems centers on the notion of *stewardship*. Stewardship represents a view that man was put on the planet as a caretaker, to see to the well-being of the earth. This "garden mentality" still relies on the concept of anthropocentrism, however, a world of hierarchical relationships in which man is the noble, managing his property, accountable only to the sovereign. Such a doctrine also requires a substantial measure of faith in a transcendent God and a belief in a reward structure in the afterlife.

Perhaps the defensiveness of the Church toward Lynn White's ideas was unnecessary. White's assertion that religion molds morals is oversimplified, and blaming one religion for our problems seems to be a unfair. In fact, during the time that the Christian religion was becoming popular, the people had many religious sects as alternatives, and the Christian ideas and ethics derived from ancient Judaic traditions seemed to fit most comfortably with the needs and existing value systems (whether

---

[4]The article first appeared in *Science* 155:1203 (10 March 1967). It has been reproduced in many books, including *Western Man and Environmental Ethics,* I. G. Barbour, Ed. (Reading, MA: Addison-Wesley Publishing Co., 1973).

[*]Or, as Mark Twain put it, "Sometimes it seems a shame that Noah and his party did not miss the boat."

[5]See for example, Scoby, Ed. *Environmental Ethics,* (Minneapolis, MN: Burgess Publishing Co., 1971), and Schaeffer, F. A. *Pollution and the Death of Man—A Christian View of Ecology,* (Wheaton, IL: Tyndale House Publishers, 1980).

active or latent). In short, the ethics existed first, and it was the people who developed these ethics into a religion.

Further, the ethics which fit the Judeo-Christian religions so well also spawned a vigorous pursuit of science, a tradition of democracy, and the capitalistic system. Although Christianity was compatible with the idea of individual worth and achievement, it was not the *reason* that these ideas fluorished. It can therefore be argued that the Christian church is not directly responsible for the traditions and ethics which promoted the destruction of nature.

According to White, the model that Western civilization has been following is

Ethics and traditions of Judeo-Christian peoples	→	Science, capitalism, technology, democracy	→	Urbanization, money, population, individual ownership	→	Environmental degradation

An equally strong argument, however, can be made for the following model:

Human nature and the search for a comfortable and compatible religion	→	Acceptance of the Judeo-Christian dogma	→	Science, capitalism technology, democracy	→	Urbanization, money, population, individual resource ownership	→	Environmental degradation

If, therefore, the latter model is equally appealing, there is little reason to blame the Judeo-Christian or any other religions for our incompatibility with nature.

Social Structure as the Cause

A second view of the roots of our ecological crisis is that our social structures are responsible. Probably the most damning piece yet written which takes this view is Garrett Hardin's "Tragedy of the Commons."[6]

Hardin illustrates his point by a story of a village which has a common green for the grazing of cattle, surrounded by individual farmhouses. In

---

[6] *Science* 162:1243–1248 (1968); and reprinted in numerous books on environmental ethics.

the beginning, each farmer has one cow, and the green is able to support the herd. It becomes apparent to each farmer, however, that if he gets another cow, the *cost* is negligible to him personally (it's shared by everyone) but the *profits* are his alone. So he gets more cows, reaping greater and greater profits, until the commons are no longer able to support the herd and the system collapses.

Hardin used this parable to illustrate the problem of overpopulation, but it applies equally well to other environmental problems. The social structure in the parable is of course capitalism, the individual ownership of wealth, and the use of that wealth for furthering one's own interest.

A lot has been written about capitalism as the major causative agent for our environmental ills, often with the implication that some form of socialism, Marxist or otherwise, is a superior system. Even mainland China has been cited as an example of a stable system.[7]

Unfortunately, the advantages of a centrally controlled (and hence totalitarian) system have not provided the answer. In fact, the environmental devastation in the USSR is substantially more serious than in the West. When *production* is the primary goal of society, the environment and human life take a poor second place. Further, the recent attempt by China to modernize attests to not only the instability of their system, but to the existence of universal human aspirations and drives to better one's condition.

The only types of socio-political systems which seemed to have developed a quasi-steady-state condition are primitive societies, such as the American Indians, the Finno-Ugric people of Northern Europe (Finland and Estonia) and the south sea islanders.

To all of these people nature holds within it spirits that are both powerful and friendly (if at times capricious). The spirits in nature do not take human forms (as in Greek and Roman religions). Yet it is possible to converse freely with the spirits. The old Estonians and Finns, for example, always explained to the spirit of the tree why it was necessary to cut it down.[8] Such a reverence for and closeness with nature is unknown in most modern societies. Imagine the difficulty in the clearing of a forest for a man-made lake if every tree required a special explanation and apology?

There was, in these primitive societies, a camaraderie with all life

---

[7] Among others, see Bryan, D. "China: A New Society in the Making," in *Notes for the Future,* R. Clarke, Ed. (New York: Universe Books, 1975).

[8] Paulson, I. *The Old Estonian Folk Religion* (Bloomington, IN: Indiana University Press, 1971).

forms. For example, old Estonians would begin the wheat harvest by cutting a shaft of wheat and placing this aside for the field mice. There did not seem to be any religious significance to this mouse-shaft (hiirevihk) and the only explanation handed down through the generations is that the mice deserve their share of the harvest.[9]

And yet these primitive societies were not all environmentally stable. The Mauri first came to New Zealand in the 1300s, and proceeded to exterminate the moa, a large ostrich-like bird which was their sole source of meat (the islands had no native mammals, except a bat).

In the face of such failures, it is difficult to argue that we should reestablish primitive societies. In addition, a society is after all the reflection of the needs and aspirations of the people. People establish societies, and thus it cannot be argued that social systems are the root cause of our environmental ills.

Science and Technology as the Cause

It has become somewhat fashionable to blame our increased knowledge of nature and our related ability to put the knowledge to work for our environmental ills. The popularity of back-to-nature communes seems to suggest that somewhere we have made a wrong turn in the road to technological capability. Numerous authors have jumped on this bandwagon and blamed technology for everything from athlete's foot to nuclear bombs.*

If technology is rightfully to blame, we must show first that other less technologically advanced societies avoided major environmental problems, and secondly, that modern technology is not value-free.

The first premise is not true. Many primitive societies were equally or even more destructive than our own. We already discussed the extermination of the moas by the Mauri. Overgrazing in Africa and the destruction of forests by the early Greek civilization are other examples. In fact, one can mount a strong argument for technology as a means of being able to *control* destructive forces and trends.

Is technology value-free? Is knowledge itself, without the application

---

[9] Oinas, Felix. Personal communication, Indiana University (1981).

* The distrust of technological advancement is not new. During the industrial revolution in England, the Luddites were people who violently resisted the change from cottage industry to centralized factories. Because the large machines threatened their way of life, they smashed a few factories to make their point, and were hanged for their trouble.

of it by humans, right or wrong, ethical or unethical? Before we jump to what might seem an obvious answer to most engineers, consider this example: You have it within your power to conduct an experiment and thus prove the feasibility of constructing a simple and inexpensive nuclear bomb. The knowledge may well be used by terrorists in untold acts of violence. Without your experiment, the construction of such a device would not be possible. Should you conduct the experiment, since you know that pure science and technology are value-free, and *you* will never put this knowledge to evil use? But does this knowledge, by its very *existence,* constitute an evil?

Strong arguments have, as you can imagine, been mounted in defense of both views. But even if science is determined to be value-laden, the use of science for curing environmental ills has an equally strong case. We cannot, in short, take the quick and dirty way out and blame science alone for environmental degradation just as we couldn't lay the blame on religion or social structure alone. And indeed, is it not better to seek a solution than to cast blame?

RESOLUTION OF ENVIRONMENTAL CONFLICTS

The development of various systems of ethics stems from the human need to create a unified and universally applicable method of "getting along." The major concern with many ethical theories is first to determine what is *good,* and second, what is *right.* In addition, it becomes necessary, if the ethical theory is to have any utility, to strive toward what has been determined to be good and right; once we decide that a system is correct, then we should live by the conclusions we derive from the system.

Many so-called systems of ethics have been proposed over the years.[10]

One of the oldest systems is *hedonism,* espoused by the sophists with whom Socrates debated. Hedonism seeks to maximize pleasure. Whatever gives you pleasure is therefore right. In isolation, this may be a workable philosophy, but not in a society which depends on the general acceptance of the golden rule.

Plato's arguments against hedonism included the concept of a person being governed by reason, and that "good" decisions can only be made

---

[10]See for example MacIntyre, A. *A Short History of Ethics* (New York: Macmillan Publishing Company, 1966), or Pearsall, G. W. "Decision Making in Hazardous Waste Disposal" in *Hazardous Wastes,* J. J. Peirce and P. A. Vesilind, Eds. (Ann Arbor, MI: Ann Arbor Science Publishers, Inc., 1981).

in the face of well-considered arguments. He extrapolated this to the idea that since not all people have the opportunity to study philosophy, decisions for the society should be made by a "philosopher-king." In Plato's system, therefore, the decisions are taken out of the hands of the masses and the power given to a benign dictator who, Plato argued, would make the "correct" decisions.

Immanuel Kant proposed a third alternative: establish a set of rules that everyone agrees to and then follow these no matter what. The *categorical imperative* commits each person to a set of consistent and universally shared required actions (e.g., do not lie under any circumstances; or "dioxin" is bad, so don't allow *any* dioxin out of a municipal waste incinerator). In this system, the focus is on the act itself, not its consequences.

An opposing view holds that consequences are the major concern. *Utilitarianism,* developed in eighteenth century England by Jeremy Bentham, John Stuart Mill and others, holds that one's actions should strive for the greatest good, defined as individual happiness. Under this system, one could construct a happiness-benefit/cost ratio for every action, and the options with the highest ratio would clearly be the ones of choice. Extending this idea to the social level, we have the concept that governmental actions should follow the dictum of "the greatest good for the greatest number." This concept blends directly into *socialism,* developed to a large extent by Karl Marx and Friedrich Engels, in which the *society* is the prime instrument and the individual is no longer very important.* Under socialism, decisions are made on the basis of maximizing production, of increasing the average wealth of the society.

There are other systems of ethics worth noting, none of which have attained the historical stature of the ones discussed above. For example:

- *legalism* – do only what the law tells you to do, and no more. If you break the law, don't get caught. (Is football played under this system?)
- *populism* – let the masses decide.
- *situation ethics* – make up rules as the game goes on, but base these on the golden rule idea.
- *altruism* – always think of the other guy first.
- *egoism* – the opposite of altruism, in that one's behavior is always governed by selfish interest.

---

*Witness the status of the medical arts in soviet Russia. A physician is paid about the same as a taxi driver or doorman, and receives only 3 years of training. In that society, it's not efficient to spend money and resources on medical care for the few workers who are ill. Productivity is enhanced by placing the emphasis on the healthy workers and ignoring the infirm.

- *existentialism* — there is no grand scheme to the world, and one's actions and one's fate are unimportant. This is a doctrine of complete resignation and hence freedom.

A major problem with the application of any of the ethical theories is that they were all developed as means of resolving conflicts among *people.* Our concerns, however, are problems between people and *animals,* people and *plants* and people and *places.* This severely complicates the issue, since now the conflict exists between two parties, each having a different *value.* As long as the actors are people, the value of a human life is a given constant. Now we have an inequality, and we must seek to place a value on nature.

The concept that nature has a value is fairly modern. With a few noteworthy exceptions,* nature has not been considered through the development of human civilization as an entity possessing value, and has thus not entered into the development of ethical theories.

In the past, only dominance and perpetual progress have shaped our relationship with the natural environment. In fact, this has been for years a substitute for an environmental ethic, and these exploitative notions have guided our behavior toward a nature which by itself did not possess value.

But simply stating that nature is valuable is not adequate. *Why* and *how* is it valuable?

This question has emerged as a major point of discussion in philosophical literature. The two primary views that seem to be emerging in this debate can be classified as *instrumental* and *intrinsic.*

The instrumental view considers nature for its real value to humans, in terms of life support, material wealth, recreational and aesthetic beauty, etc.[11] This view was first popularized by Ralph Waldo Emerson, who witnessed the wholesale destruction of nature during the nineteenth century "rape of the land," and proposed that in the use of nature's bounty, one should be grateful for the blessings and material wealth nature provides. This is, of course a utilitarian and anthropocentric view of nature, but it nevertheless represented a first recognition that nature has *value.*

---

*Such as St. Francis of Assisi. As Lynn White points out, the miracle of St. Francis was that he was not burned at the stake as a heretic. St. Francis tried to dethrone man as the center of the universe, and preached the importance of *all* of God's creatures.

[11] See for example Rolston, H. "Values in Nature," *Environmental Ethics* 3(2)(1981) or Kaufman, P. I. "The Instrumental Value of Nature," *Environmental Review* 4(1)(1981).

The second view holds that nature has intrinsic value above and beyond what we might be able to measure in dollars.[12] This, however, is a difficult argument to tie down. In a way the belief in the intrinsic value of nature represents a faith – one has to *feel* that it is right. This feeling has been even translated to legalistic terms, most notably in a work entitled *Should Trees Have Standing? – Toward Legal Rights for Natural Objects.*[13] The question we face is whether nature is here for the welfare of mankind, or does nature have its independent claims to exist.

The problem is further complicated by what we mean by nature. It's not difficult to get people upset over the senseless slaughter of whales, or of baby seals. Why? Simply because whales are too much like *humans*; they have many of our own most admirable characteristics and lead a placid, peaceful life, and baby seals are so *cute!*

Can you imagine an equal outcry raised against the destruction of the polio virus, or the rattlesnake? Do these two species also have standing, a right to be left alone and not destroyed?

The issue is of course complex. But to not face it is unacceptable. There is in the intrinsic approach to the value of nature a germ of a positive and intuitively desirable notion. By not striving to develop and foster this idea, might we be freezing our environmental ethic at too low a level? Many widely accepted principles of the past, such as slavery and the divine right of monarchs, have fallen. Albert Schweitzer noted that Europeans long considered people with darker skins to be subhuman. It was at one time stupid to think of them as equals, and to treat them humanely. And now this stupidity is a truth. Schweitzer continues:

> Today it is thought an exaggeration to state that a reasonable ethic demands constant consideration for all living things down to the lowest manifestations of life. The time is coming, however, when people will be amazed that it took so long for mankind to recognize that thoughtless injury to life was incompatible with ethics.[14]

René Dubos takes the idea further. Why Dubos argues, do we limit our concerns with animate beings? He proposes in a thoughtful article entitled "The Theology of the Earth" that everything has its place and reason for being. He suggests that there is a genius of the place, a oneness

---

[12] See for example Worster, D. "The Intrinsic Value of Nature," *Environmental Review* 4(1)(1981).

[13] Stone, C. D. (Los Angeles, CA, Wm. Kaufman, Inc., 1972).

[14] Joy, C., Ed. "The Animal World of Albert Schweitzer," quoted in "The Intrinsic Value of Nature," D. Worster, *Environmental Review* 4(1)(1981).

between man and the uniqueness of each locality, be it a city or a grove of trees.[15]

The intrinsic value of nature obviously is on a higher ethical plane than the instrumental view of nature, but it is a much more difficult idea for the mass of humanity to accept.

THE FUTURE OF THE ENVIRONMENTAL ETHIC

The birth of the environmental ethic as a force is partly a result of our concern for our own long-term survival, as well as our realization that humans are but one form of life, and that we should share our earth with our fellow travelers. It is not a religion, since the ethic is based not only on faith, but also on hard facts and thorough analyses.

One of the first to recognize the degradation of the environment and to voice a concern for nature was Thoreau. His solution to this eloquently stated concern was withdrawal, which was perhaps morally admirable but realistically ineffective. What we have experienced in the last decade has been the coupling of Thoreau's concern with activism. It is one thing to be concerned, but it's much more effective to take action in order to promote your concern.

One powerful and original attempt at defining an environmental ethic was made by Aldo Leopold, a naturalist and writer. He proposed that:

> A decision is right when it tends to preserve the integrity, stability, and beauty of the biotic community. It is wrong when it tends otherwise.[16]

Leopold argued convincingly toward the adoption of environmental values other than economic, and his writings have had a major impact on the growth of environmental awareness.

The environmental ethic is very new, and none of the doctrine is cast in immutable decrees and dogma. On the contrary, like all vital issues, the environmental ethic will undergo transformation as new data are made available and we humans are able to more rationally interpret and live with nature.

---

[15]Dubos, René. "A Theology of the Earth," a lecture reprinted in *Western Man and Environmental Ethics,* I. G. Barbour, Ed. (Reading, MA: Addison-Wesley, 1973).

[16]Leopold, A. *A Sand County Almanac* (New York: Oxford University Press, Inc., 1949).

Some critics have mistaken this maturation of the ethic for transience. The problem is so immense that to think of it as a fad is ludicrous (and fatal!). But herein is the dilemma: Unless there is public awareness as to the true nature of the problems and some realistic solutions, public concern may rapidly fade. For the impetus to survive there must be public confidence and the environmental scientist or engineer must engender this trust by analyzing and interpreting environmental problems correctly and proposing and developing constructed facilities that are compatible with our ecosystem. Sensationalism and the bandwagon tactics of some "environmentalists," politicians and other well-meaning citizens can easily destroy the public sentiment so necessary for a successful assault on our common problems.

Education of the public to environmental problems and the solutions (and nonsolutions) is of prime importance. It is necessary for people to be critical, and to be literate in the scientific, technological, economical and legal aspects of controlling environmental pollution.

But that is not enough. We must also live the environmental ethic – recognize the power of nature and feel humble in the realization that we are just one small cog in a wonderful and still mysterious system.

PROBLEMS

25.1 Read Chapter 2 of Genesis. Then study Figure 25–1. Express your thoughts relative to the "stewardship" concept of environmental ethics and the value of nature.

25.2 The three symbols in Figure 25–2 were taken from: (a) a fast food container, (b) and egg carton and (c) a paper bag.
   (a) What purposes did the respective container manufacturers have in mind in placing the symbol on their packages?
   (b) Write a critical statement of the basic honesty in the use of the recycling symbol in these three cases.

25.3 The picture in Figure 25–3 was accompanied by a story in the Air New Zealand in-flight magazine:

> As an "alarm clock," the Boeing has a "competitor." A top-dressing plane sometimes makes a habit of droning across the northern part of the city early in the morning. During the debate on Boeing noise before its introduction in 1975, Invercargill's Mayor Mr. F. R. Miller, made the comment that surely nothing could be as bad as

Figure 25-1   Drenched and shivering after nearly drowning, a calf stranded on an island gets a tender toweling from Don Frickie of the Arctic National Wildlife Range. Deposited on the stream's bank, this 2-day-old was quickly reunited with its frantic mother. [Picture and legend courtesy of the National Geographic Magazine, and the photographer, J. Rearden. The picture appeared in Rearden, J. "Caribou: Hardy Nomads of the North," *National Geographic* 146(6) (1974).]

Recyclable Package

 MADE FROM 100% RECYCLED FIBERS

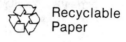 Recyclable Paper

A.   B.   C.

Figure 25-2   The recycling symbol (for Problem 25-2).

Figure 25-3   Invercargill's alarm clock (for Problem 25-3) (courtesy of Air New Zealand).

that slow-moving disturber of the peace. Further it "rings" too early for most people, is far less reliable and is one of those "alarm clocks" that jars the nerves, sends the blood pressure skywards and leaves you with the feeling that as you've just gone to bed, why are you being disturbed?

In contrast, the Boeing "alarm clock" contains the right volume of noise to arouse but not to anger and only during bad (unseasonal) weather does it forget to let you know it's time to get up. It is unreachable so it cannot be smashed to silence but perhaps most importantly, like all good alarm clocks, it can be ignored – at your own risk.

What purpose did NAC (now Air New Zealand) have in printing this article? What do you suppose the controversy was that precipitated it? What do you think of the basic integrity of the presentation?

25.4 The following paragraph appeared as part of a full-page ad in a professional journal.

Today's laws.
A reason to act now.

The Resource Conservation and Recovery Act [RCRA] provides for corporate fines up to $1,000,000 and jail sentences up to five years for officers and managers, for the improper handling of hazardous wastes. That's why identifying waste problems, and developing economical solutions is an absolute necessity. O'Brien & Gere can help you meet today's strict regulations, and today's economic realities, with practical, cost-effective solutions.

Based on a consideration of environmental ethics, why *else* might the "identifying waste problems, and developing economical solutions" be an absolute necessity?

25.5 A major purveyor of auto parts sells a book entitled *How to Bypass Emission Controls – For Better Mileage and Performance.* The ad states that the emission controls rob the car of power and waste gas, and that this book contains easy-to-follow directions for the amateur and professional on how to eliminate the emission controls. (The company, incidentally, would not allow us to reprint the ad, or to quote directly from it.) At the bottom of the ad, a statement notes that the book is not available in California, and that the buyer should check with the state motor vehicle department before removing the emission controls.

Suppose you are the president of the auto supply company selling this book. How would you ethically justify your decision to sell it? What systems of ethics would you need to adopt to make such a sale morally acceptable?

25.6 You are working as a recruiter for your consulting engineering firm, and an applicant asks you the following question:

"If I work with your firm, and I find that I object, on ethical grounds, to a project the firm has been hired to do, can I request not to work on that project, without penalizing my future career with the firm?

Give three responses you might make to such a question, and then respond to these statements according to what you believe the student/ interviewee would think. Would he/she believe you?

25.7 The Environmental Advisory Council of Canada has published a booklet entitled "An Environmental Ethic – It's Formulation and Implications," (Report No. 2, January 1975, Norma H. Morse, Ottawa). They suggest the following as a concise statement of an environmental ethic.

> Every person shall strive to protect and enhance the beautiful everywhere his or her impact is felt, and to maintain or increase the functional diversity of the environment in general.

In a short essay, discuss the validity of this statement as a useful environmental ethic.

25.8 The Dickey and Lincoln dams on the St. John River in Maine are being planned with careful consideration to the Furbish lousewort (*Pedicularis furbishiae*), an endangered plant species. An engineer writing in *World Oil*, (Jan. 1977, p. 5) termed this concern for the "lousy lousewort" to be "total stupidity." Based on only this information, construct an ethical profile of this engineer.

25.9 The selection of what science studies is the subject of a masterful essay by Leo Tolstoy, "The Superstitions of Science," *The Arena*, 20 (1898), reprinted in *The New Technology and Human Values*, J. G. Burke, Ed. (Belmont, CA: Wadsworth Publishing Company, 1966). He recounts how

> ...a simple and sensible working man holds in the old-fashioned and sensible way that if people who study during their whole lives, and, in return for the food and support he gives them, think for him, then these thinkers are probably occupied with what is necessary for people, and he expects from science a solution of those questions on which his well-being and the well-being of all people depends. He expects that science will teach him how to live, how to act towards members of his family, towards his neighbors, towards foreigners; how to battle with his passions, in what he should or should not believe, and much more. And what does our science tell him concerning all these questions?
>
> It majestically informs him how many million miles the sun is from the earth, how many millions of ethereal vibrations in a second constitute light, how many vibrations in the air make sound; it tells him of the chemical constitution of the Milky Way, of the new element helium, of microorganisms and their waste tissue, of the points in the

hand in which electricity is concentrated, of X-rays, and the like. But, protests the working man, I need to know today, in this generation, the answers to how to live.

Stupid and uneducated fellow, science replies, he does not understand that science serves not utility, but science. Science studies what presents itself for study, and cannot select subjects for study. Science studies everything. This is the character of science.

And men of science are really convinced that this quality of occupying itself with trifles, and neglecting what is more real and important, is a quality not of themselves, but of science; but the simple, sensible person begins to suspect that this quality belongs not to science, but to people who are inclined to occupy themselves with trifles, and to attribute to these trifles a high importance.

In the formulation and planning of your career as an engineer, how would you answer Tolstoy's criticism?

25.10 Re: Figure 25-4. Do fish have "standing"?

25.11 Study Figure 25-5. Comment relative to Dubos' idea of the "genius of the place."

25.12 Re: Figure 25-6. Is man adapting to his environment?

SUGGESTED READING

Barbour, I. G. *Western Man and Environmental Ethics* (Reading, MA: Addison-Wesley Publishing Company, 1973).

Burke, J. G., Ed. *The New Technology and Human Values* (Belmont, CA: Wadsworth Publishing Company, 1966).

Cahn R. *Footprints on the Planet* (New York: Universe Books, 1978).

Hardin, G. *Stalking the Wild Taboo* (Los Altos, CA: Wm. Kaufman Inc., 1978).

Leopold, A. *A Sand County Almanac* (New York: Sierra Club/Ballentine, 1966).

MacIntyre, A. *A Short History of Ethics* (New York: Macmillan Publishing Company, 1966).

Watkins, B. O., and R. Meador. *Technology and Human Values* (Ann Arbor, MI: Ann Arbor Science Publishers, Inc., 1978).

Figure 25-4   A massive fish kill in Illinois (1967) (for Problem 25–10) (courtesy of EPA-Documerica).

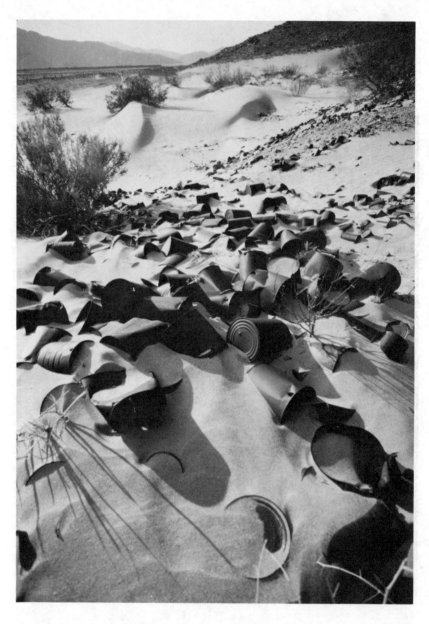

Figure 25-5   Beach at the Anza-Barego State Park, CA (for Problem 25–11) (courtesy of EPA-Documerica).

Figure 25-6   New York traffic, near 42nd St. (for Problem 25–12) (courtesy of EPA-Documerica).

Conversion Factors

Multiply	By	To Obtain
acre	0.404	ha
acre ft	1233	m^3
atmospheres	14.7	$lb/in.^2$
British thermal units	252	cal
Btu	1.054×10^3	J
Btu/ft^3	8,905	cal/m^3
Btu/lb	2.32	J/g
Btu/lb	0.555	cal/g
Btu/sec	1.05	kW
Btu/ton	278	cal/tonne
calories	4.18	joule
calories	3.9×10^{-3}	Btu
cal/g	1.80	Btu/lb
cal/m^3	1.12×10^{-4}	Btu/ft^3
cal/tonne	3.60×10^{-3}	Btu/ton
centimeters	0.393	in.
feet	0.305	m
ft/min	0.00508	m/sec
ft/sec	0.305	m/sec
ft^2	0.0929	m^2
ft^3	0.0283	m^3
ft^3	28.3	liters
ft^3/sec	0.0283	m^3/sec
ft^3/sec	449	gal/min
ft lb (force)	1.357	joule
ft lb (force)	1.357	newton meters
gallons	3.78×10^{-3}	m^3
gallons	3.78	liters
$gal/day/ft^2$	0.0407	$m^3/day/m^2$
gal/min	2.23×10^{-3}	ft^3/sec
gal/min	0.0631	liter/sec

gal/min	0.227	m^3/hr
gal/min	6.31×10^{-5}	m^3/sec
gal/min/ft^2	2.42	$m^3/hr/m^2$
million gal/day	43.8	liters/sec
million gal/day	3785	m^3/day
million gal/day	0.0438	m^3/sec
grams	2.2×10^{-3}	lb
hectares	2.47	acre
horsepower	0.745	kW
inches	2.54	cm
inches of mercury	0.49	lb/in.^{2}
inches of mercury	3.38×10^3	newton/m^2
inches of water	249	newton/m^2
joule	0.239	calorie
joule	9.48×10^{-4}	Btu
joule	0.738	ft lb
joule	2.78×10^{-7}	kWh
joule	1	newton meter
J/g	0.430	Btu/lb
J/sec	1	watt
kilograms	2.2	lb (mass)
kg	1.1×10^{-3}	tons
kg/ha	0.893	lb/acre
kg/hr	2.2	lb/hr
kg/m^3	0.0624	lb/ft^3
kg/m^3	1.68	lb/yd^3
kilometers	0.622	mi
km/hr	0.622	mph
kilowatts	1.341	horsepower
kWh	3600	kilojoule
liters	0.0353	ft^3
liters	0.264	gal
liters/sec	15.8	gal/min
liters/sec	0.0228	mgd
meters	3.28	ft
meters	1.094	yd
m/sec	3.28	ft/sec
m/sec	196.8	ft/min
m^2	10.74	ft^2
m^2	1.196	yd^2
m^3	35.3	ft^3
m^3	264	gal
m^3	1.31	yd^3
m^3/day	264	gal/day

m^3/hr	4.4	gpm
m^3/hr	6.38×10^{-3}	mgd
m^3/sec	35.31	ft^3/sec
m^3/sec	15,850	gpm
m^3/sec	22.8	mgd
miles	1.61	km
mi^2	2.59	km^2
mph	0.447	m/sec
milligrams/liter	0.001	kg/m^3
million gallons	3,785	m^3
mgd	43.8	liter/sec
mgd	157	m^3/hr
mgd	0.0438	m^3/sec
newton	0.225	lb (force)
$newton/m^2$	2.94×10^{-4}	inches of mercury
$newton/m^2$	1.4×10^{-4}	$lb/in.^2$
newton meters	1	joule
$newton\ sec/m^2$	10	poise
pounds (force)	4.45	newton
pounds (force)/in.^{2}	6895	N/m^2
pounds (mass)	454	g
pounds (mass)	0.454	kg
pounds (mass)/ft^2/yr	4.89	$kg/m^2/yr$
pounds (mass)/yr/ft^3	16.0	$kg/yr/m^3$
pounds/acre	1.12	kg/ha
pounds/ft^3	16.04	kg/m^3
pounds/in.^{2}	0.068	atmospheres
pounds/in.^{2}	2.04	inches of mercury
pounds/in.^{2}	7140	$newton/m^2$
tons (2000 lb)	0.907	tonne (1000 kg)
tons	907	kg
ton/acre	2.24	tonnes/ha
tonne (1000 kg)	1.10	ton (2000 lb)
tonne/ha	0.446	tons/acre
yd	0.914	m
yd^3	0.765	m^3
watt	1	J/sec

Index

It's what you learn after you
know it all that really counts.

Harry S. Truman

Vesilind   **Peirce**

P. Aarne Vesilind is Professor of Civil and Environmental Engineering at Duke University, Durham, North Carolina, and Director of the Environmental Engineering Center. Born in Estonia, he holds BS and MS degrees in civil engineering from Lehigh University, and an MS and PhD in sanitary engineering from the University of North Carolina. Dr. Vesilind has also been a Fellow at the Norwegian Institute for Water Research in Oslo and has received a Fulbright-Hayes Senior Lectureship to study in New Zealand.

Among his many honors and awards are the Duke University Outstanding Professor Award for 1971–1972 and the 1971 Collingwood Prize from the American Society of Civil Engineers.

Dr. Vesilind is the author of two Ann Arbor Science publications, *Environmental Pollution and Control* and *Treatment and Disposal of Wastewater Sludges,* is the series editor of the *Design and Management for Resource Recovery* series, and is a member of the Ann Arbor Science Publishers Editorial Advisory Board.

J. Jeffrey Peirce holds a bachelor of engineering science in engineering mechanics from The Johns Hopkins University. He received his MSCE and PhD in civil and environmental engineering/economics at the University of Wisconsin, Madison. As a faculty member at Duke University, Dr. Peirce lectures on topics ranging from air pollution abatement technologies to resource recovery systems, and conducts research within the Department of Civil and Environmental Engineering. Dr. Peirce's current research develops and compares hazardous waste disposal options for Duke University, analyzes data on municipal sludge to provide a basis for future disposal options, and studies less costly alternatives for regional solid waste disposal and resource recovery. He has more than 20 publications in such journals as *Water Resources Bulletin* and the *Journal of the Environmental Engineering Division of ASCE*. Dr. Peirce is a full member of Sigma Xi, the American Society for Engineering Education and the Association of Environmental Engineering Professors.